PLANTS
FOR PEOPLE

PLANTS FOR PEOPLE

Anna Lewington

eden project books

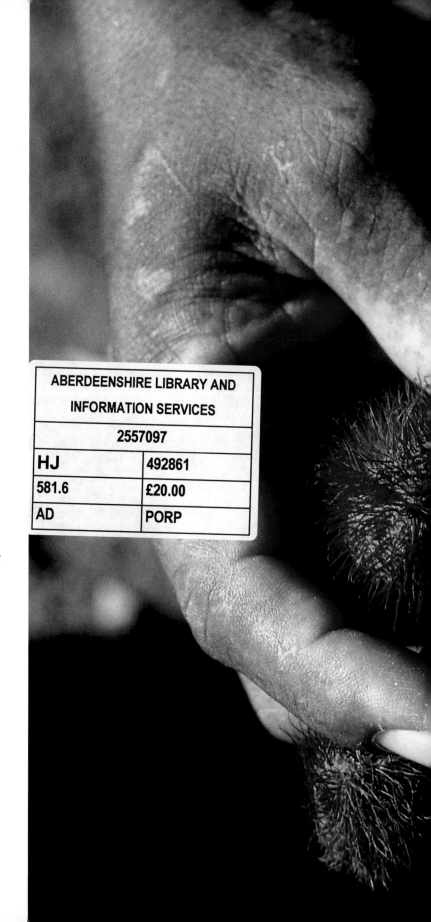

TRANSWORLD PUBLISHERS
61-63 Uxbridge Road, London W5 5SA
a division of The Random House Group Ltd

RANDOM HOUSE AUSTRALIA (PTY) LTD
20 Alfred Street, Milsons Point, Sydney,
New South Wales 2061, Australia
RANDOM HOUSE NEW ZEALAND LTD
18 Poland Road, Glenfield, Auckland 10,
New Zealand
RANDOM HOUSE SOUTH AFRICA (PTY) LTD
Endulini, 5a Jubilee Road, Parktown 2193,
South Africa

Published 2003 by Eden Project Books
a division of Transworld Publishers

Printed in Germany by Appl, Wemding

1 3 5 7 9 10 8 6 4 2

Papers used by Eden Project Books are natural,
recyclable products made from wood grown in
sustainable forests. The manufacturing processes
conform to the environmental regulations of the
country of origin.

■ The publishers would like to express their
gratitude to WWF-UK for their generous support in
the preparation of this book.
WWF-UK registered charity number 1081247
www.wwf.org.uk/research/plants

CONTENTS

Introduction

Today, in the 21st century, the survival of everyone on earth, from businessmen to bushmen, depends on plants. Countless raw materials for our food, our clothing, our buildings and homes, our medicines, the vehicles that transport us, the ways in which we communicate and keep ourselves entertained are provided by a multitude of plants. Who would have thought that their clothing might be made from eucalyptus wood or that cork helps insulate the fuel tanks of vehicles sent into space?

Though their origin, indeed their existence, may seem peripheral to our modern lives — modified or disguised as they often are — plant products are as strategic as oil. In the coming decades, as reserves of our fossil oil diminish, this importance must surely rise.

The idea of *Plants for People* arose nearly 15 years ago, when I was invited by the Royal Botanic Gardens, Kew, to produce a book that would show a non-specialist audience just how important plant materials are to our lives. Rather than focus on the critical role that plants play in maintaining all life on earth, on the vital environmental services that they perform, the idea took shape of revealing the largely hidden array of plants used to make countless products we depend on — as relatively affluent consumers — from toothpaste to powerful anti-cancer drugs.

As I carried out the research for that first edition of *Plants for People*, I became increasingly excited by the discoveries I was making: the fact that plants are involved in so much of what we use and take for granted. But I was concerned that from the viewpoint of the general public this was a subject that remained essentially hidden. I felt strongly, as I still do, that this situation should change, that people should know the origin of what they use, that products should be properly labelled and that we should

all somehow acknowledge the fundamental link we have with plants from all around the world.

But what I became aware of too, of course, was our general ignorance, not just of the material origin of so many things that support our lives, but of the people and environments that provide them. I was concerned in particular about the serious repercussions for the environment of our industrial usage of plants and about the accelerating threats to the survival of so many species (plant and animal alike) including traditionally cultivated plants. Above all, however, I have been concerned about people: the millions of people around the world, whose labour we use, whose land we take, whose health we affect, whose knowledge we trade and whose lives we determine by our insatiable consumption of resources from plants.

In this new edition of *Plants for People*, although it has been impossible to cover every aspect, I continue to present the tremendous diversity of plant use — largely from the perspective of Western consumer goods — with many examples of traditional plant use from around the world. But I also reveal aspects of the wider social and environmental contexts or 'footprints' of what we perpetuate, often unwittingly, as consumers of mass-produced plant products. The routine accidental poisoning of banana

Page 2: Ferns are a very versatile group of plants with a wide range of uses, including foods, medicines, fibre, dyes, compost and fertilizer. Some ferns are also able to remove pollutants from contaminated soils.
Pages 4–5: The ripe seed capsules of *Bixa orellana* reveal the source of annatto, a major natural colorant used by the food industry — in dairy products, confectionery, ice cream and many other processed foods — and in cosmetics.
Opposite: Woman plaiting coconut fibre for making into sleeping mats, Mafia Island, Tanzania.

workers with agrochemicals in Costa Rica, for example, or the destruction of vast areas of rainforest in Indonesia to make way for African oil palm, which provides cheap oil for margarine and soap (amongst many other things), are part of the real price of products that is never printed on our till receipts.

I mention many of the more positive industrial initiatives that are taking place too – from the increasing use of hemp and other fibres to make composite panelling for cars and energy-efficient building materials for our homes to the use of coffee waste in El Salvador to make environmentally friendly paper. Had space allowed, countless examples could have been given of small-scale projects – which attest both to the ingenuity of people working together in communities and the versatility of plant materials – to find sustainable local solutions to problems of environmental degradation and to alleviate poverty. In the Udzungwa district of Tanzania, people have begun to use a mixture of dried leaves, wood ash, sand and water to make 'vegetable charcoal' – thereby helping to ease the firewood crisis. In the Brazilian Amazon, a number of community projects are making use of abundant native palm species – with oil from the babaçu palm for soap and cosmetics, for example, and crafts and furniture from the buriti palm.

At the same time, 'Western' scientific and technical developments continue to push back boundaries, to discover or create new ways of using plants or plant materials undreamt of 50 years ago. Recent research has shown the existence of 'antifreeze' proteins in carrots, for example, that – in binding to and preventing the growth of ice crystals – could have future applications in frozen foods and in the storage of human transplant organs. Maize is now the source of new kinds of plastic used in

some high-tech electronics, such as laptop computers, while potatoes are being turned into biodegradable food packaging and carrier bags. Plant breeders have produced crops – our major staples – whose yields and characteristics have made possible a highly mechanized agriculture and surpluses that sustain economies as well as the diets of more and more people in the developed and developing worlds. Such diets, however, are increasingly processed, and many crops, such as potatoes, are 'designed' and grown purely to meet the specifications of producers of homogeneous 'fast food'. Around 60% of our processed food contains soya derivatives, for example.

The controversy that surrounds genetically modified (GM) plants has given a new, sharper focus to the long-standing concerns, felt by many, about an increasingly privatized global agriculture. It has raised issues not just about the technology involved (regarded by many as insufficiently understood in terms of possible repercussions for the environment and human health), but about who owns the (hybrid) seeds that produce the crops on which so many depend. Many believe that genetically modified crops cannot be justified and that, above all, they are unnecessary. In 2000, a Food and Agriculture Organization report concluded that the world could feed itself without the need for genetically modified organisms.

The importance of diversity

Whatever our opinion about GM crops, it cannot be denied that they present physically and conceptually a uniformity that is the very antithesis of the diversity of nature. It is at last being acknowledged that much of the world's biological diversity (biodiversity) is a reflection of – or has evolved alongside – cultural diversity. It is cultural diversity, the

great variety of different peoples that grace our world, that has held the key to the development of so many of the world's useful plants. But everywhere this diversity is being undermined. As 'free trade' policies and the privatization of resources extend and continue to widen the gap between rich and poor, more and more people (in 'developing' countries) are obliged to abandon their lands, and their plants, to become impoverished consumers in urban slums.

Analysts have shown that the model of further economic growth through the expansion of the free market is not sustainable. There are simply not enough resources for the 4 billion or so people in the developing world to match US per capita consumption levels (supported by a fossil-fuel-based, throwaway economy). Yet this model is still perpetuated by governments and the world's most powerful financial institutions. As more forests are destroyed, as habitats are degraded and grasslands overgrazed, as wild plants are over-harvested and wetlands drained, as urban areas increase, no wonder that so many – about 25% – of our plants are believed to be threatened with extinction in the next 50 years. And this mirrors what has been happening to so many of the world's indigenous peoples, who question 'progress' based on an economic model that devalues or expropriates the land, labour and goods of others.

A sea-change is needed so that we alter the focus of our world view to one that encompasses an economic policy that respects the earth's natural systems and what they can sustain, as well as its people.

We are learning, albeit slowly, about limits. Alongside the recognition that it is in all our interests to consume less and to recycle more of what we have already used, is the growing movement to produce food – and other goods – locally and to appreciate regional and seasonal diversity.

Organic agriculture promotes sustainability, it restores respect and gives real value to what we consume, to the land and to the people who work it for us. For far too long we, as affluent consumers, have been discouraged from ever considering the real price, in environmental or ethical terms – of the products we use – from our bowl of cornflakes to the pairs of jeans we own; from garden furniture to an electric guitar (perhaps made from mahogany) – and that make our lives so comfortable.

With the developing initiatives of fairer trade, organic food and farmers' markets, and of environmentally friendly and ethically produced clothing, and with numerous other items available today – from locally produced charcoal for the barbecue, to reclaimed timber for our homes – it is now possible for each of us to make a difference, however small. This book attempts to show something of the extent to which we all use plants in the products that support our lives, and I hope that it will also encourage us to consider the people and environments from which they have come.

Anna Lewington

9

Plants that **care** for us

Most of us use plants each day to help us keep clean and to enhance our looks. Almost everything we use to wash and beautify ourselves involves plant extracts of some kind. These could be gums, resins or waxes, cellulose or any of a great number of compounds obtained by a variety of methods. Often, however, these are oils, which are incorporated in their 'raw' state or, most commonly, in the form of derivatives (whose technical names disguise their origin) obtained by complex processing. By the time we have opened the door in the morning we will almost certainly have used dozens of different plants.

Long before the invention of the bar of soap it was the natural properties of the oils produced by various aromatic plants that led people to use them to banish body odours and unpleasant smells. During the Middle Ages, Britain and the rest of Europe relied heavily on the scented oils from herbs to alleviate the generally malodorous conditions that existed. While scented waters were sometimes sprinkled onto clothes, floors were often strewn with aromatic plants (some of whose essential oils have germicidal properties) both to mask bad smells and to deter fleas, lice and vermin. Such plants included meadowsweet, sweet rush, lavender, sage, thyme, camomile and hyssop, which released their fragrance as they were trodden on. Since it was thought that foul air caused disease, posies of fragrant flowers were also carried as a means of protection.

Before this use of herbs in Europe, however, the ancient Egyptians were using the oils from plants and fragrant flowers as part of their ablutions. The Greeks and Romans similarly anointed their bodies with perfumed oils, and used fine sand as an abrasive, scraping this off with a metal implement – the 'strigil' – to cleanse their skin. Well known for their public baths, the Romans placed lavender in their bathing water, not only to scent it but to act as a disinfectant. In fact, the scientific name for lavender, *Lavandula*, is derived from the Latin *lavare*, meaning 'to wash'.

Most of us use soap in one form or another to start the day. In Britain alone £203 million was spent on bar and liquid soaps in the year 2000. The base for British soap manufacture was traditionally tallow. Now, however, plant oils, particularly from the African oil palm (*Elaeis guineensis*) (see page 16) and coconut palm (*Cocos nucifera*) (see overleaf), are extensively used in the industry to produce the vast array of soaps that are on sale in supermarkets and stores around the world.

Today's household soaps are generally complex formulations which are likely to include synthetic detergents, hardening agents, anti-static agents, lather-boosters, water-softeners, preservatives, antioxidants, pigments and perfumes, some of which are made from petrochemicals.

The base of most toilet soaps, however, is provided by a mixture of animal fats and vegetable oils or their fatty acids, which are reacted with inorganic, water-soluble bases such as sodium or potassium hydroxide. A soap's characteristics traditionally depended largely on the proportion of one to the other, but technical developments and the discovery of a readily available plant oil with similar characteristics to tallow have blurred this distinction.

Previous page: Rosemary.
Opposite: Fragrant rose petals produce a scented oil used in perfumery and aromatherapy.

13

Tropical plant oils for soap

As a starting material, animal fat – mostly in the form of tallow obtained from beef and mutton – is still a major component of many soaps, but plant oils have become increasingly important. Two palms provide us with the bulk of the plant oil currently used: the African oil palm and the coconut palm. Several other plant oils can be utilized instead of or as well as these, among them soya bean, cottonseed and peanut. Which oils are chosen will depend on where the soap is made and on its price and availability.

The African oil palm is a particularly valuable tree since its fruits yield two different types of oil: palm kernel oil from the seeds and palm oil from the fibrous pulp that surrounds them. This palm is now one of the world's most important providers of edible and soap-making oils, yielding more oil per year per hectare than can be obtained from any other vegetable or animal source. Coconut oil – which also has other major industrial and edible uses – is extracted from the edible white flesh of the coconut, which is first dried either in the sun or artificially to form copra, the tree's most valuable product.

Coconut palm

■ The coconut palm (*Cocos nucifera*) can be found growing along almost all coasts of the tropics where the rainfall is greater than 1.5 m (60 in) per year. Tolerant of salty, sandy soils, it will grow on strips of land where most other crops could not survive. While many local varieties exist, which may take different forms, mature trees reach a height of up to 25 m (82 ft), developing slender, erect, but often curving, trunks, with a distinctive feathery crown of leaves at the top.

■ The fruits of the coconut palm – from which coconuts are extracted – appear from about the fifth to the seventh year onward (varying according to variety and growing conditions) and may be produced until the tree is about 80 years old. Green at first, they turn yellow as they ripen. The tall, branchless trunk of the palm can make harvesting very difficult, but skilled barefoot climbers often use ropes to help them ascend the trees and in some Southeast Asian countries monkeys are trained to do the job, throwing the fruits to the ground as they reach them. Alternatively, the fruits may be left to drop of their own accord or are cut from the trees by knives attached to long bamboo poles.

■ All parts of the fruit are useful. The thick layer of fibrous husk that lies beneath the outer skin, for example, can be combed out and sold as coir, an important material used for making ropes and matting. The object that we call the coconut lies within this husk and is actually the nut of the fruit.

The first soaps from plants

The use of soap in tablet form for washing the body did not become widespread in Europe until the mid-19th century, but its **manufacture** has a long and interesting history. Though their role has changed somewhat, plants have been essential to the soap-making process since the earliest times.

The first soaps, as we know them, appear to have been developed not for personal hygiene but for **cleaning wool and cloth**. Over 5,000 years ago, it is reported that the Sumerians were making soap mixtures by boiling fats and oils with alkaline solutions, obtained by steeping wood ashes in water. The Egyptians also made a kind of natural **washing soda** involving wood ash and animal fat. Though the Romans are known to have used a particular clay for cleaning fabrics, and possibly learnt the art of soap making from the Celts, the remains of what seems to have been a **soap factory** were discovered beneath the lava and ash of the eruption that smothered Pompeii in AD 79.

Little is known of the use of soap in Medieval Europe, but its manufacture appears to have grown at the end of the first millennium. One of England's **earliest centres** of soap making seems to have been Bristol, where, in 1192, the monk Richard of Devizes noted the number of soap makers and the unpleasant smells associated with their trade.

Soap making also became established early on in Marseilles, Venice and Savona, where later, some of the most sophisticated soaps, incorporating

Coconut palms like these in the Solomon Islands provide oil for edible and industrial use.

Although olive oil from the Mediterranean has been used for centuries, the large-scale move from tallow to tropical plant oils did not come about until the beginning of the 20th century, when the world experienced an acute shortage of animal fats. At this time a means of hardening or hydrogenating plant oils previously considered too soft or liquid for use in soap manufacture was invented in Germany, and the new vegetable sources were a great success. In comparison with tallow, plant oils produce soft, clear-textured soaps, which lather well and are less prone to cracking and have therefore been used in much higher proportions in the finer, more luxurious toilet soaps. The wide availability and cheapness of the major plant oils used in soap making have also made them very attractive to soap manufacturers.

Inside the nut's shell is the coconut 'meat', the edible, white, fleshy layer (the endosperm), and within this, a cavity partly filled with a watery liquid containing sugars, which is absorbed by the fruit as it ripens. 'Coconut milk' is made from the endosperm.

■ The coconut palm has a great number of uses, and has become a vital component of many local economies. Its trunks provide building timbers and the leaves material for thatch. Wood and leaves are also used as fuel, whilst the fruit provides food, drink and fibres. The most important commercial product of the coconut palm, however, is copra, the dried coconut flesh, which is the source of the oil widely used for the manufacture of soap. Coconut oil has many other industrial uses, including the production of numerous processed foods, such as margarine.

15

■ In Thailand – which produces over a billion coconuts a year – it has been reported that a small amount of kerosene mixed with coconut oil (1:20) is a good substitute for diesel fuel.

■ In 2000, nearly 3.3 million tons of coconut oil were consumed worldwide. The Philippines – where the coconut industry employs about 30% of the population – is the world's largest exporter of coconut oil. The biggest producer of coconuts was Indonesia, followed by the Philippines, but Malaysia, India and New Guinea are also major exporters. Though produced in small quantities, copra and coir are particularly important to many Pacific Island communities.

exotic perfumes and aromatic powders, were to be made.

By the 14th century, the famous Castile soap was being exported from Spain, manufactured using olive oil (instead of animal fats such as goat tallow) to give a distinctive, hard white soap. By the 17th century, different kinds of soap were being produced, and might incorporate fish oils, 'train' oil (from whale blubber) or tallow, or, for 'sweet soaps' fine enough to wash the skin, olive oil. Using soap to wash the body, however, was still a luxury. Some fortunate households made up wash-balls from a mixture of plant oils, shavings of Castile soap, distilled water and various herbs and spices. Ground almonds, raisins, breadcrumbs and honey were also sometimes added.

By this time, the ash from burnt wood (often beech), which had formed the alkali for many soaps, had been replaced by that obtained by burning sea plants, especially the spiky-leaved saltwort (Salsola kali), which grew commonly on seashores and salt marshes. This ash, known as barilla, was widely used by soap and glass makers until the late 18th century, when it was discovered that alkalis could be made from common salt. This discovery now gave the soap maker access to an unlimited supply of one of the basic raw materials. It was not until the Industrial Revolution, however, that increased wages together with mass-production methods gave ordinary working people the chance to buy a bar of soap with which to wash themselves.

THE REAL PRICE of a bar of soap

A product as unassuming as a bar of soap seems an unlikely candidate for association with human and environmental tragedy on a massive scale. The African oil palm (*Elaeis guineensis*) is now the source of the world's best-selling vegetable oil and is widely used in the food industry and for the production of oleochemicals in soap and cosmetics. In 1996 it accounted for 52% of exports of all plant oils and now represents 40% of the total global trade in edible oils.

Until the 1960s, the cultivation of African oil palm was largely confined to West Africa, where it is native and where it has long been used in the production of food, medicines and fibres. By the early 1990s, however, it was being grown widely in industrial plantations in the tropics, since its oils (palm and palm kernel oil) had been transformed from a commodity scarcely traded internationally to a lucrative agribusiness supplying an insatiable global demand.

With a host of different applications, from that of a lubricant in oil drilling, to the production of coffee whitener and face cream, the global palm oil industry (referring henceforward to both kinds of oil) has been extraordinarily successful.

Increasing consumption

Between 1993 and 1998, consumption of palm oil products increased by 32% while production increased by about 1 million tons a year. By 1997, around 6.5 million ha (16 million acres) had been converted to plantations, producing 19.6 million tons of oil (17.5 of palm and 2.1 of palm kernel); by 2002 this figure had risen to 23 million tons per year on some 10 million ha (25 million acres) worldwide. Major consuming nations include the UK, Germany, Netherlands, Spain and Italy, and India, Pakistan, the Middle East and China.

This demand has fuelled the rapid expansion of oil palm plantations using specially selected and cloned varieties in non-native regions of the world – chiefly Malaysia and Indonesia, which now produce 80% of the world's palm oil, and Papua New Guinea, the world's third largest exporter. Thailand is also a major producer and ambitious plans exist for the Philippines, Cambodia, India and the Solomon Islands. In Latin America, palm plantations are covering ever increasing areas in Ecuador, Colombia, Brazil and many other countries. As recent reports show *so far all this has only been possible through the massive destruction of both natural ecosystems (mostly tropical rainforest) and the livelihoods of thousands of people.

In numerous cases it is the direct theft of ancestral lands – mostly by setting fire to the forest –- that has enabled new plantations to be set up. In Indonesia many of the devastating fires of 1997, which shocked the world, destroying nearly 10 million ha (25 million acres) of forest are thought to have been started deliberately by major plantation owners. Such deeds were perpetrated to make it easier for land to be expropriated from locals.

Many Dayak communities, such as those in Kalimantan, have been obliged to witness the destruction of their forests and homes and also their traditional way of life. Relocated in inappropriate settlements, having perhaps been coerced with money or the promise of facilities that don't materialize, they have been obliged to become labourers, working under harsh conditions on vast new plantations. Complaints are reported to be either ignored, met with violence or provoke the mysterious outbreaks of yet more fires destroying traditional lands. In East Kalimantan, village leaders who tried to halt plantations in 1996 were

Many Dayak communities have been obliged to witness the destruction of their forests, homes and traditional way of life.

stripped, beaten and burned with cigarettes by Indonesian security forces. Other protesters have been shot. Linked to Indonesia's widely condemned 'Transmigration' project, and dominated by business conglomerates linked with the ruling elite, huge areas of Kalimantan and Irian Jaya are scheduled for conversion to palm oil plantations.

A valuable crop

Oil palm is a lucrative crop for domestic and foreign businesses and is particularly attractive to governments burdened by external debt. Cheap labour, cheap land, a short growing-cycle and a lack of effective environmental control add up to big profits. The promotion of the crop on a vast scale, however, reduces world prices (which may mean that some producers are financially ruined along the way), but stimulates consumption and ensures profits for those involved in marketing and reprocessing, often in other countries. African oil palm also attempts to compete with other oils whose prices are kept artificially low by American and European subsidy schemes, thus helping to ensure that the North will have a continuous supply of oil while producers in the southern hemisphere face economic risk.

With the IMF and the World Bank encouraging foreign investment in oil palm, many Western financial institutions and private banks are supporting this human and environmental tragedy, often stating that the plantations offer a solution to unemployment and even benefit the environment.

Diversity destroyed

Rather than the tree itself, it is, of course, the model under which it is grown that has caused so many problems. The replacement of rainforest with monocultural plantations is destroying an enormous diversity of plant and animal species. Studies in Indonesia and Malaysia show that 80–100% of rainforest animals cannot survive in oil palm monocultures. Those that do survive are often classified as pests as they feed on the trees. Species already endangered, such as the orang utan, Sumatran tiger, Asian elephant and rhino (in Indonesia), now face even greater threats.

Some good news

In January 2002, Switzerland's largest retail chain, in collaboration with WWF, became the first European retailer to commit itself to obtain all its palm oil from plantations that were not set up at the expense of tropical forests. WWF's Forest Conversion Initiative is working to ensure that expansion of palm oil (as well as soya) plantations is not a threat to 'high conservation value' forests.

* *The Politics of Extinction*, Environmental Investigation Agency (1998)

* *The Bitter Fruit of Oil Palm,* World Rainforest Movement (2001)

17

African oil palm plantation replacing rainforest in Cameroon.

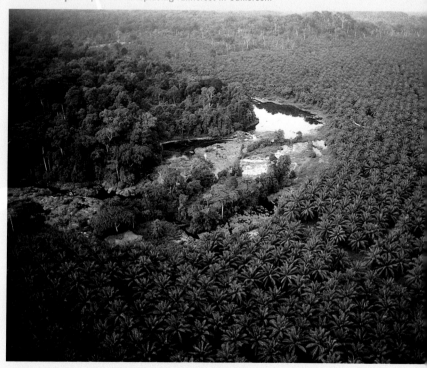

Dangerous suds?

Toiletry products – especially soaps – that look and sound good enough to eat, containing plant extracts often advertised as vital aids to a healthy skin, do not encourage us to consider what **dangers** might be lurking within them or how their ingredients may have been produced.

An area of increasing concern is the toxicity of some of the main ingredients, due to the involvement of a **huge range of chemicals**. Over 3,000 chemicals are currently registered in the EU for use in cosmetics and toiletries, some 15 or more being routinely included in any one product. It is reported that each year over **100,000 tons** of chemicals are incorporated into so-called 'natural cosmetics'.

Of particular concern are substances such as diethanolamine (DEA) and triethanolomine (TEA), which are commonly used in **cosmetic and toiletry products** in association with plant-derived and other ingredients.

Debate continues, too, about the use of sodium lauryl sulphate (SLS) and the related sodium laureth sulphate (SLES). A mainstay of the **shampoo industry**, SLS is one of the most common surfactants to be used in any product requiring suds (including bubble baths, shaving creams and toothpaste).

Used on their own, however, soaps made from plant oils dissolve very quickly and lather a great deal, so tallow has been essential in the production of mass-produced household soaps to give hardness. But palm oil has been found to be chemically very similar to tallow and can now replace all or part of it with only minor adjustments. Curiously, palm oil and palm kernel oil are very different in physical and chemical characteristics, though the latter is so similar to coconut oil in its fatty acid formulation that the two can be interchanged. Of the 17 major oils and fats in world trade, only palm kernel oil and coconut oil – known as lauric oils – have been considered the 'aristocrats'. This is because lauric acid is the major fatty acid in their composition (about 50%), which is more than 80% saturated. Good quality soaps require specific blends of fatty acids for the right balance of solubility, detergency and lather. 'Superfatted' soaps are rich in such fatty acids and these stabilize the lather, giving a creamy texture. Typical formulas for some household soaps are: 25% coconut oil or palm kernel oil and 75% tallow; or 20% palm kernel oil, 20% palm oil and 40% tallow. In recent years, however, rapeseeds have been genetically engineered to produce an oil high in lauric acid, which is already used in soap and cosmetics. This may become a major source in the future, ousting the 'tropical' oils.

Although glycerine is a by-product of soap manufacture, it is often washed away or removed (for resale to the cosmetics and food industry) during processing. It is, however, sometimes re-introduced to the soap to enhance its softness and appeal.

Plant oils that produce soft, clear-textured soaps, are used in much higher proportions in the finer, more luxurious products.

Glycerine (also referred to as glycerol) is a very versatile compound and has a wide range of industrial uses – as a humectant (moisture enhancer) in food, cosmetics and pharmaceuticals and as a lubricant, plasticizer and thickener. In the form of nitroglycerine it is the key ingredient in explosives such as dynamite (see page 259).

Plant oils such as rosewood, cedar wood and rose are often added to soaps for the scent they impart or to offset unpleasant smells in soaps that may still be labelled 'fragrance free', but many aromachemicals are now synthesized from turpentine oils taken from various species of pine tree.

Alternative soaps

Saponin is a generic term applied to a range of organic compounds that produce soapy, frothy solutions. Many plant species contain saponins and will produce appreciable amounts of **cleaning lather** when rubbed or boiled in water. In Europe, the plant most commonly used in the past was soapwort (*Saponaria officinalis*), also known as fuller's herb, latherwort, crow soap and soap root.

In Sudan and Chad, a plant oil used by the Egyptians some 4,000 years ago has been traditionally used as a type of soap. It is produced by crushing the **bark and seeds** of a small spiny tree, *Balanites aegyptiaca*, which yields a yellow oil, rich in saponins.

In Arabia and Somalia, some desert peoples clean their bodies by crouching over a **smouldering pot** of charcoal and aromatic herbs, gums and spices, their robes spread around them to catch the smoke and promote heavy sweating. The inhalation and fumigation of the body with smoke from the burning of fragrant plant materials is, in fact, one of the oldest uses of plants by man and thought to cleanse, regulate and heal the body.

Saponins derived from other plant sources find a major use in the food industry. They are **particularly useful** in the manufacture of beer and lager, helping to produce and maintain a good head of foam.

Soapwort (*Saponaria officinalis*)

This herbaceous perennial, which grows to a height of 1.5 m (5 ft), was once **widely used** in Europe for washing, as soap was rare and costly. One early herbal describes the 'great scouring qualities' of the leaves, which 'yield out of themselves a certain juice when they are bruised which scoureth almost as well as sope'. When **rubbed or boiled** in water, the leaves and root of soapwort produce a green soapy solution, which is still used for the cleaning of very old and delicate fabrics such as tapestries. Soapwort is incorporated today in some gentle skin-cleansing products.

Balanites aegyptiaca

Indigenous to the Nile valley, all parts of the small tree *Balanites aegyptiaca* can be **soaked in water** to extract saponins and the resulting solution used for washing clothes. The edible oil contained within its large, fleshy fruits can also be processed to make a soap with **good foaming** and detergent qualities. This oil was used as a base for cosmetics and perfumes in ancient Egypt, but the main use of the tree was, and still is, as a food – its nutritious fruit, eaten fresh or dried, contains 20–30% protein and 30–60% oil. The flowers, seeds and leaves (used as nutritious **cattle fodder**) can also be eaten.

B. aegyptiaca has great potential for the people of Africa's Sahel region, since the tree yields an impressive range of foods, raw materials, and chemical and medicinal extracts for domestic use and commercial production. The most important future use of *B. aegyptiaca* may be **medicinal**. The root, bark, fruit-pulp and seeds have all been found to be lethal to fish and fresh-water snails, which act as an intermediate host for bilharzia parasites. The presence of an extract of the tree in water can also kill minute free-living forms of these **parasites**, which affect millions of people in tropical countries each year.

The soap tree (*Quillaja saponaria*)

A tree native to Chile is one of the major sources of industrial saponins. The foaming properties of quillaja bark were first recognized by the **Mapuche Indians**, who used extracts in water for washing hair and clothes. For decades, however, the bark has been **commercially harvested** and its extracts used, among other things, as foaming agents in drinks and toiletries, and emulsifiers in foods.

The tree is now **widely grown** in southern California, but in central and southern Chile, over-exploitation of the bark has caused the serious depletion of what were once extensive forests. Having been felled – in recent times – at the rate of 50,000 trees a year, mature individuals, which may reach 30 m (98 ft), are now rare in the most accessible areas, and much bark is taken from **shrub-like trees** reaching about 10 m (33 ft) in height, which grow back from old stumps. In an effort to harvest this resource sustainably, saponins are now also being extracted from pruned trees.

Quillaja has a wide range of industrial applications, from improving the texture of **'slush' type ice drinks** (food uses account for 70% of production) to helping distribute the silver halide crystals evenly in photographic film.

In shampoo, quillaja can be used as a secondary surfactant to improve foam stability, and is said to **help combat** dandruff and hair loss. An infusion of the bark applied to the scalp is also claimed to assist hair growth.

Sunflower seeds yield an oil with many uses, from toiletries to margarine and bio-diesel fuel.

Concern about the chemicals involved in the manufacture of soap and a large number of cosmetics has led to the production – sometimes by hand – of more 'natural' soaps, generally by small companies. Such specialist, luxury soaps are often based on saponified, cold-pressed olive and other plant oils such as almond, avocado, sunflower or castor. They also contain natural colours and fragrance and entice us with an enormous range of fragrant essential oils and plant products, from orange flowers and seaweed to hemp oil, added for their various properties. These soaps are generally much less drying to the skin, as the glycerine that is produced during saponification is stirred back in.

Despite the increasing popularity of soap-free synthetic detergents and gels (which are made from petroleum derivatives), the appeal of the traditional bar of soap looks set to stay.

Shampoo and conditioner

Just as most households never lack a bar of soap, shampoo, in some form or other, is rarely absent from our bathrooms. More time and money is probably spent caring for our hair than any other part of the body. Once again, plants are indispensable. Though their involvement is disguised by technical terminology, plant products generally provide the base for the cleansing and foaming agents of the shampoo itself and add thickness as well as fragrance to the product.

Shampoo is a more recent introduction than the bar of soap, but aromatic plants and flowers have been used since the earliest times to beautify the hair. Aristocratic Chinese ladies, like many Eastern women, are said to have oiled their hair and bound into it strong-scented flowers such as frangipani and jasmine, which perfumed it long after the flowers had died. A commonly used hair oil in Polynesia today is monoi tiare, which is made of coconut oil scented with fresh gardenia flowers (*Gardenia taitensis*), to produce a fragrant shine.

Appealing plant names and the promotion of extracts – from nettle to nasturtium – and oils, such as jojoba or apricot, which claim to beautify and tame our hair often encourage us to buy shampoos. But it is not these well-advertised ingredients – often making up less than 1–2% of the volume of the product – that are generally responsible for cleaning it. Although a number of European plants, including rosemary and camomile, have a long history of use for conditioning the hair, their efficacy in most modern shampoos is likely to be eclipsed by the other ingredients.

The main ingredients of today's shampoos are water (80–90%), detergents and perfume, to which are added thickeners, lather-boosters, antistatic and detangling agents, colour and preservatives. Since soap would leave a scum on the hair and also dull it, the cleansing agent in most shampoos is usually a soluble synthetic or non-soap detergent. Although the starting materials of this detergent, fatty alcohols, may be derived from petrochemicals or tallow, they are often made from natural plant oils. Coconut and palm oils are likely to be used. To improve detergency and the stability of the foam, 'secondary surfactants' such as sodium lauryl sulphate are also present in most shampoos. Chemically, these are fatty acid derivatives, which are again mostly taken from coconut and palm oils. Some more recent additives to shampoos are compounds – generally referred to as AHAs (alpha-hydroxy acids) – which occur naturally in fruits. Helping loosen adhesion between skin cells on the scalp and acting as useful humectants, they are said to be good for people with dry, flaky scalps.

The smell of our shampoo is an important part of its appeal. Since, for example, a strong mental association exists between lemon juice and the removal of grease, shampoos formulated for greasy hair are often given a 'lemon' or other 'citrus' smell. Whilst many of these fragrances are chemically synthesized, the starting materials used to make them are, as in the case of soap, often plant-derived.

Thickeners

The colouring and preservatives in most shampoos are almost entirely synthesized from minerals or mineral oils, but some very important components of shampoos and liquid soaps – the thickeners or viscosifiers – are usually derived from plant materials. Coconut fatty acids are the starting materials for an ingredient such as 'cocamide DEA', which acts as a thickener and foam stabilizer, but cellulose, the chief component of the cell membrane of all plants, has become a major raw material for this purpose. Plant cellulose is the raw material behind a huge, worldwide industry that produces a range of thickeners and suspending agents essential to a large number of cosmetics, toiletries and detergents, and also to the manufacture of a great range of convenience foods and pharmaceuticals as well as many other industries. Many hundreds of thousands of tons of cellulose are used each year to help make products as diverse as fish fingers and hair mousse.

The lac tree

■ In India the lac tree (*Schleichera oleosa*) is grown for its seeds, which contain the edible fat that was the original source of Macassar oil. This fat is still used today for hair oils and is burned as a source of illumination. The fruit pulp can be eaten raw, the unripe fruits are consumed as pickles and the young leaves as a vegetable.

■ The hard red timber and bark also have local uses. Though the oil is no longer used in Britain, rectangular cloth coverings for the backs of chairs, originally designed to protect them from macassar oil, and thus called 'antimacassars', are still found in many homes.

As an ingredient of shampoo, cellulose, which has been chemically treated to produce a substance that will dissolve in water (a water-soluble polymer), is added in the form of colourless granules. Though present in relatively small quantities in relation to the total weight of the product, polymers can be made to give a range of consistencies, from very thin liquids to thick gels, which help bind the ingredients together. They do not affect the shampoo's clarity but help make the foam produced heavier, creamier and more stable. By effectively wrapping themselves around dirt particles, they also help dirt to be washed away.

Bath oils, skin fresheners and cleansers, aftershave lotions, deodorants and anti-perspirants may all contain cellulose as a thickener. In colognes and perfumes, water-soluble polymers often act as binders, holding the odour components to the skin to lengthen the time that they will be effective.

To produce these polymers (such as carboxymethyl cellulose [CMC]), plant cellulose in the form of wood pulp and/or cotton linters (the short stiff hairs that remain on the cotton seed after the removal of the longer cotton fibres) is first reacted with an alkali and other chemicals and then purified, dried and powdered. Natural cellulose does not dissolve in water, but these changes to its chemical structure will make its derivatives soluble in either hot or cold water or alcohol. Whilst wood pulp generally produces the lower-viscosity polymers, cotton linters are almost pure cellulose and therefore produce a higher viscosity.

Sources of wood pulp are extremely diverse and involve a great range of hardwoods and soft woods from plantation and natural forests around the world, but are likely to include a high percentage of eucalyptus (see page 83).

Hair conditioners

A great number and variety of plant materials – and some highly unpleasant animal ingredients – have been used throughout the ages to strengthen, thicken and perfume the hair. Besides relatively innocuous gums and woods, all kinds of alcohol and the ashes of burnt bees and goats' dung were once used. Puppy fat and bear grease have also been recommended in the past. In contrast, the modern use of lanolin, the fat extracted from sheep's wool, in conditioners and shampoos seems a fairly civilized alternative. Most famous in Victorian times for the well-groomed gentleman's hair was Macassar oil. It was based on the oil obtained from the seeds of the lac tree (see left), also known as the Ceylon oak, kussum tree and Malay lac tree, which is native to India and parts of Asia.

The primary active ingredients in many modern hair conditioners may be extracted from tallow or plant oils. They often contain

Eucalyptus plantation, Brazil.
Woodpulp, often made from eucalyptus, is a major source of cellulose used, after pro-cessing, as a thickener in a great variety of products. These range from shampoo and toothpaste to glue and household paint.

water-soluble polymers made from cellulose, as do many hair gels and mousses used for grooming and styling, which leave a very thin, clear, flexible and non-sticky film on the hair shaft after drying.

Selected plant oils such as galbanum, distilled chiefly from the aromatic gum resin produced by *Ferula gummosa*, which is native to Iran and able to give 'shine and hold', are present in some conditioners in small quantities. Protein too is present and is said to give a strong protective coating to the hair shaft and add body and lustre. The protein may be derived from animal fat but it is also often made from soya bean oil (see page 147), one of the cheapest and most widely used oils in the cosmetic industry.

Alginates from seaweeds are present in some conditioners and styling products as they impart a smoothness and gloss to the hair, while carrageenan (a natural gelling substance) is used as a fixing agent in some hair lacquers and balms.

Bathing with plants

Lavender

■ Familiar as a garden plant, lavender (*Lavandula* spp.), with its distinctive spikes of fragrant flowers and aromatic leaves, is a sun-loving shrub, native to the Mediterranean region. Some 30 lavender species are recognized but *L. angustifolia* (English lavender), *L. latifolia* (spike lavender) and the hybrid *L.* x *intermedia* (lavandin) are those most widely grown commercially, chiefly for their aromatic oils, much used in perfumery and aromatherapy, but also for their flowers, which retain their fragrance when dried. Hardier than English and spike lavenders and able to produce a larger quantity of oil, recent years have seen an increase in the cultivation of lavandin.

■ A cheaper, lower-quality oil with a more camphorous smell disdained by the devotees of true lavender oil, lavandin has been steadily replacing lavender oils in many household toiletry products. Global annual production – mostly within Europe – is currently estimated to be about 1,000 tons of lavandin oil and 200 tons of true lavender oil.

A range of plant oils and extracts are also available at bath-time to soothe and relax us. As in the case of soaps and shampoos, coconut and palm oil often provide the cleansing and foaming bases of bath and shower gels and, in the form of fatty alcohols, act as solubilizers for numerous perfumed bath oil constituents. Citric acid obtained from sugar cane and sugar beet may be added to help the product match the pH value of the skin's outer surface, and molasses is sometimes the source of substances used to keep the skin supple and moisturized. But one particularly useful plant for bath products is the castor oil tree (see page 260).

Very few oils disperse quickly in water, but castor oil, sometimes sold as Turkey Red oil, not only disperses but does not mark the bath. Native to India and tropical Africa, but now widely distributed in many tropical and temperate regions, this shrub was known to the Egyptians, and the oil pressed from its seeds was used as an unguent base as well as a fuel for lamps. Today, castor oil is an important ingredient of many bath preparations, helping to soften and lubricate the skin.

Of all the plant oils and extracts that are used for the bath, lavender (see left) is one of the best known and loved. The centre of a large and very valuable industry, lavender oil or extract has been popular as a bath ingredient since the days of the Romans and continues to make its way into many of the most widely used toiletries and scents. At one time Britain was famous for its lavender fields and Mitcham in Surrey was the centre of the industry until the early 20th century. Lavender is still grown commercially in Norfolk and in other parts of the UK, but most of Europe's supply currently comes from Bulgaria, France (in particular Provence) and countries of the former USSR.

A fragrant oil is to be found in all parts of the lavender shrub, but the essential oil of lavender is extracted only from the flowers and flower stalks, usually by steam distillation. Lavender is often sold dried as an ingredient of potpourri and in sachets for scenting clothes and bath water. Besides scenting the water, it is said to stimulate the action of the skin's pores, relax muscles and soothe the joints. As the Romans discovered, lavender's antiseptic properties also make it a useful addition to the bath.

Long valued as a perfume ingredient and in medicine, the impressive list of therapeutic properties attributed to lavender oil have promoted its wide use today in aromatherapy. Indeed, considered the most versatile

Familiar as a garden plant, **lavender** (*Lavandula* spp.) can be dried easily at home for use in sachets to scent clothes as well as bath water.

25

Sponge gourd

■ Two species of *Luffa* are widely cultivated, but *Luffa aegyptiaca* is the one most widely used in commerce. The genus name and our own term 'loofah' are derived from an Arabian name for the plant, which is thought to have originated somewhere in Asia.

■ *L. aegyptiaca* has deeply lobed leaves and large bright yellow flowers, which open in the morning sun. Its fruits are cylindrical and smooth, growing to some 50 cm (20 in) in length. Green when young, they become a yellow or tan colour when mature.

■ In Japan, which has been growing the gourds commercially since the 1890s, the vines are trained over trellises to give maximum exposure to the sun, and 20–25 fruits are allowed to mature per vine.

■ To extract the loofah 'sponges', the fruits are placed in tanks of running water until the outer walls disintegrate, and further washing removes any seeds or pulp still clinging to the skeletal fibres. Other processing methods include freezing or using boiling water. The sponges are then left in the sun to dry and bleach – a process often completed with hydrogen peroxide – before they are graded and packed for export.

■ Various varieties of sponge gourd are grown and in India and parts of Southeast Asia some are eaten as vegetables when young and tender. In some areas, an edible oil is also extracted from the seeds during the drying process.

of essential oils, and said to be able to make any other blend more dynamic in its action, some value it above all others. The beneficial properties of lavender oil include helping to calm as well as stimulate the mind and body (aiding sleep yet increasing alertness), and balancing the mental and physical energies. Acting as a natural anti-depressant, a cup of lavender tea is recommended by some to help relieve anxiety. One of the few oils that can be applied neat to the skin, a drop of lavender oil applied to the temples and hair line is said to be beneficial to headache and migraine sufferers, while it is also highly valued for its ability to help heal scars.

Lavender is said to boost lymphatic circulation while firming and tightening the skin's surface cells, and is now incorporated in a plethora of skin products designed to help alleviate the stresses of modern life – from sun-block and shower gel to shaving cream and body lotion.

Loofah

Commonly found residing in bath racks all over Britain and looking like the dried form of some strange sea sponge, there is one plant – or rather plant fruit – that helps us with the practicalities of bathing. This is the loofah (see left), the bleached skeleton of a tropical fruit – the sponge gourd – a large climbing plant native to Asia. Removal of the papery rind of the mature fruit reveals the familiar fibrous network, which we can use to scrub ourselves. The main commercial supplier of these gourds is Japan, where the juice extracted from the plants' stems is highly prized for softening the skin. It has also been used as an ingredient of toilet water and medicinally for respiratory complaints.

Sponge gourd skeletons have also been used for cleaning cars, glassware and kitchen utensils and as a packing material. Until the Second World War, the USA imported large quantities from Japan, some for cleaning purposes, but 60% for use as filters for steam engines and diesel motors. With the outbreak of war, production was started in South America, and Brazil now features among the main producing countries.

Talcum powder

If we use talcum powder we are likely to be using plants again. Though the traditional ingredient of talc (from the Persian *talq*, a soft greasy powder first introduced into Europe in the 16th century) is powdered hydrated magnesium silicate, or mica, the term talcum 'powder' is used rather loosely today and a number of other substances may be included. Calcium carbonate (precipitated chalk) is commonly used and talcum powder may also contain finely powdered maize or rice starch. Because talc in its natural form may be carcinogenic, a move away from this traditional ingredient in favour of vegetable powders has led to the development of various alternatives, especially for babies.

Toothpaste

Having bathed or showered with the help of plants, how do you fancy brushing your teeth with wood pulp? If you use a standard toothpaste, you will probably be using wood pulp twice a day already. Sodium carboxy-methyl cellulose or cellulose gum, derived, as we have seen, from wood pulp and cotton linters (page 22), is used in greater quantities in toothpaste manufacture than in any other cosmetic or toiletry item. Though it may form only 1–2% of the total weight of the paste (much of which is comprised of abrasives), it holds the other ingredients in a free-flowing gel that will dilute easily in water. Cellulose also suspends colour, helps the toothpaste keep its shape when squeezed onto the brush, yet disperse evenly in the mouth, and enables it to be rinsed free of the toothbrush after brushing.

This is not the only plant-derived ingredient in toothpaste. Humectants, which prevent the paste from drying out and help preserve its flavour, are traditionally made from glycerine, which may be derived from plant oils, or sorbitol, made from wheat or maize starch. Xylitol, found in many fruits and vegetables, but often produced commercially from the bark of silver birch trees, is a less common, but very useful, humectant, said to help reduce the incidence of decay. These humectants, which often comprise about one-quarter of the toothpaste's weight, also act as sweeteners. Xylitol, the sweetest of all bulk sugar substitutes (and now widely approved for use in foods, pharmaceuticals and oral hygiene products), but with 40% fewer calories than sugar, also produces a cooling sensation as it dissolves in the mouth. The foaming agents or detergent (such as sodium lauryl sulphate) will often have plant oils, especially coconut or palm, as their base, whilst alginates and carrageenans from seaweeds (see page 23) may be added as binding agents and thickeners, helping the product to retain its moisture content.

The presence of detergents and other ingredients requires some strong flavourings to mask their taste. If one set of plants is widely associated with toothpaste, it must be the mints. Though some synthetics are used, the minty flavour of most toothpastes is derived from natural mint oils – chiefly peppermint and spearmint. Four different species of mint are commonly used: *Mentha X piperata* (peppermint); *M. arvensis* var. *piperascens* (Japanese mint); and two varieties of spearmint, *M. spicata* and *M. gracilis* (see right).

Much of the peppermint and spearmint oil used in the world today comes from large-scale producers in the USA, and in particular the

Mint

■ The mint family (*Labiatae*) contains a number of important herbs, many of which have a very long history of use. The flavour and aroma of mint plants (*Mentha* spp.) come from aromatic essential oils, which are secreted by glands in the leaves and stems.

■ Extracted by steam distillation, peppermint oil (from *M.* x *piperata* – a hybrid of *M. aquatica* and *M. spicata*) and spearmint oil (from *M. spicata* and *M. gracilis)* are used as flavourings for toothpaste and chewing gum. Peppermint oil is also often found in mint-flavoured sweets, chocolate and liqueurs, while spearmint leaves are commonly made into the jelly or sauce often served with roast lamb.

■ More bitter and astringent than peppermint oil, which it is sometimes sold as, Japanese mint oil – extracted from *M. canadensis* – is used for the production of menthol.

Vanilla

■ The vanilla plant (*Vanilla planifolia*) is a tropical climbing orchid native to Mexico and Central America (where it was used by the Aztecs) as well as the Caribbean, but is now grown commercially in Madagascar, Indonesia, Mexico, the Comoros Islands, Tonga, Reunion and French Polynesia. Of the 4,403 tons of vanilla beans produced in 1999, 2,102 tons came from Indonesia and 1,815 from Madagascar.

■ One of the most labour-intensive agricultural products, it takes 18 months–3 years, from planting an orchid cutting, for it to produce flowers. These flowers, which would otherwise die within a few hours, are then pollinated by hand. The resulting seed pods – which are the vanilla beans – must mature on the vine for nine months before harvesting. It takes several more months for the pods to undergo a curing process during which time the distinctive scent and flavour compounds are formed. Over 250 organic components – including vanillin, the main substance responsible for vanilla flavouring – create the distinctive flavour and fragrance that we know as vanilla. These compounds are extracted by percolating an ethanol mixture over the chopped, fermented pods, thereby dissolving them out.

■ This very time-consuming production process, as well as its scarcity relative to demand, has made vanilla, which is the world's most important food flavouring and essential for many perfumery purposes, the second most expensive spice after saffron.

states of Idaho, Oregon and Washington (these states accounted for 65% of world spearmint oil in 1992), though China and India are also major producers of these oils. India is the largest producer of Japanese mint oil – about 75% of world production. Though this oil is often sold as or mixed with peppermint oil, it is regarded in the trade as a low-quality substitute.

Mint accounts for over half of the flavouring materials used in toothpastes in the Western world each year, but it is not the only natural flavouring from plants. Vanilla extract is very common in many toothpastes (see left) and others are flavoured with aniseed, fennel, sage, eucalpytus or cloves as well as citrus oils. Some alternative toothpastes, which avoid the use of fluoride, detergents, synthetic fragrance, colours and preservative, incorporate ratanhia extracts (from the root of a *Krameria* species) for their astringent properties, as well as calendula, myrrh and camomile.

Before leaving the subject of toothpaste, it is worth remembering that our toothbrushes may also owe their existence to plants. Many moulded plastic toothbrushes are made from cellulose acetate derived from wood pulp reacted with various chemicals.

If, after all our chewing or brushing, we still need dental care in the shape of false teeth, plants may still be helping us. Gum karaya tapped from the trunk of the large deciduous tree *Sterculia urens*, native to India and Sri Lanka, is an important dental fixative in the West. Its special adhesive properties, as well as its resistance to enzymatic and bacterial breakdown, have led to its wide use in dentistry, helping to keep false teeth, veneers, plates and crowns in place. Gum karaya has other important uses as a laxative and as a stabilizer in a variety of processed foods, including frozen desserts, salad dressings and aerated dairy products.

Wood for teeth

Not everybody cleans their teeth with a toothbrush and toothpaste. In parts of **East Africa**, chewing sticks are used to keep teeth clean and help prevent mouth infection. At least two of the *Diospyros* species involved (*D. usambarensis* and *D. lycioides*) have been shown to contain **potent** antifungal compounds. Also in Africa, as well as parts of India and the Middle

East, the woody stems of *Salvadora persica*, the **toothbrush tree**, have been used for centuries to help clean teeth and gums. When chewed, the stems release juices that appear to have a protective anti-bacterial action, helping to offset decay and **gum disease**. An extract from the toothbrush tree is now found in some toothpastes sold in Britain.

■ Vanillin is now synthesized for commercial use on a large scale, sometimes from turpentine or wood pulp. A bacterium found in a British soil sample has recently been discovered to be able to transform a chemical (ferulic acid) found in bran as well as fruit and beet pulps into vanillin.

■ An unrelated study, meanwhile, has found that vanilla-scented patches attached to the skin can significantly reduce the appetite for chocolate and sweet foods and drinks.

29

Vanilla: The delicious aroma and subtle flavours produced by natural vanilla pods are still unmatched by man-made substitutes.

Cosmetic creams and lotions

Almonds

■ Almond oil is one of the most important of the fine cosmetic oils, and is pressed from the kernels of the ripe fruit of two varieties of almond. *Prunus dulcis* var. *amara*, the bitter almond, is the main source of the almond oil, which is used for flavouring as well as cosmetics, but the oil from the sweet almond (*P. d.* var. *dulcis*), which is grown for its delicious nuts, is also used in skin preparations.

■ The oils are light and make very good emulsifiers and emollients, softening and smoothing the skin. Though native to southwest Asia, almond trees are grown extensively in Mediterranean countries, especially Italy and Spain, and cultivated on a large scale in California, China and Iran. The distinctive almond smell and taste is largely due to the presence of prussic acid in the almond kernels.

30

In the race to win customers from a rushed and polluted world, the idea of good health and correspondingly beautiful skin is often sold to us with the powerful imagery of plants. In a huge and fiercely competitive market, a vast array of complex cosmetic preparations, often in expensive packaging, is available to help us to enhance our looks, and in particular to smooth, protect, enliven and – miracle of miracles – 'de-age' our skin. Most of these products, including those, of course, with simpler or more 'natural' ingredients, produced with minimal processing, rely very heavily on raw materials from plants and promote them as active beauty aids.

Some of the basic plant-derived ingredients used in cosmetic creams and lotions today – in particular oils, gums and waxes – as well as disparate items from other sources, such as beeswax, honey and lanolin, have been used for centuries, but originally in much simpler ways. The first recipe for cold cream, for example, as recorded by the Greek physician and teacher Galen in the 2nd century, involved beeswax and oil of roses. Both of these are used today for various cold cream formulations, though the very expensive oil or attar of roses (see page 44) is often replaced with cheaper vegetable oils.

Plant oils certainly have a very ancient pedigree where beauty treatments are concerned. Olive oil perfumed with saffron and other aromatic plants was used by the Egyptians to keep their skin supple, and later the Romans used pumice stones to rub their bodies with this much-prized lubricant. Though olive oil is still found as a cosmetic base today, cheaper oils generally replace it.

Sweet almond (see left) and grapeseed oils are probably those most widely employed in cosmetic preparations today but natural oils suitable for cosmetics are constantly being discovered. Mineral oils are also widely used but plant oils, which may be refined and deodorized to give better colour, odour and stability, provide the starting materials of many modern creams and lotions, often as the source of a great range of derivatives.

The idea of good health and correspondingly beautiful skin is often sold to us with the powerful imagery of plants.

Many vegetable oils are triglycerides, that is they have three fatty acids attached to their glycerol molecule. The fatty acids most commonly occurring in vegetable oils are palmitic, stearic and oleic, and these can be separated or further converted into a number of other compounds. Reacted with fatty alcohols, they form esters, important in the manufacture of perfumes and flavourings.

Fatty alcohols obtained from plant oils such as soya bean, linseed, safflower, rapeseed, peanut, palm and coconut are often added to emulsions to help adjust viscosity. The esters of these fatty alcohols also act as emulsifiers, combining substances that cannot normally be mixed, such as water and oil, and evenly distributing solid particles in a liquid medium.

A large number of other, more expensive, plant oils can be added (in their whole forms) to cosmetic preparations for their own particular characteristics. Some of the most popularly used include almond, avocado, apricot and peach nut. Even the humble carrot may be pressed – quite literally – into service, yielding a yellow oil that is said to soothe chapped skin and combat dry hair. Other oils recommended for their therapeutic properties for sore, damaged or bruised skin include sea buckthorn, St John's wort, marigold, rose hip, rice bran and wheatgerm.

Avocados for sale in Brazil: A food and source of cosmetic oil.

Argan oil: saving trees, helping people

A project that is helping to protect the endangered argan tree (*Argania spinosa*) of Morocco is also helping meet the needs of **Berber families** in rural Morocco by promoting and developing the oil obtained from its seeds.

Found only in the arid Souss region in the southwest of the country, the long-lived **argan tree** has for centuries been the source of a wide range of valuable materials. A precious oil is pressed from the **seed kernel** within the fruit; its wood is used for joinery and – along with argan nutshells – for fuel; its leaves and fruit pulp are used as forage, and the residue from pressed kernels as a food for livestock. Practically the only natural resource for the region's local people – and playing an important **ecological role** – it is estimated that the production of argan oil, as

well as other activities connected with the tree, provides support for around 3 million people.

For food use, **argan oil** is an important ingredient of Berber cuisine and is pressed from the roasted kernels, while the oil pressed from the raw kernels is highly valued for its **moisturizing, healing and tonic properties** for the skin, hair and nails.

The overexploitation of the argan forests, however – with much damage being done by goats who eat leaves, branches and shoots – has led to the loss of one-third of the original area in the last century. Threatened also by **extreme drought** (a facet of the desertification of the Mediterranean region as a whole), the area was designated a World Biosphere Reserve in 1999. While programmes of reafforestation

and environmental protection have also been initiated, a plan first drawn up in 1995 has led to the formation of a number of **cooperatives** dedicated to the sustainable production of argan oil.

One such cooperative – Amal, which particularly benefits widowed or divorced Berber women, for whom life is particularly hard – processes about 300 tons of fruit each year. As well as providing a reliable income, the cooperative is helping to demonstrate the importance of protecting the trees and provides an alternative source of food for **domestic animals** in the form of the fruit pulp and processed kernels. While 80% of the coop's output is sold locally, a proportion of the **cosmetic oil** is sent to the main Moroccan cities and is also shipped to the cosmetics producers of Europe and the USA.

Moisturizing

With our desire for an ever more youthful appearance, a large proportion of the skin preparations that we buy are intended to revitalize the skin by moisturizing it and replacing lost elasticity. Whilst most plant oils will help soften it, macadamia nut oil (which is extracted from the fruit of *Macadamia tetraphylla* and *M. integrifolia* trees native to northeast Australia) has been of particular interest in the past. With a fat content of around 80%, the macadamia nuts contain very high levels of palmitic and oleic acids, components of the skin's epidermal fluids, said to be responsible for its softness, suppleness and water-retaining capacity.

The amount of lipids and palmitic acids decreases in women rapidly after the age of 20, and so this nut oil is one of those that enjoyed a spell of publicity as a suitable component of creams intended to improve the appearance of ageing skin, promoting extra elasticity. Other 'exotic' oils now being incorporated into some skin preparations for their emollient and other properties include baobab and oysternut from Africa, cohune from Guatemala and shikonin and artemisia seed oil from China.

Cocoa butter, a cream-coloured waxy solid pressed from the shells or husks of the seeds contained within the cocoa pod (*Theobroma cacao*), is another useful substance, often employed as a lubricant and moisturizer. Melting at body temperature, cocoa butter gives the skin a silky finish. Artificial cocoa butter, which is often made from palm and palm kernel oils, is used for the same purpose and a traditional, cheaper substitute is sheanut butter. With very similar chemical and physical properties to cocoa butter, and widely used in confectionery and bakery products, sheanut is extracted from the dried roasted nuts of the West African tree *Vitellaria paradoxa*, which is harvested from the wild.

Also derived from plants and acting as a humectant, especially in cosmetic creams, is glycerine (see page 18). Lecithin, mostly derived from oil seeds, is another humectant and emollient. About 95% of the world's lecithin comes from soya beans, in a process removing the phosphorus salts from their oil. Peanuts and maize are also used and sunflower and rapeseeds are being developed as new sources. Lecithin is added to cosmetic creams, to act as an emulsifier, foam stabilizer and anti-oxidant, stopping the plant oils from turning rancid. It is also widely used in edible products for this purpose. Liposomes, promoted during the 1980s as the fashionable de-ageing ingredients of some face and body creams, are also made from lecithin. Acting as carriers for other molecules and therefore assisting in the dispersal of active ingredients to the skin, they have been claimed to help slow down the ageing process.

Kelp forests

■ Amongst a number of very large brown algae found off the Pacific coast of North America and generally known as kelps, *Macrocystis* species grow together in dense stands called kelp forests and are amongst the largest of all the seaweeds. Able to grow as much as 50 cm (20 in) a day, these giant seaweeds, such as *M. pyrifera*, can reach up to 65 m (213 ft) in length.

■ Rooted to the seabed from a depth of 6–30 m (20–98 ft), leaf-like 'blades', as they are known, grow continuously along the length of the plant. Gas-filled bladders called pneumatocysts, located near the base of each blade, provide buoyancy and keep the kelp plant vertical in the water.

Thickening and stabilizing

Not to be forgotten, and playing an indispensable role in skin preparations and most other cosmetics, are the thickeners. As we have already seen (page 21), plant cellulose provides the raw material used to make an important range of polymers, which have a thickening function, and it is frequently present in the beauty products we buy.

Other naturally occurring plant materials are also widely used. They include gums, tapped from various trees or produced from seaweeds, and starch, often provided by maize or potatoes.

Though diminishing somewhat in their use today, gum arabic (see page 280) and tragacanth may both be found in cosmetics, as well as locust or carob bean gum and guar gum (see right). These thicken and stabilize creams and lotions, adding spreading properties to the product and a smooth, protective coating to the skin. They are also used in a large number of other industries, from food processing to oil drilling.

Gum arabic, the most widely used of all the natural gums in industry as a whole, and still important in the food, printing and adhesives industries, makes an excellent adhesive in face masks, helping to keep the liquified ingredients from dripping down the face.

As well as these terrestrial plant gums, which have some of the properties of water-soluble polymers, gums and alginates from seaweeds are often used. The red seaweeds are particularly valuable as they yield a group of natural gelling substances (or hydrocolloids), most of which cannot be replaced by synthetics. One of these is carrageenan, which has gelling and viscous properties that are very useful for 'oil-in-water' emulsions. Different kinds of carrageenan, with different properties are derived commercially from a wide range of red seaweeds including *Chondrus crispus* and *Gigartina stellata* (both known as Irish moss), *Euchemia* and *Ahnfeltia* spp. Although mostly used in the food industry, carrageenan is a useful cosmetic thickener for a variety of products from sunscreens – in which it provides a cooling film, active against the sun's rays – to face masks, in which it forms an absorbent base. Carrageenan and agar (which are often derived from gelidium and gracilaria seaweeds) may also form a base for perfumed deodorant sticks.

The larger, brown seaweeds yield alginates, a wide range of products derived from alginic acid. For cosmetics these are very useful seaweed derivatives as they can form a base for creams, gels, hair sprays and colorants – to name but a few – and are often added to stabilize emulsions or foams. Able to form protective films, alginates can be mixed into barrier creams and have a softening effect. In cosmetic creams they are also very useful for holding the ingredients together so they can be absorbed by the skin. Brown seaweed species that are commercially harvested for the production of alginates include

Guar gum

■ Guar gum – also called guaran, and often labelled E412 on product packaging – is obtained by processing the seeds of the leguminous shrub *Cyamopsis tetragonolobus*, in which it acts as a food and water store. The endosperm (plant tissue that provides food for the plant's developing embryo) is milled into a flour, which is then reconstituted for use.

■ Sold in the form of a white to yellowish powder and in various grades, guar gum is an emulsifier, thickener and stabilizer that has been approved for use in cosmetics and a wide range of foods and pharmaceuticals.

■ The plants are grown extensively in parts of India and Pakistan, and in hot, dry areas of the USA, with smaller areas in Australia and Africa. Jodhpur City in Rajasthan is currently the world's most important processing centre of guar gum, contributing around 40% of the world supply. While much of this is used in textile sizing, finishing and printing, paper production, processed foods, cosmetics and pharmaceuticals, guar gum is also used in the explosives industry, helping to maintain the explosive properties of the various materials used, even in wet conditions.

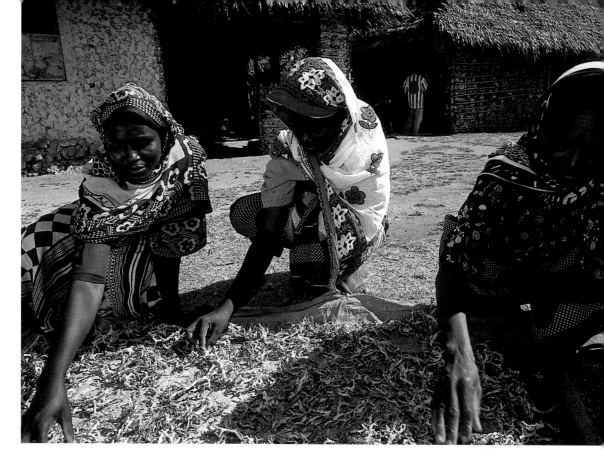

Seaweed being dried in Tanzania.

Laminaria, *Ascophyllum* and *Macrocystis*. A typical extraction process comprises many different stages, following an initial treatment with dilute sulphuric acid. Around 40,000 tons of alginates are produced each year in various locations around the world, including the UK, France, Norway, Chile, California, Japan and China.

Anti-ageing and protecting

There seems to be no limit to the range of skin products – from anti-ageing creams to sun-tan lotion – in which plant extracts appear. The range and nature of these extracts, providing many chemical compounds mostly concerned with rejuvenating the appearance of the skin, seem to be ever more complex.

Selected for a variety of alleged benefits – often moisturizing or with a high vitamin content – plant extracts can be taken from almost anything from the plant kingdom, from oak bark or hazelnuts to guava fruits or passion flowers. From a commercial point of view, the only restriction seems to be whether the plant matter is available in a suitable form and in sufficient quantity. Alongside essential oils, we may find lipid-rich oils, tannins, enzymes and fatty acids, for example. An entire range of skin products currently available incorporates an enzyme produced by grapevines to protect them from fungal attack. Advertised as 'grapeseed extract', this is claimed to 'revitalize the cells and promote the synthesis of collagen and elastin' when applied to the skin. Wheat, oats and soy protein are the source of various compounds said to have a skin-tightening effect, while active ingredients from seaweeds are used as sun-tan accelerators and anti-wrinkle agents and oryzanol from rice bran may be included as a natural antioxidant and UV filter.

The highly technical names of many of these compounds mean that ingredient lists may be incomprehensible without a reasonable know-ledge of chemistry. The consumer is unable to tell what he or she is actually buying and whether the product contains extracts in sufficient quantities to do the job claimed and promoted through persuasive advertising. Increasing numbers of products, however, are basing their appeal on the inclusion of ingredients that are organically grown, which have eliminated or reduced chemical additives or processing. One example of this is the use of hydroxybenzoates, which take their name from benzoin trees, which are natural antibacterians found in the tissue of many plants. These have been incorporated as natural preservatives in some organic cosmetics.

New uses for plant materials are constantly being made. Sometimes it seems that our appearance might be improved by eating them rather than applying them to our skin. One recent candidate is acerola (*Malpighia glabra*). The cherry-like fruits of this shrub or bushy tree, which is native to tropical America, are eaten in various forms or used as a commercial source of vitamin C (in which they are very rich). But acerola extracts are being hailed as a new weapon against anti-ageing. Researchers in Brazil have developed a range of products, including moisturizing cream and shampoo, which incorporate them. Because of their antioxidant properties, they are said to be able to combat free radicals and therefore slow down the ageing of the skin.

Another tropical extract – buriti oil obtained from *Mauritia flexuosa*, a palm that grows abundantly in the Amazon floodplain of Brazil – has long been appreciated by Amazonian Indians for food and other uses. Recently it has been found to have the ability to filter and absorb cancer-causing UV rays from the sun and the large-scale production of skin-moisturizing products containing it has begun.

Soothing and healing

One plant whose natural extracts have maintained their popularity since they were first widely adopted for use in the toiletries and skincare industries in the last 30 years, with claims that elevate it above many others, is *Aloe vera* (see right and overleaf). The remarkable extracts of this plant have been soothing and healing the skin of peoples around the word for a very long time. Two separate substances have been singled out for special attention: a pale, very pure gel that is contained

Aloe

■ *Aloe vera* is just one of over 360 aloe species that flourish in the hot, dry regions of the world. With its long, spiny, pointed leaves that grow in a rosette from the centre of the plant, this succulent is often mistakenly referred to as a cactus. When mature, its yellow flowers may tower some 6 m (20 ft) above the tips of the leaves, on a long central stalk.

■ *A. vera* is believed to have originated in Africa or Southern Arabia, but it is now distributed throughout the tropical and subtropical regions of the world and may also be found as a house-plant. Today, plants are being grown commercially in large plantations in Mexico, the USA, the Dominican Republic and Venezuela (the four largest producing countries) as well as Spain, Thailand, Australia, China and India.

■ While the annual sale of raw aloe extracts to product manufacturers is estimated to be as high as $85 million, the value of products containing them is thought to be some $110 billion or more.

35

within the central pulp of the leaves and which is used to heal and moisturize the skin, and a darker juice, which is extracted from the crushed leaves and is used, among other things, in sun-screen preparations. There has been some debate over the precise definition and description of the active ingredients – more than 200 different compounds are found within the leaves – and over the efficacy of the aloe vera juice or gel (which is often highly adulterated as a commercial raw material), once processed into creams or lotions for the skin. But there is little doubt that in its raw state aloe vera can work wonders.

Aloe uses

Although its use in cosmetics is relatively recent, aloe vera has been valued for its **medicinal properties** for at least 2,000 years. The Greeks, Romans and Chinese used the plant for the healing of wounds, skin irritations, rashes, sunburn and headaches, as well as stomach complaints. **Alexander the Great** is said to have conquered the island of Socotra in order to obtain supplies of aloes for his soldiers' wounds, while the **Sonora Indians** of Mexico still continue their age-old use of aloe vera for healing burns and preventing scarring. When cut, the leaves exude a bitter yellow sap or juice, containing a mixture of phenolic compounds – the source of **aloin** (used for many years as a laxative ingredient) and **aloe extracts**. Said to help protect the skin from UV rays, they are incorporated into some modern sun-screen lotions.

The mucilage or semi-solid gel contained within the leaves' central pulp, and which consists of a mixture of many **different compounds**, has become renowned for its moisturizing, soothing and anti-inflammatory properties, highlighting it for use in a range of modern cosmetic creams and moisturizers as well as in shampoos. The effectiveness of these compounds and particularly their **healing attributes** (which have extended its use to the treatment of radiation burns, chronic skin and mouth ulcers, eczema and poison ivy rashes) are now the subject of research which aims to understand and specify the **precise nature** of their activity. A further use of aloe vera gel is as a base for various nutritional drinks.

Face cream with pesticides

An array of **undesirable** chemicals are finding their way into the beauty products we use every day. The great majority of the plants used for the supply of raw materials are grown with the help of chemical fertilizers, herbicides, pesticides, defoliants and fumigants – many crops receiving numerous **chemical treatments** – residues of which remain in the end product. Nearly one-third of all pesticides are reported to be **possible cancer agents**, while another third of all pesticides may disrupt the human nervous system. Since our skin absorbs up to 60% of what it touches, the chemical cocktail we apply to our skins should be a cause for concern.

An additional concern to some is that key starting materials for toiletries and cosmetics may be derived from **genetically modified** (GM) plants. These include soya, cotton seed and rapeseed oils, as well as maize, also used for oil, protein and starch.

Aloe vera: Often mistaken for a cactus, *Aloe vera* is a succulent, now grown on a large commercial scale in many countries.

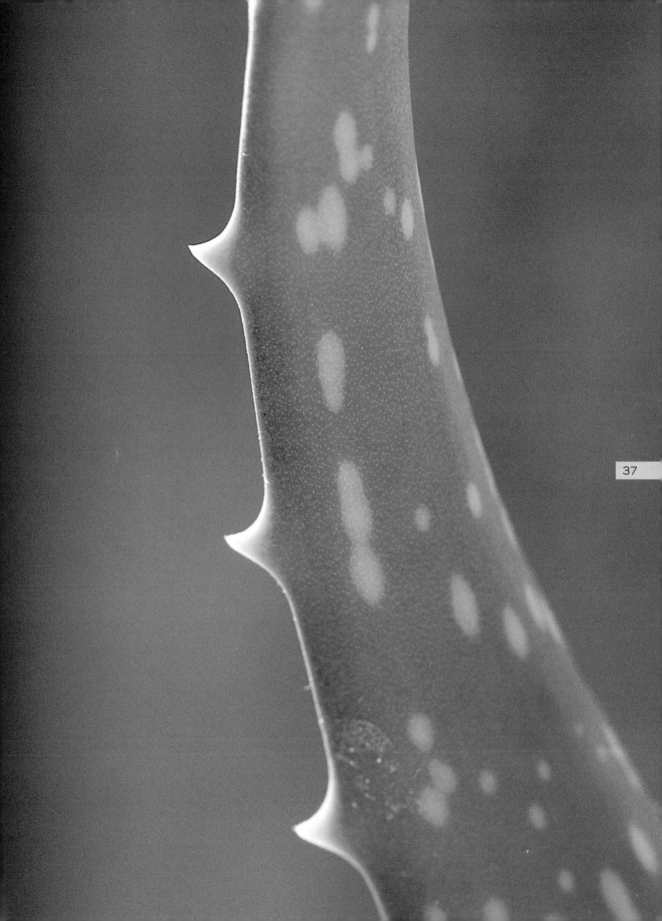

PLANTS THAT CARE FOR US

Colour from plants

Jojoba

■ Jojoba (*Simmondsia chinensis*), a low-spreading evergreen bush that grows wild in the Sonoran Desert of Mexico and parts of the southwestern USA, has gained great popularity in recent years. The dark brown seeds that develop within the fruit capsule contain up to 60% of a light yellow, odourless oil (technically a wax). This was reported to be an almost perfect substitute for the sperm whale oil needed for the production of ballistic missiles. The millions of dollars invested in its future were, it has also been claimed, essentially to provide a natural raw material for military use.

■ However, production difficulties associated with growing this only partly domesticated plant have meant that supplies have been uncertain, and the price high. Shampoo, hair conditioners, gels and mousses, and a range of cosmetics (including moisturizers, lipstick and nail products) account for 80% of its current usage. Jojoba oil may also be used for the manufacture of pharmaceuticals, engine lubricants, printing inks, plastics and varnishes. In solid form it can act as a polish, protecting many different surfaces.

■ North American Indian peoples of the Sonoran and Baja California regions traditionally used jojoba seeds and oil for cooking, hair care and for healing purposes. A recent commercial development, however, is the extraction from jojoba meal (subsequently fed to cattle) of compounds known as simmondsins, which have been discovered to suppress appetite. A

If concerned with the texture or softness of our skin, we may also be concerned with enhancing its colour. A huge array of cosmetic colours entice us to enhance our looks by emphasizing lips, cheeks and especially eyes. More coloured waxes, powders, crayons, creams and gels are available to help us do so than ever before.

Though we may not use colour in such striking ways as some of the peoples of New Guinea, for example, or the Woodabe men from Niger for whom colourful face or body designs often represent important, cultural messages about status, many people feel undressed without it. Our messages are generally much simpler, usually concerned with attracting others or representing an opinion, but whichever way we view it, colour, or the lack of it, is often used to make a point.

Fashion has, of course, largely dictated the appearance of our faces. Roman ladies were much concerned with whitening the skin and used extraordinary mixtures including pulped narcissus bulbs in the form of face packs. Other face-whitening and highly poisonous ingredients used by European women in the 17th and 18th centuries included ceruse (white lead), ground alabaster, sulphur and borax.

The rather less dangerous rice flour mixed with plant oils was used as a night-time face pack by aristocratic Chinese ladies, and a face powder based on ground white rice was used during the day. Today most modern face powders are made from complex mixtures including talc, silk cocoons, petroleum derivatives, lanolin, synthetic pigments and various resins. Some, however, may include maize and rice starch in powder form.

Very few of the colouring materials found in modern make-up are plant-derived. Most, like those used by the tribespeople of New Guinea, are based on aluminium or iron oxides, which give a range of colours from pinks and reds to yellows, browns and black. Pressed face powders often comprise mineral rather than plant oils, but most coloured creams and gels, as well as crayons and mascaras, are likely to contain the ever-useful cellulose ethers from pulped wood or cotton linters. As with the other products, cellulose acts as a thickener, helping to give a smooth, stable, creamy texture and, in the case of coloured crayons, helps prevent the crayon itself from shattering on application. Many rouges contain plant cellulose, gum arabic or algin from seaweed as a suspending agent, though the Romans used certain fucus seaweeds (whose generic name is derived from the Latin for rouge) to provide the red pigment for this use. A similar extract mixed with fish oils was traditionally used by women in Kamchatka, in the extreme east of the former Soviet Union.

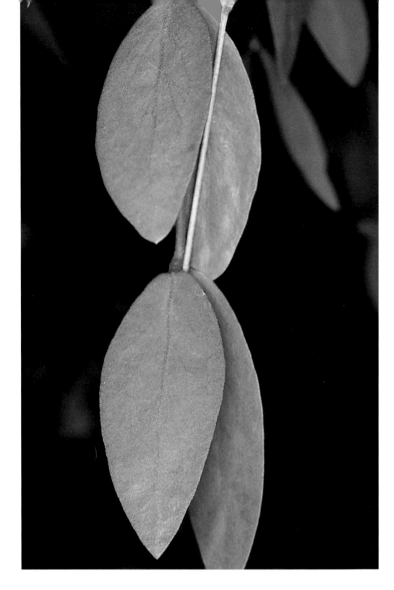

Jojoba's leathery leaves help protect it from moisture loss in its arid desert home.

nutritional supplement containing these compounds (now patented) has been developed for human use.

■ Jojoba plants are cultivated on around 8,000 ha (19,800 acres) of commercial plantations in arid or semi-arid regions of the world, including Argentina, Israel, Australia and Peru, as well as being collected from the wild in the USA. Over 4,000 tons of seeds have been harvested each year since 1997 and production is set to rise.

Lipstick

Lipstick, one of the most popular cosmetics, depends very much on plant materials for its effectiveness. Most lipsticks are based on complex blends of mineral and vegetable oils and waxes, plus some animal products such as lanolin and beeswax. Plant oils feature in both their whole and fractionated forms: myristic acid, commonly from coconut or palm kernel oil, for example, is mixed with alcohol to form a colourless liquid softener used to smooth the lips and make them glossy. Fatty alcohols, also from coconut, soya or linseed oils, are also often present. Castor oil is an important ingredient of many lipsticks, since as well as having a reasonable taste, it can disperse colour pigments easily, is also fairly resistant to oxidation and forms a tough, shiny film when it dries after application. Castor and jojoba oils (see opposite) also provide 'slip', helping to spread the lipstick evenly as it is applied. Cocoa butter is often incorporated as a moisturizer.

Brazilian wax palm

■ The Brazilian wax palm (*Copernicia prunifera*) is a long-lived tree thought to be able to reach some 200 years. It is found growing along rivers and on the edges of marshy or saline land in parts of northeastern and southern Brazil as well as the Chaco region of northern Argentina and Paraguay, where it may form dense forests.

■ As they form, the young leaves of the palm become coated with a thin glossy film of carnauba wax, which helps protect the leaves from moisture loss. To extract the wax, a number of leaves are cut from the palm twice a year, then graded according to age. The best-quality wax is produced by the most tender, unopened leaves at the centre of the palm. After drying, some wax is chipped off before the leaves are beaten and leaf fragments processed to remove more. This wax will be melted, purified and filtered before moulding into blocks. Some 200–300 leaves are required to make just 1 kg (2.2 lb) of 'average' wax.

■ The hardest natural wax commercially available (harder than beeswax and many synthetic substitutes), carnauba wax is used mostly with other waxes, improving their lustre and durability and increasing their melting point, while decreasing stickiness. It may be found in a variety of products, including cosmetics, floor, furniture, leather, shoe and car polishes, glazes for pills and various foods, including confectionery and fruit, as well as a coating for paper.

Plant waxes also play a major part in lipstick manufacture. Two in particular appear repeatedly: carnauba wax, produced by the leaves of the Brazilian wax palm (see left), and candellila wax, produced by *Euphorbia antisyphilitica* and *Pedilanthus* spp., tough shrubs native to Mexico's Chihuahua desert region. Carnauba wax is a regular if not essential ingredient in lipsticks. A hard wax, it helps to give them firmness and rigidity and is available in various shades and blends. Four different categories of wax are produced, depending on the age of the leaves. 'Prime yellow', the grade used widely by the cosmetic industry, is obtained from the younger, fresher palm fronds, whilst the older, tougher leaves give much darker waxes of a greyish green or brown colour.

Carnauba wax is exported only by Brazil, and a shortage some 20 years ago led to the introduction of candelilla wax, which continues to be used together with or instead of carnauba wax. This wax, which has slightly different properties from carnauba, is obtained by boiling the succulent stems of the shrubs in water to which sulphuric acid has been added and then skimming the wax from the surface. Large blocks of candelilla wax gathered in this way are remelted, filtered and bleached before commercial use.

Plant waxes also play a major part in lipstick manufacture. Two in particular appear repeatedly: carnauba wax and candellila wax.

The brilliant pigments associated with lipsticks come mostly from iron, aluminium or titanium oxides, though 'natural' red may also be carmine, derived from the female beetle *Coccus cacti*. Plants, however (particularly as cell cultures), are increasingly being investigated. In Japan, the bright red pigment traditionally extracted from the roots of *Lithospermum erythrorhizon* is now being produced by biotechnology, for use, among other things, in lipsticks.

Though much of this work is commercially sensitive and details are hard to obtain, research continues in the field of 'new', natural colorants from plants. Some that have been investigated for red pigments include safflower, red cabbage and strawberries, all of which are rich in anthocyanins. These are the largest group of water-soluble plant pigments, responsible for the attractive colours of many fruits and vegetables. Research has also been done on the natural blue pigments in red microalgae.

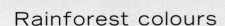

Rainforest colours

For face and body decoration, many groups of **Amazonian Indians** have traditionally used the juice pressed from the green fruits of *Genipa americana*, a tall, stately tree reaching some 18 m (59 ft) in height. When exposed to the air, the juice from these fruits turns a deep blue-black. Txucarramae women in Brazil's Xingu National Park paint some of the most **complex designs** to be found among Brazilian Indians, creating the impression of garments clinging to the body. Charcoal is also used for drawing designs.

In traditional Amerindian societies drabness is generally disliked and associated with death. Worn ceremonially, for religious purposes and to show status, **colour and design** are integral aspects of such indigenous cultures. Red pigments are especially important. In the northwestern Amazon, a red powder obtained by **boiling the leaves** of the vine *Bignonia chica* is used by both men and women and is an important item of trade. Chilli pepper juice is sometimes used to help designs stand out.

The best-known **red pigment** from the Amazon is that obtained from the tree *Bixa orellana*. Described for commercial use as E160(b) or 160b, **annatto, bixin or norbixin**, a rich paste made from the waxy red coating of the seeds, is used throughout Amazonia to decorate the face and body. While we may eat it every day as the colouring used to liven up our butter, cheese or margarine, the Mehinaku Indians of Brazil apply it to their **hair and bodies**. Two sorts of urucu, as the paste is known locally, can be distinguished: a deep scarlet used by the men and a paler orange exclusive to the women. Oil from **pequi fruits** (*Caryocar brasiliense*), eaten as a staple food by peoples of the Xingu during the rainy season, is used to lubricate the body before urucu is applied.

The Tapirapé, who along with most Amerindians find facial and body hair unattractive, have traditionally used the fruits of **streptogyna grass** as tweezers, while other wild grasses are used as razors.

The Baka pygmies of southeast Cameroon have also traditionally used a red pigment to decorate and protect their bodies. Made from the vivid heartwood of the ngélé tree (*Pterocarpus soyauxii*), **ngélé paste** is said to give them courage. It is also used to relieve aches and fevers, give strength to dancers and special protective powers to pregnant and newly wed women.

Plants for hair dyes

It has never been much fun putting benzine or hydrogen peroxide mixtures on our hair to boost or change its colour. But these and a range of semi-permanent solutions comprising combinations of red, blue and yellow dyes made soluble in water and alcohol have become the standard colouring materials for hair. Apart from the detergent bases, almost all the ingredients are made from mineral oils and other inorganic materials, but plants are still there to offer us alternatives, in subtle though generally less permanent combinations.

Perhaps the best-known natural hair dye in the West is henna. Available commercially as a green-brown powder, it is made from the crushed, dried leaves of the henna plant (see left), and gives a variety of deep red tones when mixed with hot water and applied to the hair in the form of a paste. The colouring material responsible is lawsone or hennotannic acid.

For commercial purposes henna may be altered chemically to provide a base for synthetic hair colorants, but it is also sold in its natural state, sometimes mixed with several other plant materials. These may include indigo, the famous blue-black dye obtained from various plant species (see page 85), ground coffee-beans, or catechu, a reddish-brown dye extracted from the heartwood of the Indian tree *Acacia catechu*. Two other plant sources have a long history of use to darken the hair: the black walnut (*Juglans nigra*) and oak galls. Both were recorded by Pliny in about AD 50. One recipe required walnut shells to be boiled with a portion of lead parings and mixed with such exotica as earthworms. Extracts from the leaves of walnut trees are added to some commercial hair colours today.

Perhaps the best-known natural hair dye in the West is henna, made from the crushed, dried leaves of the henna plant.

Though not found in contemporary hair colours, oak galls, sometimes known as oak apples, were used by the Greeks and Romans, and by European women in the 17th century, to darken hair. Formed by the action of cynipid wasps on the leaf buds of various oak species (mainly harvested from *Quercus infectoria*) and, in the Near East, sumac trees (*Rhus* spp.), they contain relatively high levels of tannin. It was the presence of this tannic acid that first promoted their use not just as dyes for hair or cloth but for the manufacture of inks and medicines, and for tanning leather (see page 89). For blond hair, rhubarb and camomile are the traditional herbal lighteners. Extracts of Roman and

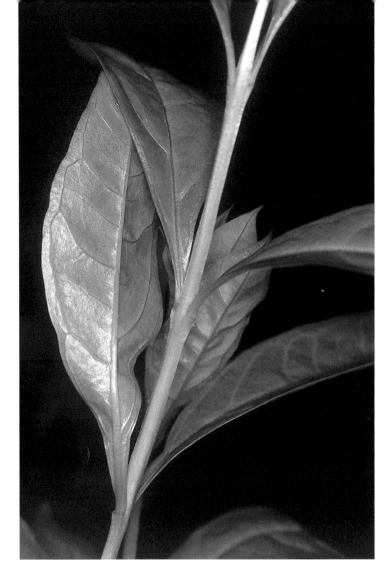

Henna leaves, when dried and mixed with hot water, produce a variety of red shades.

Henna

■ A shrubby perennial plant, with narrow grey-green leaves and reaching about 3 m (10 ft) in height, henna (*Lawsonia inermis*) is believed to be indigenous to the region between Iran and northern India. Having been culti-vated for a very long time, its range includes many other (mostly Muslim) countries, especially in North Africa, where it is also grown commercially. India, Pakistan and Iran are the main suppliers of henna to some Middle Eastern and North African countries (where local production requires imports to meet demand), but it is also exported, for European use, from Morocco, Iran, Sudan, Syria and Oman.

■ The henna powder sold for use as a hair dye and for body art consists of the dried powdered leaves of the plant. Powders can vary in composition and quality, depending on which varieties and parts of the plants are used, and are also affected by methods of storage and transport.

■ Derived from the Arabic name *Al Kenna*, henna is one of the oldest known hair and body dyes. The prophet Mohammed is said to have used it to redden his beard and it is still commonly used by Eastern women to colour their cheeks, hands, nails and feet. Henna has a special religious and mystical significance for the Berbers of North Africa, who believe it represents both fire and blood, and links man with nature.

German camomile (*Chamaemelum nobile* and *Chamomilla recutita* respectively) are those that are widely advertised in all sorts of modern hair preparations, often in shampoo, but their power to dye the hair is not as strong as that of rhubarb. Containing oxalic acid, crushed rhubarb roots yield a yellow dye that can be applied to the hair in paste form to lighten it. Other plant-derived colours some-times found in 'colour-enhancing' shampoos and conditioners include red from annatto and madder root. Mallow extracts from *Malva pusilla* are sometimes added for their moisturizing properties.

Once we have washed or dyed our hair, we may well want to hold its style in place. Most hair sprays once used plant resins in a solvent base to do this job. Although these are mostly now replaced with synthetic polymers, shellac, a dark red transparent resin that is produced by the action of lac insects on the twigs of the Indian tree *Ficus benghalensis* and other species (see page 55), may still be used in hair sprays.

Fragrances from plants

Rose

■ Amongst the oldest and most famous of perfume ingredients, oil or attar of roses (*Rosa* spp.) continues to play a major role in perfumery and aromatherapy. Its particular properties make it one of the most antiseptic of the natural oils and one of the least toxic. Rose oil has long been valued for its soothing action on the nerves, which may induce sleep, as well as for its powers as an anti-depressant. It is also said to tone the vascular and digestive systems.

■ According to legend, rose oil was discovered in Persia (now Iran) when a princess at her wedding feast noticed the oil released from rose petals, which had been thrown into a pool and warmed by the sun. Persia was once very famous for its rose water, but the production of the rose oil that gave rise to it is no longer commercially significant. India was also a major supplier of rose oil in the past, but Bulgaria and Turkey have taken its place, producing the best-quality and therefore most expensive oil, from the damask rose, *R. damascena*. Other rose species used for this purpose include *R. gallica*, cultivated widely in Morocco, and *R. centifolia*, grown in the Grasse region of southern France.

■ Rose oil is not red in colour but an orange-green. Over 9,000 roses are needed to make just 25 ml (1 fl oz) of oil by steam distillation.

Universally associated with desire, we have used perfume to make ourselves attractive since the earliest times. It is said that Cleopatra did not win over Anthony with her looks alone, but with delicious perfumes, which he found himself unable to resist. This same association is used very effectively today to sell several hundred million dollars' worth of perfumes every year.

Alongside this widespread use of scent and its addition to most toiletries and cosmetics, perfumes may well help us to decide which writing paper, disinfectant, disposable handkerchiefs or even shoe cream to buy, as they have become essential components of hundreds of familiar household items.

The ingredients of many are now made synthetically from various organic compounds, but all perfumes were, until relatively recently, derived directly from the natural sources (chiefly plants) suggested by their different aromas.

Perfume history

It is impossible to say when people first discovered that they could extract and use the very varied fragrances of plants to scent the skin. One of the earliest recorded methods – which linked healing with religious ritual – was by burning scented herbs or incense and inhaling or standing in the smoke. Believed to have special powers, such purifying smoke was thought to be able to drive out evil spirits and associated illness or disease. Evidence of this ancient therapeutic use of plants comes down to us in the word perfume, derived from the Latin *per*, meaning 'through', and *fumus*, 'smoke'. The link between fragrant substances from plants and medicine continues today in the form of aromatherapy, which uses plant oils to correct imbalances in the body.

Like many other early civilizations, including those of India and China, the ancient Egyptians were familiar with the medicinal properties of a large number of aromatic plants and burned plant materials in religious purification ceremonies. Perfumery had become so closely associated with religion in ancient Egypt that each god was allotted and its image anointed with its own particular fragrance – priests being those who dispensed such unguents. Plant oils such as olive, castor and sesame, scented by the infusion of aromatic herbs and gums, were used to anoint the hair and body, while aromatics including cedar wood, bitter almond, juniper, coriander, henna, spikenard, cinnamon, cassia and calamus as well as myrrh and frankincense were blended and used as potent perfumes. Many different raw materials were imported from other countries.

For the Egyptians, as for the Greeks and Romans, who were to learn a great deal from them, the use of perfume was not restricted to women. Egyptian men would place a cone of solid perfume on their heads and let it melt, slowly covering the rest of the body. The Greeks – whose use of aromatic medicines and cosmetics was as prevalent as the Egyptians', and who attributed the origin of fragrant plants and perfumes to the gods – used different scents for the different parts of the body. Mint, marjoram and thyme as well as fragrant flowers such as rose and hyacinth were favourite ingredients of some cosmetic oils and scents. The Romans, even more lavish in their use of perfumes than the Greeks, scented not only their bodies and hair, but their clothes, bedding, military flags and the walls of their houses. Saffron, cardamom and lupin seeds, costus, melissa and narcissus flowers were just some of the plant ingredients used.

Whilst the trade in gums and plant oils continued to flourish in the Holy Roman Empire, it was not until the late 10th century that the Arabian doctor Avicenna perfected the art of distilling essential or volatile oils to make essences and aromatic waters, first discovered by Aristotle in the 4th century BC. The fame of Arabian rose water in particular spread and reached Europe along with other exotic perfumes from the East at the time of the Crusades.

By the end of the 12th century Europe had its own perfumers and by the 13th, when lavender was already being grown at Mitcham in Surrey for this purpose, it was developing distinctive fashions of its

Plant scents used for religious purposes

Two of the three gifts presented by the Wise Men to the infant Jesus, myrrh and frankincense, were **sweet-smelling gum resins** produced by plants.

A reddish-yellow in colour, **myrrh** is produced by a number of small, tropical, thorny trees of the genus *Commiphora,* native to Africa and southwest Asia. Some confusion exists over the precise botanical origin of the myrrh so widely used for **incense and perfumery** in ancient Egypt and the East, but *C. myrrha* and *C. abyssinica* are important sources. Myrrh is still used in some cosmetics and perfumes as well as pharmaceuticals.

Frankincense is the gum resin produced by certain *Boswellia* trees, including *B. sacra*. It was recommended by Ovid, the Roman love poet, as an excellent cosmetic and was used widely by the Egyptians, Greeks and Romans as an ingredient of **incense**. When burned, the volatile oils contained within the resin are released and diffused into the air.

Both frankincense, which was offered to the sun god Ra, and myrrh, offered to the moon, were used to **anoint the dead** in Egypt and formed part of the process of mummification. **Spikenard** was similarly important for religious ceremony amongst the Greeks, Romans and Egyptians. The term appears to have been used to refer to more than one species of plant, however. While spikenard has been identified as a perfume distilled from the **desert camel grass** (*Cymbopogon schoenanthus*), the 'ointment of spikenard' that Mary used to anoint Jesus's feet before the Last Supper is thought to have been derived from the stems of the **Himalayan valerian** (*Nardostachys grandiflora*).

Ylang-ylang

■ One of the key ingredients prized by the best perfumers, ylang-ylang (*Cananga odorata*) oil has a voluptuous scent and is distilled from the beautiful yellow flowers of a tall tropical tree, *Cananga odorata* subsp. *genuina*, native to Southeast Asia. Today the tree is found in various tropical countries and for commercial use is generally grown on plantations. The very similar but more stable cananga oil, extracted from *C. odorata* subsp. *macrophylla*, and produced almost exclusively from uncultivated trees in Java, may also be processed to yield various grades of ylang-ylang oil. Though almost identical botanically, trees growing in different locations – such as Java, Fiji and Samoa – produce oils with quite different characteristics. The Comoros Islands have traditionally been the world's primary producer of ylang-ylang oil, but Indonesia has also now become a major supplier.

■ The production of oil is a very delicate business. The flowers, which are hand-picked in the early morning, must not be bruised or they will turn black and start to ferment. Requiring expert handling once extracted, oils of different qualities are produced according to the length of distillation. Top-quality oils are used in perfumes and those of lower quality are used to give fragrance to soaps. Some 9 kg (20 lb) of fresh flowers, harvested from mature trees, are said to yield just 30 ml (1¼ fl oz) of oil each year.

own. The process of distilling oil from plants and flowers in alcohol gave rise in the 14th century to Hungary and, later, Carmelite waters – the latter including angelica, lemon peel, lemon balm and coriander in a base of orange flower water and alcohol. By the 16th century, most large houses in Europe possessed a stillroom, where herbs and flowers were processed into simple perfumes.

Eau de Cologne from Germany, one of the most famous of perfumed waters and still popular today, was invented at the beginning of the 18th century. Its success was due to a combination of rosemary, neroli (orange flowers) and bergamot oils distilled in a grape spirit.

Essential oils

None of the plants or flowers so valued in the past or those prized by the perfume trade today would have been used at all if it were not for their essential oils. Also known as volatile oils because of their ability to diffuse into the air, they are the essence of a plant's fragrance. Of course, not all plant oils smell as good to us as others, though very different plants will sometimes release quite similar odours. In nature, the perfumes given off by flowers are an important means of attracting pollinating insects. Some, such as that produced by the maroon and mustard coloured flowers of the succulent *Caralluma speciosa*, which mimics the smell of rotting meat, are attractive to flies. Those with paler flowers – since their perfume replaces and is derived from

Extracting essential oils

Various methods are used to extract essential oils, depending mainly on their tolerance both to heating and to the different extraction processes. Some oils, such as citrus, can be obtained by **pressing** (in this case, the skin of citrus fruit), but the most common method is the ancient technique of **steam distillation**. This involves the heating of plant material by forcing steam through it under pressure so that the essential oils are vaporized and carried away in the steam. Once this steam has condensed back into water, the oil can simply be **skimmed** from its surface. Also fragrant, this water may have a commercial use, as in the case of

rose or orange blossom water. Though the idea is simple, the process is a **highly skilled** and delicate one. Most essential oils have an extremely complicated chemical make-up – several hundred different types of molecules might be present in each one. The heating process may alter the chemical composition of the fragile oil and may change its **odour**, so only the tougher flowers such as roses or lavender, or the leaves, bark, seeds or roots of certain plants may be suitable. **Delicate flowers** such as jasmine and tuberose must receive gentler treatment. In the recently revived process known as enfleurage, these blossoms are carefully

Oranges: Both bitter and sweet oranges are sources of citrus oils used in perfumery.

Sandalwood

■ Once common in the dry regions of peninsular India, the sandalwood tree (*Santalum album*) has played a prominent part in Indian culture for over 2,000 years and was formerly thought to have originated there. Sandalwood is now considered to have originated in Indonesia, where it is still found, the species being widely scattered in China, the lesser Sunda islands, the Philippines and Australia.

■ Known in commerce as East Indian sandalwood to distinguish it from other *Santalum* and unrelated species, which are used as substitutes, the tree may grow up to 18 m (59 ft). It is evergreen and semi-parasitic, using nutrients from the roots of other trees to help it grow. One of the few woods that is weighed correct to fractions of a kilo and formerly classified into 18 different grades, sandalwood timber is extremely valuable.

■ Great quantities have been used for buildings and for carving ornaments and both the scented red wood and the oil are traditionally used in incense and medicines. In India, Hindu women apply a paste made from the ground wood to their foreheads, combining religious and cosmetic uses. The most famous use of the oil, however, is in perfumes.

placed in layers on trays coated on both sides with purified fat, or in cloths soaked in olive oil, and left to impart their **fragrance**. The flowers may be removed and replaced with fresh ones several times until the fat or oil is **saturated**. The essential oil is then extracted with alcohol, which must now be boiled from the solution to leave the essential oil.

A gentler heat is employed in the process known as **maceration**. Here successive batches of fresh flowers are stirred in warm liquid fat for several days until it becomes impregnated with their essential oils. After extraction with alcohol, this solution is, again, concentrated by boiling.

Solvent extraction seeks to minimize the use of heat. Plant material is first **mixed with a solvent** such as hexane or petroleum ether, and the filtered solution is then boiled to evaporate this solvent, leaving the oil behind. The product (known as a 'concrete') thus formed also contains other **compounds** from the plant, however, such as waxes, fatty acids and resins. Refining with alcohol isolates the essential oil, now known as an **'absolute'**. Hydro fluorocarbons (HFCs) have more recently been found to be ideal solvents for the extraction of natural oils, since their use yields absolutes directly, without the need for **further refining**.

Brazilian rosewood

■ Demand for the unique oil produced by Brazilian rosewood (*Dalbergia nigra*), used in some of the most expensive perfumes and extracted until now only from its wood, has brought this magnificent tree to the brink of extinction in its native Amazonia. Now, however, Brazilian researchers from the University of Campinas have shown that the oil – rich in linalool, which helps fix the aroma to the body – can be extracted from the leaves, meaning that the trees need not be felled.

Rosewood seedlings are grown in Brazil for their oil and are also used in soaps.

pigments similar to chlorophyll – often have delicious scents that we have come to cherish. Lily of the valley, honeysuckle and ylang-ylang (see page 46) are just three examples of these.

Essential oils may be present in different combinations and concentrations in different parts of a plant. While orange flowers contain their own distinctive oil, known as neroli, orange peel produces a citrus oil. Lemon grass oil, on the other hand, comes from the leaves and stems of a lemon-scented grass, cinnamon oil comes from a bark, and sandalwood (see page 47), pine and cedar wood oils are extracted from the wood of these trees.

Produced mostly by steam distillation of the chipped wood (from the roots and heart of the tree), sandalwood has a heavy, sweet scent and a very lasting odour. Since it has no overwhelming top notes and has the advantage of a pale colour, it is very valuable as a fixative and can be blended with other oils, such as rose, to make attars. This warm, spicy oil is in great demand for use in soaps and perfumes, cosmetics and soaps as well as medicines and incense.

In India almost all sandalwood oil is produced from wild trees and here, as in Indonesia, over-harvesting, plus the impact of fires and cattle grazing, amongst other things, has led to a serious decline in wild populations. With a ban in place on the export of logs from India, smuggling has become a serious problem. Logs are, however, exported by Hawaii, Fiji, Indonesia and Western Australia. Though substitutes from other parts of the world are used, *Amyris balsamifera*, for example, known as West Indian sandalwood (chiefly used as a perfume fixative in soaps and cheaper cosmetics), and *Osyris tenuifolia* from tropical and South Africa, nothing is said to compare with the fragrant qualities of East Indian sandalwood.

Making perfumes

Good perfumes consist of mixtures of essential oils, chosen for their compatibility, with many other ingredients playing a supplementary role. Very few perfumes have less than 20–30 ingredients and some have over one hundred.

As well as the natural plant odorants, animals provide some very important substances. Civet from the civet cat, castoreum from the beaver and musk from the musk deer are often used as 'fixatives', helping to slow down the evaporation of the volatile oils and so 'fix' their odours. Some plant resins also perform this function.

Though the more expensive perfumes rely to a great extent on these natural plant and animal sources, which are often very difficult to imitate exactly, many fragrant materials used today are made synthetically and may account for 90% of some perfumes. While many chemical additives are derived from petroleum, a large number of synthetic aromachemicals are derived from turpentine oils, extracted from the balsams or wood of several species of pine tree. Flavours and fragrances as diverse as spearmint, nutmeg, lavender and lime can all be made from the pinenes in turpentine, as well as valuable compounds such as linalool, an important component of certain natural and synthetic oils. Individual components of some essential oils are also chemically isolated to make separate substances, such as geraniol, from citronella and palma rosa oils.

Putting all these sources together, the modern perfumer will have over 3,000 plant-derived materials from which to create a masterpiece.

The French, whose perfumers were granted a special charter as early as 1190, are still considered leaders in the perfume trade, despite much competition from America. Of the several million dollars now needed to launch a major perfume, however, only a very small part will relate to the raw materials used.

But whether using turpentine to manufacture the smell of fresh limes for commercial use in aftershave, or rose petals at home to make a simple scented water, plants leave us smelling, and often therefore feeling, a great deal better than we would without them.

The Atlantic cedar (*Cedrus atlantica*) is one of the sources of cedarwood oil.

51

Plants that **clothe** us

If we believe the Bible, man's first item of clothing – a fig leaf – was simply plucked from a plant. However, the steady observation of plant materials and subsequent experimentation appear to have led our ancestors to a much more skilful and ingenious use of plants very early on in our history. Many thousands of years ago, numerous societies seem to have discovered that it was the fibres, separated or processed in various ways, which were generally the most useful parts of plants for making clothes. Cotton, for example, which came to be cultivated in the Old and New Worlds, was being grown for use in South America at least 6,000 years ago.

Some societies developed quite different ways of adorning or covering their bodies, but they show a sophisticated understanding of the properties of the materials being used. The Yagua Indians of the northwestern Amazon, for example, have traditionally prepared fringed items of dress (with important cultural significance) from the split fronds of the chambira palm (*Astrocaryum chambira*). The petioles or leaf stalks of the freshly gathered leaves are soaked in water and then split to separate the fibres into long strands, which are skilfully knotted or threaded into place.

Many other palm species native to tropical America have provided fibres for body apparel, including arm and leg bands, necklaces, amulets and bracelets. Chambira palm fibres are generally regarded as the finest for weaving, but those obtained from the buriti palm (*Mauritia flexuosa*) and various *Bactris* species are also important

and are commonly made into hammocks, nets and multi-purpose cords. Some Amerindians have traditionally used penis sheaths made from woven palm leaflets, whilst the nuts produced by *Astrocaryum* palms are still carved into earrings and beads for necklaces. Many tropical seeds and sometimes carved, scented woods are also used as beads, often accompanying animal teeth and feathers in very beautiful and eye-catching combinations.

On the island of New Guinea, many of the approximately 750 different tribal peoples use the stems and seeds of pit-pit grass (*Coix lacryma-jobi*) for decoration. The Huli men wear the stalks through their noses and the Henganofi women cover themselves in 'necklaces' made of the seeds to show that they are widows. Sometimes weighing up to 23 kg (51 lb), the strands are removed one by one over a span of several months until the official period of mourning is over.

Over a very long period of time, the world's peoples have developed a vast knowledge of the uses of plants for different kinds of clothing or body ornamentation – in many traditional societies still reflecting important practical considerations and socio-religious beliefs.

A range of plant materials is presented in this chapter, from those – such as bark cloth – that are little known outside their regions of origin, to others – such as cotton – that are now the basis of global industries.

Previous page: Mangrove tree, Indonesia.
Opposite: Raffia fibres form part of a ceremonial 'Ekpe' costume, at a religious masquerade in Cameroon.

Bark cloth: recyclable and cool

The paper mulberry

■ Cultivated since the earliest times in Japan, China and Polynesia, the paper mulberry (*Broussonetia papyrifera*) is perhaps the best-known source of fibre used for making bark cloth. Unlike cotton and kapok, which are 'surface' fibres, the paper mulberry is used for its internal stem or 'bast' fibres. These fibres are very strong and made up of long strands of cells, held together by gums and pectins.

■ In China around AD 100, fibres from the paper mulberry's bark were separated and mixed with those of flax and hemp to make a form of paper. The Mayans independently developed a way of making paper-like sheets by pounding the bark of various trees, but they used these not as the Polynesians did for clothes, but for recording their customs, traditions and cosmology.

One of the most interesting and extensively used plant materials for clothing, but which never became an item of world trade, is bark cloth. Captain Cook's first voyage to Tahiti and the Pacific Islands (1768–71) brought back news and examples of a cloth 'almost as thin as muslin' and often dyed beautiful shades of red, made from the beaten inner bark of certain trees. Joseph Banks, naturalist on this voyage, recorded the materials and processes involved in the making of this *kapa* or *tapa* cloth (from *ka*, the, and *pa*, 'beaten' – meaning 'the beaten thing').

In Tahiti, he observed the making of various sorts and thicknesses of cloth from different trees that were grown and carefully tended for the purpose. The most important included the paper mulberry (see left), the breadfruit tree (*Artocarpus altilis*) and certain species of fig (see opposite). The bark of these trees was stripped from trunks having grown to only 5–8 cm (2–3¼ in) in diameter and placed in running water for several days. The outer layer was then scraped with pieces of shell until only the fine inner fibres remained and these would then be spread out in layers for beating.

To do this, the women used rectangular wooden batons with grooves of differing widths cut into each of the four faces. The cloth would be beaten until it reached the desired thinness, sometimes being doubled over and splashed with a glutinous liquid often containing Polynesian arrowroot starch (*Tacca leontopetaloides*) to help the fibres adhere, then bleached and often dyed.

Tapa making using a number of different tree barks was once common all over the Pacific, from Hawaii to Papua New Guinea. Sometimes made in pieces that were hundreds of feet long, red and yellow plant dyes were much favoured for its decoration and the cloth was often glazed or varnished with a special vegetable gum, then stamped with abstract or leaf designs. Both men and women wore *tapa* cloth, particularly for ceremonial use, styles ranging from elaborate wraps and poncho-like garments to simple loin cloths or 'skirts', according to region and rank. *Tapa* was also used to make mosquito curtains and fine screens as well as bedding. Banks, impressed with the cloth's softness, which he said resembled that of 'the finest cottons', often slept in it, finding it 'far cooler than any English cloth'.

The powerful influence of Western culture in its various forms has meant that *tapa* making has now been abandoned in many areas, although a number of islands, including those that make up the South Pacific states of Tonga and Fiji, continue to practise this very ancient art for ceremonial occasions and for sale to tourists. The ultimate in cool clothing for hot climates, one of the special features of bark cloth is that, once dirty, or worn out, it can simply be washed, re-pulped and beaten again to make clothes that are as good as new. Each year in Britain alone, about 500,000 tons of the clothing we no longer want ends up in landfill sites.

Plant dyes for bark cloth

Turmeric (*Curcuma longa*) is traditionally valued in many parts of Asia as a yellow dye and is still used in Papua New Guinea for dying bark cloth. The roots of the plant are scraped and mixed with water and the cloth is left to soak in this mixture. Brown dyes can be obtained from mangrove bark, rich in tannins, while cloth is still dyed black in parts of Southeast Asia by burying it in mud beneath plots of taro (*Colocasia esculenta*).

Bark cloth from the Colombian Amazon, painted with natural pigments, depicting rainforest palm and wildlife.

Fig trees

■ Around the world all manner of bark cloth garments have been made – from jackets worn by the Dayak peoples of Borneo to the trailing robes of Samoan chiefs. The fig trees that Captain Cook saw growing on Tahiti, along with the paper mulberry and the breadfruit, all belong to the *Moraceae* family. Within this family, the genus *Ficus* is particularly interesting. It contains 750 different species of fig and many of these have supplied peoples living in the tropics with fibre for cloth as well as raw materials for dyes.

■ Once commonly worn by Polynesian islanders, bark cloth made from fig species is still used by various ethnic groups of the Philippines, Malaysia, Indonesia and Papua New Guinea. Some Amerindians make items of clothing and slings for carrying babies from fig bark and its use has been traditional in Madagascar and parts of central and eastern Africa, along with fibre from the baobab tree (*Adansonia digitata*).

■ The dye material produced by some species of fig comes not from the bark but from the sap. A red dye noted by Joseph Banks in Tahiti was made by mixing the milky sap exuded from the fruit stems of tiny figs (*Ficus tinctoria*) with the leaves of various tropical trees and plants such as *Solanum latifolium* and *Convolvulus brasiliensis*, which were first placed in 'cocoanut water'. The action of the fig's sap on the leaves produced a beautiful dye, described by Banks as 'a more delicate colour than any we have in Europe, approaching, however, most nearly to scarlet'.

New Zealand flax: *harakeke*

Another very fine cloth made not from stem but leaf fibres was worn traditionally by the Maoris of New Zealand. Greatly impressed with the quality and softness of this clothing, Captain Cook noted that it was made from 'a grass plant like flags, the nature of flax or hemp, but superior in quality to either'. The plant was in fact New Zealand flax (*Phormium tenax*), or *harakeke* (the Maori name), not a grass but related to the onion and iris families. By the time of Cook's arrival, *harakeke* had become an essential part of the Maori economy. Although the strap-like leaves could be plaited to make simple clothing and sandals (as well as containers, floor coverings and fishing nets), the discovery of the strong, thick fibres within them prompted the development of new weaving techniques and increased the range of uses.

The many different varieties of the plant – which were cultivated and named by the Maoris – offered fibres suitable for many purposes. Among these were the fine, soft cloaks admired by Cook as well as other fabrics and cordage of many types.

Maori weavers continue to grow selected *harakeke* varieties today. To obtain the fibres, the upper surfaces of the leaves are first scraped

New Zealand flax in flower.
The long, slender leaves of this plant contain strong, versatile fibres that can be used for making textiles and cordage of many kinds as well as mats, baskets and paper. Various ornamental varieties are also commonly grown.

56

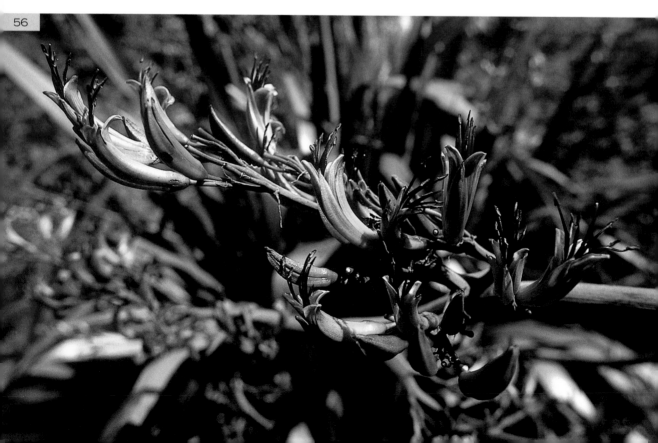

with mussel shells (still regarded as the best instrument for the job). For the finer fabrics the fibres are next soaked in water before drying and rolling into plied threads. These are then 'hanked', soaked again, beaten, and then rubbed with the hands until they are soft, white and crinkly before being dyed, if desired, and finally woven on a simple frame.

Captain Cook noted 'a grass plant like flags, the nature of flax or hemp, but superior in quality to either'. The plant was New Zealand flax.

The past 30 years have seen a huge resurgence in weaving and plaiting *harakeke* as part of the Maori cultural renaissance. While traditional techniques and patterns have been maintained, the fibre is also being used in innovative ways in contemporary arts and crafts, including paper making. Many items, such as baskets, are being enjoyed by New Zealanders and tourists alike.

Today, there is also a renewed interest in the potential of *harakeke* to make fine textiles on a large commercial scale. Coarse, machine-stripped fibre – used for rope making – had become New Zealand's major export crop by the early 20th century, but by the 1940s commercial production was restricted to the domestic market. At that time the fibre was being used to make the sacking for wool bales, upholstery padding, baler twine, carpets, mats and underfelt. Competition from cheaper, imported fibres and synthetics, and other factors, brought commercial production to an end – the last factory closing in 1985. But a new era may now be dawning for New Zealand flax, as techniques are being investigated that may allow commercially produced *harakeke* fibres to achieve the silky quality of the best of those produced by hand, for the finest fabrics.

Clothes fit for a king

Neither the soft bark cloth of the Polynesians nor the silky fabrics made by the Maoris were adopted for use in Europe when they were first observed, but the delicate inner fibres of the **lacebark tree** (*Lagetta lagetto*) met – for a limited period – with greater success. Held in several layers beneath the outer bark of the tree, this natural network of fibres can be stretched apart to resemble the finest **hand-made lace**. It was traditionally used by the peoples of Jamaica, Cuba, Haiti and Hispaniola, where the tree is native, but with the arrival of the Spanish, and at their instigation, it was made into bonnets, capes and entire **lace suits** for Europeans. The most famous use of lacebark was a cravat, frill and pair of ruffles sent from Jamaica in the late 17th century and worn in Britain by King Charles II. Reported to have been commonly used in Jamaica, for example, in the early 19th century for **a variety of items**, including ropes, caps, suits, bonnets and whips, and for cordage at least until the mid-20th century, the local use of lacebark appears now to have ceased.

Raffia for keeping cool

Raffia palm

■ There are 28 species of raffia palm (*Raphia* spp.) worldwide and all of them are found in the wetter areas of tropical Africa, with the exception of one species that is also found in South America and one that is also found in Madagascar. Raffia palms have the biggest leaves in the entire plant kingdom. The stemless *R. regalis*, which grows in Cameroon, can produce leaves that are up to 25 m (82 ft) in length. One of the most widely used palms in West Africa is *R. vinifera*, whilst in Madagascar, *R. farinifera* is of supreme importance to local people for a range of domestic purposes.

58

In many parts of tropical and sub-Saharan Africa, the leaf fibres of the raffia palm (see left) have been used for centuries to make traditional clothing. Though this use has now greatly decreased, raffia is still worn in West Africa and on the island of Madagascar, especially for ceremonial occasions. The most splendid and elaborate raffia costumes have traditionally been reserved for people of status, such as elders and ritual specialists, for use at masquerades and dances. The power and prestige of the wearer is denoted by their colourful design.

Slightly different methods are employed in different regions to produce the supple, straw-coloured strands still coveted by European gardeners, but the principles are the same. Young, tender palm leaves are first detached from the petiole of the palm frond and a small incision is made on their underside with a knife, close to the end. The raffia fibre, a colourless membrane produced by the upper epidermis of the leaflet, is then quickly pulled away by hand or by running the leaflet across a knife blade. The membranes are next tied together at one end – about ten at a time – and left to dry in bundles in the sun, during which time they acquire their familiar creamy yellow colour. For good-quality fibre that does not roll up at the edges to look like string, the drying process must be carried out carefully. The fibres, which can be split into very fine strands, are now ready for knotting, knitting or weaving together.

In the Zaïre Basin area, raffia was the only plant fibre traditionally woven into cloth, but its production has now decreased. However, the fibres from the raffia palm remain an important source of rope, fishing tackle and all sorts of cord, and the midribs of the leaves are often used as roofing poles. An alcoholic wine is also made from the sap, which is tapped from the trunk of the palm.

Lime for the Ice Man

One of the organic materials put to expert use by the prehistoric peoples of Europe was **lime tree bark**. The small-leafed lime (*Tilia cordata*) in particular, which, like its large-leaved sister *T. platyphyllos*, appears to have played an important **spiritual role** to the peoples who preceded us, offered a valuable 'bast' fibre from its inner bark, used for cordage, clothing and matting. The Bronze Age Ice Man who died in 3300 BC and who was discovered in 1991 on the Italian/ Austrian border was found to have been **wearing a cloak** that had been carefully constructed from the stems of grasses, some 90 cm (36 in) long. This cloak, which on reconstruction proved to be lightweight, hard-wearing and warm, was held together and fastened at the front with cords made of **'bast' fibre** taken from the small-leaved lime. The upper framework of the **Ice Man's shoes**, which had soles of brown bearskin, was also made from and fastened with cords of this strong fibre, which keeps its strength when wet. Rope from lime bast was still being used in Scandinavia until the mid-20th century.

Cotton: king of plant fibres

The chances are that each of us is wearing at least one garment made from cotton, the most important and most popular natural fibre in the world, and the raw material used to make 40% of all our textiles.

The fibres come not from leaves, stems or bark but from the seeds of the cotton plant (see right). Attached to the outer surface of these seeds, each cotton fibre has the appearance of a fine hair. Beneath these long fibrous hairs lies a second layer of shorter, fuzzy fibres, known as linters. Removed from the seed, these are used industrially for making water-soluble polymers and paper.

The biological function of the cotton seed's silky hairs – also known as 'surface fibres' – is to help the seeds disperse in the wind. The strength and flexibility of these fibres, however, plus a gradual spiral in their length, which causes the strands to interlock as they are twisted, have made them very suitable for spinning into thread.

Archaeological evidence has confirmed that people have been using cotton fibres to make textiles for a very long time. The discovery of seeds at Chilca, near Lima, show that cotton was being grown in coastal Peru 6,000 years ago, while other Peruvian finds of cotton yarn, fabrics and fishing nets, as well as seeds and fruit dating back some 5,000 years, are also common. Cotton seeds that are some 6,000 years old have also been discovered in coastal Ecuador (though it is not known if these are from domesticated plants), as well as ancient cotton textiles in northern Chile and cotton bolls in Mexico. The oldest cotton finds in Africa, meanwhile, are 4,500 years old and in India 5,000 years old. Tufts of cotton of a similar date found in plaster in eastern Jordan may well have been imported from the Indian subcontinent.

The ancestry of cotton and the history of its cultivation present a fascinating if complex picture. There are 50 wild species of cotton, but only four have been domesticated. It appears that they were domesticated and that spinning and weaving techniques developed independently – though probably repeatedly – in different parts of the world: two species in the Old World and two species in the New. *Gossypium arboreum* is believed to have been domesticated in India (though this is not certain), and *G. herbaceum* in Africa. The origin of *G. hirsutum* is Central America and that of *G. barbadense* is the coast of northwest South America (Peru and Ecuador), expanding later into Central America, the Caribbean and Pacific islands.

A subject that has continued to intrigue scientists is the parentage of these domesticated cottons, since both the New World species have one genome (set of chromosomes) that is similar to genomes found in

Cotton

■ Cotton (*Gossypium* spp.) belongs to the same family (the *Malvaceae*) as hibiscus and the hollyhock. About 50 species exist and are native to the tropics and subtropics of Asia, Africa, South and Central America and Australia, preferring hot, dry conditions. Amongst these species and the many varieties developed from them there is great variation in size and form and whilst some are shrubs, others grow into trees of 5.4 m (18 ft) or so. The flowers of the cotton plant also vary and may range from creamy white to beautiful reds and purples.

■ After pollination of the flowers, a fruit or boll develops, containing the seeds. The seeds of all cotton species are covered with hairs, revealed when the ripe boll bursts open, but these too vary greatly in the wild – from short and straight to long, dense and tangled. Wild species have only one type of hair – but the seed hairs of the cultivated species have become differentiated into two types. Short 'fuzz' hairs or linters, up to 1 mm ($\frac{1}{10}$ in) long, have become an important source of industrial cellulose, while longer 'lint' hairs – up to 25 mm (1 in) long – are the cotton fibres used for textiles. As these fibres dry, they collapse into flat, convoluted ribbons that enable them to be spun into yarn.

■ It has been calculated that 1 kg (2.2 lb) of cotton contains roughly 200 million individual seed hairs – each one a single cell, 3,000 times longer than its width.

Cotton weaving

Some of the finest and most beautiful of all **woven cotton** was produced in Peru. A succession of coastal and highland cultures brought spinning and weaving to such a fine art that weaving continued to be the most prized of possessions and most sought-after trading commodities up to the time of the Spanish conquest. Almost all the **weaving techniques** known today were practised by the ancient Peruvians. They produced an extraordinary variety of fabrics, from simple plain weaves to intricate gauzes, fringed brocades, open-work, tapestries and cloth that could be tie-dyed, printed on or painted. **Mummies** preserved in the sand of the coastal desert were sometimes wrapped in thousands of square feet of cotton cloth.

the Old World and one similar to those found in the New. Although this has led to much speculation as to the possibility of trans-Pacific contact, genetic evidence shows that this hybridization occurred several million years ago. Wild cotton seeds remain viable after prolonged immersion in sea water, enabling trans-Atlantic drift of wild cotton seeds from Africa to the Americas. Following domestication, people began to plant seeds beyond the native ranges of the four domesticated species; bursts of agronomic development were then followed by a much wider dispersal of 'improved' plants. The development of these domesticated cottons is the result of human selection and dispersal, which has taken place over very long periods of time. Wild perennial shrubs and trees with small seeds, sparsely covered with coarse, poorly differentiated seeds hairs, have become short, annual plants producing copious quantities of long fibres on large, easily germinated seeds.

Gossypium hirsutum

Because of their superior yield and fibre qualities, it is the New World species that supply us with almost all the cotton grown commercially today. Over 90% of this is produced by varieties of *G. hirsutum*. The large native range of the wild form covers much of Central America and the Caribbean and it is thought likely that it was domesticated more than once in different parts of this range at different times by various indigenous peoples. The beautiful cotton textiles that have been discovered at archae-ological sites, with their complex, colourful designs and intricate weaving techniques, show the extraordinary sophistication and ceremonial significance of much cotton cloth production. The Mayans so revered the art of weaving that it was given its own patron: the moon goddess Ixchel. Only Mayan nobles and their servants were allowed to wear the finest cot-ton clothes, however; ordinary people wore garments made of maguey fibres, extracted from the tough leaves of *Agave pacifica*. The Aztecs were also accomplished weavers. The last Aztec emperor, Montezuma, collected beautiful and distinctive cotton blankets and also warriors' costumes, including quilted cotton armour, by way of tribute from different parts of his empire.

It is one of the two major cultivated forms of *G. hirsutum* that is the progenitor of the modern, highly improved 'Upland' cultivars which currently dominate world cotton production. These Upland lines, as they are known – developed from old Mexican lines – have made the most important contribution to commercial cotton cultivation during the last 200 years. The historical development of these highly produc-tive modern cultivars is extremely complex and has involved many stages and numerous genetic additions.

Cotton seed hairs form the cotton fibres used for textiles.

61

Other cotton varieties

Although they account for less than 10% of the world's cotton production, it is modern improved cultivars of *G. barbadense* (extra long staple, **Egyptian and Pima** varieties) that supply the fine, long fibres used for much high quality 'luxury' cotton cloth. These are all thought to owe their origin to **Sea Island cotton** (an annual *G. barbadense* type that arose in the offshore islands of southeast USA in the 18th century). This was further developed, by crossing with old **perennial forms** of *G. barbadense* growing in Egypt, to give the 'Egyptian' cottons.

Of the Old World cottons, *G. arboreum* is still an **important** crop plant in India and Pakistan, while *G. herbaceum* is cultivated on a small scale in several regions of Africa and Asia, from Ethiopia to western India.

Ripe bolls reveal cotton seed hairs, ready for picking.

Cotton production today

The People's Republic of China is currently the world's biggest producer of cotton. In 1999/2000 (according to the United States Department of Agriculture) 3,832,000 tons were grown there and the forecast for 2001/2002 was for 4,899,000 tons. The USA has been the world's second largest producer for several years. Other major producers, in order of current output, are India, Pakistan, Uzbekistan, Brazil and Turkey. The regions importing most cotton in 1999/2000 were Southeast Asia (including Indonesia, Malaysia, the Philippines, Singapore, Thailand and Vietnam) and the EU (which imported 871,000 tons in 1999/2000). In total, nearly 21 million tons of cotton were forecast to be grown for world trade in 2001/2002.

The modern cultivation of cotton by the world's major producers is far removed from the way in which it has been grown for thousands of years. The indigenous peoples of South and Central America, for example, intermixed it with other crops, probably within 'garden' plots of different sizes and within varied botanical and ecological environments.

In general, cotton cultivation has become a mechanized monoculture requiring heavy chemical inputs to keep production levels high. In America, huge machines have been developed to plant multiple rows of identical seeds, and also harvest the cotton bolls after the plants have been defoliated by chemicals. Whilst the cotton is growing, the fields are heavily fertilized (with phosphate and ammonia), and additional fertilizer is also added to the irrigation water. Pesticides (including two that have been classified as 'highly hazardous' by the World Health Organisation [WHO]) are liberally applied, often sprayed by plane.

It has been estimated that cotton cultivation currently accounts for one-quarter of all the insecticides used on earth. This chemical deluge is responsible for appalling health problems now being endured by many of the world's estimated 140 million cotton workers – who are largely poor people, obliged to work in cotton fields or who are involved in the processing of cotton. It is very difficult for cotton workers to avoid either breathing in the chemicals as they are being sprayed, or absorbing them through the skin. Drinking water in cotton-producing areas is also often heavily polluted by pesticide run-off. One recent study in Pakistan reported that of 88 female cotton pickers only one had acceptable levels of pesticide in her blood – the rest had moderate or dangerous levels of blood poisoning. The Pesticide Action Network has estimated that agrochemical use causes about 20,000 accidental deaths and 200,000 suicides a year. Stories that show how modern cotton is damaging the lives of real people are truly shocking.

Naturally coloured cotton

Vast quantities of chemical dyes are currently used each year to colour our clothes. Most of these are **harmful** not only to the environment but to the people who work with them and, in varying degrees, to many of us who wear them.

But an alternative has existed for thousands of years in the naturally coloured cottons that are **native** to South and Central America. There, varieties of native cotton with pigmented seed hairs, including various shades of mauve, dark brown, 'rust' and beige, were developed and used by the peoples of ancient civilizations such as the Mochica in northern Peru (which flourished between AD 0 and 700) and others, in the production of a range of **exquisite fabrics**.

Relatively recently, in regions selected for commercial cotton cultivation, many of these native cottons were supplanted by **'commercial'** white-fibred varieties giving higher yields and stronger fibres that were easier to spin. In Peru, farmers were fined for cultivating native varieties where Pima cotton, for example, was being grown for export, for fear of **cross pollination**. But coloured cottons continue to be grown by indigenous peoples (for their own use) in small family gardens in parts of Central and northern South America.

Peru's native cottons

Though commercial plant breeders have used the native cottons as a source of desirable characteristics, such as disease and pest resistance, the fibres were **too short** and weak to be spun using modern machinery.

In an attempt to revitalize Peru's ancient cotton-working tradition, whilst providing a valuable source of income for local people, and with the traditional spinning and weaving by hand of **coloured cotton** on the decline, research and development began in the 1980s. This prompted a revival in the cultivation and weaving of the coloured cottons, though mostly for tourist consumption and export.

Much in demand in the textile industry, blue-coloured cotton is being developed.

At about the same time, a traditional breeding programme began in America to select coloured cotton with **improved fibre** and other traits for organic production. The selection and crossing of varieties of wild, coloured cotton with longer-fibred white cottons led to the commercial availability in 1989 of cotton of various hues: red-brown (with a natural flame resistance), bronze and green.

To date, some two dozen countries are reported to be **researching** coloured cotton. Of these, China is said to be working on the greatest scale in conjunction with several different companies and organizations (including the Chinese Academy of Agricultural Science). Rather than using traditional breeding methods, however, the **Chinese** – and many others – are using biotechnology and genetic engineering to boost yields, for example, and to work on the development of **blue cotton**.

Much in demand in the textile industry, chiefly for the production of denim, blue-pigmented cotton has proved particularly difficult to develop. One American company is said to be using genes from an **'indigo-producing'** species for this purpose, and plants being used in China by another to produce naturally coloured cotton (worth $2 million in 1999) are similarly unknown in nature.

Ownership concerns

Such developments have raised important questions and concerns about the ownership and control of naturally coloured cotton as well as lack of **compensation** for the knowledge and germ plasm taken from the south. Breeders and biotech companies claim that **coloured cottons** are their inventions, resulting in the award of various patents (see page 236).

Naturally coloured cotton has **huge benefits** for manufacturers, consumers and the environment, for example by reducing or eliminating the need for toxic and costly chemical dyes, bleaching and the disposal of toxic waste. But part of our acceptance of green, **environmentally friendly** fibres must surely be the acknowledgment and recompense of all those involved in its development.

63

Dying for cotton in Andhra Pradesh

The momentum dictating this use of chemicals is the privatization and globalization of the seed supply system largely for the benefit of the world's major agrochemical companies. In 1997, nearly 500 farmers in the Indian state of Andhra Pradesh were reported to have committed suicide because of the situation they found themselves in. The farmers had incurred serious debts because of the high cost of the hybrid cotton seeds they had bought and the pesticides needed to enable them to flourish. The following year, despite heavier than ever use of pesticides, insects developed high levels of resistance and large swathes of the cotton crops were eaten away; more farmers, in over 200 villages, committed suicide.

An important crop in India for over 5,000 years, cotton has played a significant economic and symbolic role in the transformation of India to an independent nation. Mahatma Gandhi actively promoted the cultivation and hand weaving of local cottons, both as a symbol of India's indigenous capabilities, and as an economic weapon against imports from Britain's cotton mills. Since independence (in 1947), the Indian government has promoted the cultivation of cotton – particularly

Cotton in Europe

Europeans first came to hear of cotton cloth not from expeditions to the New World – where **Columbus** and later explorers were amazed to find exquisite fabrics and other artefacts made from cotton – but from travellers to the East. There, on the Indian subcontinent in particular – where abundant **archaeological evidence** shows cotton was being grown about 4,000 years ago – the cultivation and weaving of cotton using species native to the Old World had also reached an extraordinary **sophistication**. A great range of fabrics, including fine calicos and muslins, came to be traded overland and by sea and some made their way to northern Europe.

Since Europeans had become accustomed to coarser garments made from linen or wool, stories of fine **cotton textiles** from the East and their mysterious origin fuelled the feverish imaginations of many. In the 5th century BC, the Greek historian Herodotus had described how, in **India**, 'wild trees bear wool more beautiful and excellent than the wool of sheep' and similar descriptions appear to have given rise to the northern European belief, still current in Britain in the mid-14th century (by which time cotton was already being **imported** for use as padding in clothing, embroidery thread and candle wicks), that cotton fibres came from a kind of tiny 'vegetable lamb' which grew on trees.

The sites of the first domestication of the **Old World cottons**, thousands of years ago, are not precisely known. But it is thought the Indus Valley was an important centre of diffusion for the *G. arboreum* cultivars that came to dominate Old World cotton until the introduction of the New World cultivars there in about the 1950s.

Arabic peoples had also been growing and trading cotton for a very long time (the word 'cotton' comes from the Arabic *qutun*) and it has been suggested that the cultivation of *G. herbaceum* began in the region of **Ethiopia**. The term 'muslin' comes from the Arabic name Mosul, a town in Mesopotamia, where this smooth and delicately woven cloth was made.

Increasing cotton trade
Having spread to **China** by the 7th century, cotton was being grown in almost every region of the Islamic world, including southern Spain, by the 10th century and in Italy by the 12th century. By the early 16th century, England was **regularly importing** 'cotton wool', as it was known.

The manufacture of cloth in **Manchester** was underway by 1641, with raw cotton imported on a larger

American varieties, which account for 67% of plantings today. While this policy has resulted in a successful textile industry, and a three-fold increase in cotton production, the shift to industrialized agriculture has exacerbated economic problems in some areas. American cottons have proved highly vulnerable to attack by the cotton bollworm, and in the 1980s pesticides such as DDT and synthetic pyrethroids were widely applied. The development of resistance to these pesticides by the insects, however, led to higher and higher applications. Caught in a spiral of needing ever greater doses of chemicals without any increase in production, and with many crops failing because of insect attack, over 10,000 people are reported to have killed themselves in India since the mid-1980s.

Dr Devinder Sharma of the Forum for Biotechnology and Food Security in New Delhi has suggested that it is the agricultural scientists whose work supports the commercial interests of the seed and agrochemical companies who should be held responsible for this massive human tragedy – 'perhaps the greatest in the history of independent India'.

scale from the eastern Mediterranean region and Syria. By the early 1700s, however, the **British East India Company** was importing large quantities of cotton textiles cheaply from India – a trade which was to flourish and expand, creating a significant import and export market, and playing a **major role** in shaping Britain's economy and position of world power.

By 1800, the British had set up cotton plantations in the **southern USA** with seeds from Indian plants (to avoid importing cotton from India), having also introduced the 'Upland' cultivars which were to form the basis of all cotton grown there.

With the inventions of the **Industrial Revolution** in the second half of the 18th century, new machinery for the carding, spinning and winding of fibres was provided. Furthermore, a cheap and plentiful supply of cotton from America (where the invention of the cotton gin had dramatically increased output) meant that the British industry **boomed**, helping the nation to gain dominion over a huge empire. But all this bore an immense price in human misery, both at home and abroad. In England, the rapid mechanization of the cotton industry led to massive urbanization and **rural depopulation**, to cruel exploitation (especially of small children, who were obliged to work long hours) and appalling living and working conditions. With Britain importing vast quantities of American cotton in the mid-19th century, **slaves** were to become an essential labour force for plantation owners. Unimaginable misery and suffering were imposed upon millions of Africans. America's 4 million slaves were not freed until after the **Civil War** (1861–5).

The British cotton industry
By the early 1800s, **Liverpool** had become the centre of the world trade in cotton, and by the 1830s, 50% of all British exports were of cotton cloth. The **inequalities** that had developed between the northern and southern states of America, however, with southern farmers constantly indebted to the northern bankers who marketed their crops, and disputes over trade policy, led eventually to Civil War, temporarily cutting off supplies of cotton to Britain. Although the British cotton industry continued in the 19th century, it was seriously affected by the **First World War**. The economic depression of the 1930s, and the **Second World War**, the loss of former colonies and the increasing industrialization of other European countries all played a part in the subsequent collapse of the British cotton industry.

Genetically modified jeans?

The cultivation of cotton plants as vast monocultures has increased the vulnerability of these crops to failure. As we have seen, modern cultivars are the result of very complex breeding and exchanges of genetic material from cotton species and cultivars in both ancient and recent times. In the search for ever-better yields and other characteristics (such as improved fibre quality in *G. hirsutum*), this process is continuing, using traditional genetics and breeding methods.

Today, 12% of the world's commercial cotton (over 70% in the USA) is currently harvested from seeds that have been produced by a much less understood modern technology: genetic modification (GM). In the USA, farmers currently have over 25 transgenic cotton varieties to choose from.

Herbicide tolerance and insecticides active against a wide range of cotton pests have now been engineered into cotton and many other staple crops. Claiming that they will be resistant to attack from cotton bollworms, the biotechnology company Monsanto, for example, has produced strains of cotton using a toxin gene from the bacterium *Bacillus thurigiensis* (BT) – the source of the insecticides that are most commonly being engineered into plants. The rationale behind this is that fewer pesticide applications should need to be applied externally to crops; instead, when insect larvae begin to feed on any part of the plant they will soon die.

But, as is discussed in more detail on page 150, the outcome and side-effects of the radical interference with the sensitive and complex natural functioning of plants (and other recipient organisms), which is in itself highly controversial, are very uncertain and plants don't necessarily behave according to scientists' plans. When Monsanto's NuCOTN was first grown by farmers in southern USA in 1996, an unusually hot dry summer appeared to prompt the plants to produce less BT toxin than under 'normal' conditions. The plants then needed heavy doses of pesticides. In 1997 some US farmers reported that cotton bolls were malformed. In China and Australia, cotton bollworms have developed resistance to BT crops, whilst it is known that the BT toxins are, conversely, able to kill insects that are beneficial.

Experience shows that leaving the eggs of certain insects ('pests') to hatch and be dispatched by natural predators – of which the American bollworm has many – in integrated pest management schemes involving the careful rotation of crops can be more effective than the use of disruptive insecticides. The insecticides only seem to stimulate a critical proportion to hatch, become larvae and cause damage, whilst also acquiring an increased pesticide resistance.

Despite the very considerable fears of many scientists and individuals about the repercussions of such technology and doubts about the alleged

Cotton picking near Piura, northern Peru. Here, local people gain work from cotton, but the crop is causing environmental problems and contributing to the overuse of water in an area of limited supplies.

benefits of using it (including concerns about the quality of cotton fibre produced from GM seed), many countries (from south Sulawesi, to Mexico and Australia) have been persuaded to grow genetically engineered cotton. Other countries, including India, which has the largest area under cotton cultivation in the world, have come under intense pressure to follow suit. In March 2002, India lifted a four-year ban on growing GM crops and authorized production of three types of GM cotton. By September, however, in three major Indian states, the BT cotton crop was reported to have been destroyed by diseases and pests – including the bollworm, to which the new engineered cotton was claimed to be resistant – leaving already poor farmers in a situation of crisis. Having spent on average four times as much in buying the GM cotton seeds, the farmers have demanded compensation.

A sustainable future – organic cotton

A logical alternative to the massive human misery and environmental damage currently being caused by large-scale commercial cotton cultivation for the clothing industry is its production by organic methods. The adoption of integrated pest management techniques is an important step in the right direction, together with a commitment to consume less.

Producing clothes that are socially acceptable, environmentally sound and economically viable is perfectly possible. We should surely pay a price for our clothes that reflects their real cost – to those who provide them – rather than support the sweat-shop conditions endured by large numbers of people in developing countries, including 1.6 million Bangladeshi garment workers, many of whom are children. Improved conditions for textile workers are required alongside responsible agricultural methods.

In the UK, the movement to end this sort of exploitation has been promoted, to date, by a small number of organizations and companies selling organic cotton (and some hemp-fibre clothing), largely by mail-order. But some well-known brands available from stores have begun to incorporate organic cotton in their products. Elsewhere in Europe, higher levels of public awareness of the social and environmental issues linked to clothing and a long-term commitment to environmental well-being have created a greater demand for 'green' clothing.

Around 95% of the clothing featured in the catalogues produced by one German company (with an annual turnover of DM123 million) is now organic and all products are labelled to show where and how they were made. In the USA, some clothing brands are now blending organic with conventional cotton, while some pioneers have become fully organic. A 'long life' guarantee has also been introduced in Germany to challenge the trend towards throw-away fashion. Currently, 6 million tons of post-consumer clothing is thrown away each year in Europe.

THE REAL PRICE OF A PAIR OF JEANS:
the disaster of the Aral Sea

The production of cotton is largely responsible for one of the most appalling human and environmental disasters the world has seen – the destruction of the world's fourth largest inland body of water, the Aral Sea. It is also the cause of chronic health problems affecting more than 1 million people who live in the Karakalpakstan Republic (formerly the Soviet Republic of Karakalpakia), which borders the Aral Sea.

'You must remember, the Aral Sea is not only an ecological catastrophe ... It is above all a human one.' Abdikrim Tleyov, 1995

In order to irrigate vast plantations of cotton as well as rice, water from the two rivers (the Amu Darya and the Syr Darya), which naturally fed the Aral Sea, has been systematically diverted over some 40 years. The result is that what was previously a vast stretch of water has been largely transformed into desert, with just three small,

highly saline lakes remaining. According to Don Hinrichsen, who travelled through the region in 1995 and subsequently wrote a detailed report (see page 296), the former seabed had become a salt pan desert, 'a virtual graveyard littered with the rusting remains of fishing trawlers and barges and the bleached bones of cattle, which died from eating salt-poisoned vegetation'. What looked from above like snow was salt, blown by the winds as far as the Himalayas.

Now incorporated into Uzbekistan – the world's fifth largest producer of cotton – the Karakalpakia Republic was a region selected by the former Soviet Union for the development of huge cotton plantations. Since cotton would not grow here naturally, in 1958 vast canals specially constructed for the purpose began to siphon water

from the Amu Darya river for the irrigation of the new cotton fields and rice paddies in Uzbekistan, Turkmenistan and Tajikistan. The Aral Sea began to shrink.

An environmental tragedy

Between 1960 and 1995, the surface of the Aral Sea shrank by over half, from 64,500 sq km (24,900 sq miles) to less than 30,000 sq km (11,600 sq miles), dropping in level by 19 m (62 ft) and tripling the water's salinity. Over 50 lakes in the Amu Darya delta dried up. By 1995, the town of Muynak, formerly on an island in the delta with the waters of the sea lapping against its shore, was 70 km (43 miles) away from the sea.

Whilst this disaster was occurring, the heavy use of pesticides and other treatments for the cotton being grown in the region meant that drainage water became polluted with agricultural chemicals and, along with untreated industrial and municipal waste, was entering the Amu Darya river, poisoning drinking water. The 3 million ha (7.4 million acres) of seabed now exposed to weathering has increased soil salinization and desertification, and whilst the sea formerly regulated the local climate, frequent dust storms now scatter salt and pesticides for miles around. The irrigation of cotton with water from the Aral Sea has raised the water table, bringing salt formerly held by

Americans are said to own, on average, seven pairs of jeans each.

68

No clean water – lives destroyed

Health problems on an unprecedented scale are the most shocking result of the abuse of the waters of the Aral Sea.

■ In 1995, it was estimated that over 1 million people living in the republic were under 'immediate and continued health threats from toxic pollution'.

■ Drinking water containing high levels of heavy metals, salts and other toxic substances and vegetables contaminated with the pesticides were routine.

■ Women and children were the most severely affected by the pesticides and other pollutants that have accumulated in the sea. Many had kidney diseases, thyroid dysfunctions and high levels of lead, zinc and strontium in their blood.

■ Over 80% of women in Karakalpakia were suffering from anaemia. Most women haemorrhaged while giving birth and the region had the highest levels of maternal and infant mortality in what was the USSR.

■ Diseases of the liver and kidneys, especially cancers, had increased 30–40-fold, arthritic diseases 60-fold and chronic bronchitis 30-fold since 1965. Tragically, the region is desperately under-equipped in terms of facilities and people to deal with this disaster. In 1995, there was only one children's hospital in the entire republic.

the soil to the surface, such that the whole of the Karakalpakia region is either too overloaded with salt or waterlogged to make normal agriculture possible.

No more fish – an economy destroyed

Until the late 1960s, about 3,000 fishermen earned a living from the rich waters in the vicinity of Muynak, catching 22 species of fish. But by 1982 all commercial fishing had ceased as the sea itself had become too distant and too salty to sustain enough fish to catch. Now only four species survive in the polluted water of the delta and practically nothing can be harvested.

Any conversion from fishing to agriculture has also failed for most people, as the soil is too salty for most crops to grow. The collapse of the economy means that the majority of the people have no income and have been obliged to live a hand-to-mouth existence by bartering. Their problems have been compounded by the government of Uzbekistan, which, with its own chronic water shortages, has ignored the plight of the people who are ethnically and culturally distinct from the Uzbeks.

Whilst some international efforts have been made to address these problems, very little has actually changed. Recent modifications in farming practices are reported to have begun to slow the further shrinking of the Aral Sea, but the human tragedy remains.

69

Potatoes and cereals for size

Before any of our cotton and some of our synthetic clothing can be made, half of the threads involved will have come into contact with some important substances produced by other plants. Once they have been spun, the fibres that make up the warp threads in most of our cotton and some synthetic clothing must be 'sized'. Sizing provides a protective coating to the fibres, preventing the warp fibres (placed lengthwise in the loom) from being pulled apart during the rapid weaving processes without losing elasticity, thus increasing their tensile strength and improving abrasion resistance during weaving. The size solutions used are often made from natural starch.

Though it is present in almost all plant parts, starch tends to be stored in the largest quantities in seeds and tubers. In America, cereals such as wheat and maize (mostly waxy maize, whose starch has a different chemical make-up) are often used, whilst in Europe these are joined by potatoes (of which Holland is a principal producer). In the Far East, rice and tapioca are important sources of starch, though the different qualities and applications of each source mean that the starches produced in a particular region may, in theory, be used any-where in the world. Other plentiful sources of starch that have been used commercially include cassava tubers, arrowroot, taro tubers, sorghum grains and the pithy stems of the sago palm.

The starch itself (a natural polymer made up of glucose molecules) is present within the plant cells in the form of insoluble granules. When the cells are crushed – for example, when a potato is cut or peeled – and then placed in water, the starch is simply washed out and the water takes on a characteristically milky appearance. If the water is then evaporated, a solid starch residue is left behind, which can easily be dried and powdered. Cornflour, arrowroot and potato starch are all sold in this form for domestic use. For use as a fibre size, most of the

Corn for clothes

Once used to stiffen collars, corn (maize) starch is the raw material behind a **new kind of plastic** now being turned into a range of biodegradable products, including clothes. Known as PLA, polylactic acid is obtained from the dextrose derived from corn starch after a process of **fermentation**. It can be adapted to form a wide range of products from plastic cups to car-pets. As a **clothing fibre**, PLA is said to be very versatile, combining the feel of natural fibres with the performance of synthetics. Both biodegradable and coming from an annually renewable resource, the promoters of PLA claim that its use has **environmental benefits**.

Critics, however, have drawn attention to the fact that the makers of PLA are leading breeders of GM maize, which will be used in its products.

starch used today is in modified form (a 'starch derivative'), having been physically and/or chemically modified to lower the level of viscosity by treating with heat, acids, oxidants, enzymes or chemicals.

The type of size selected depends on its price, the kind of yarn to be treated and the use to which the woven fabric will be put. While wheat and maize starch are currently cheaper than potato starch, the films formed by potato starch derivatives are tougher and more flexible than those from cereals. Cotton is often sized with potato starch derivatives. Two other sizes – sodium carboxymethyl cellulose (SCMC), produced by treating wood pulp or cotton linters with alkalis (see page 22), and polyvinyl alcohol – are also used on various yarns.

Once the fabric has been woven, and before it can be worn, the size must be removed. Since potato starch, for example, is not strictly soluble in water, the sized material is usually treated with hot water, enzymes (which break down the starch into sugars or smaller molecules), or acids, or a combination of these.

Natural starches and starch derivatives are also used for 'finishing' some textile fabrics. This involves passing the fabric through a dilute solution of a natural or modified starch and then removing the excess before drying. This improves weight, smoothness, stiffness and strength of the cloth, though it will be removed on washing.

Sizes and finishing agents are also made from natural plant gums – including guar gum (from *Cyamopsis tetragonolobus*) (see page 33) – and seaweeds, such as *Macrocystis pyrifera* and *Laminaria* species, which are important sources of alginates used both for sizing and the thickening of printing inks in very large quantities.

Native potatoes growing at 3,000 m (10,000 ft), in the Ecuadorian Andes. The potatoes that supply starch for textile sizing and numerous other uses are – unlike these – mostly grown in highly mechanized industrial systems in many countries worldwide.

Linen: strength and sophistication

The main purpose of sizing cotton yarns (see previous page) is to give strength to the warp threads that form the cloth's backbone. Until the early 19th century, these warp threads were generally replaced in the hand-woven cottons of Europe and America by the much stronger fibres of another plant: flax (see left).

Flax was, in fact, the principal source of plant fibre for all the clothes worn on both these continents until the time of the Industrial Revolution. A slender, herbaceous plant with striking sky-blue flowers, flax was domesticated in the Near East about 10,000 years ago, spreading later to Greece and reaching Germany about 6,500 years ago.

The world's most ancient textile is a fragment of linen, discovered in southeastern Turkey and which has been shown by radiocarbon dating to be 9,000 years old. In Europe, the richest finds of prehistoric linen have been made among the remnants of Neolithic lakeside villages in Switzerland, which date from about 3200 to 2600 BC. The Egyptians wore fine linen clothing, and also used the cloth to wrap the bodies of their dead. The oldest surviving garments from Egypt are linen shirts or dresses dating back to 2500 BC. Long before the Christian era, linen manufacture had become part of domestic life around the Mediterranean basin and both the Greeks and the Romans appreciated its lustrous quality and texture. Some early linen fabrics were so fine, with up to 500 threads per square inch, that a piece the size of an overcoat could be pulled through a ring the width of a £1 coin.

Despite the ancient ancestry of flax in Switzerland and its use in other parts of northern Europe in prehistory, the Romans are the people credited with promoting the development of linen manufacture in France and Belgium. The Emperor Charlemagne established centres of linen weaving in several Flemish cities in the 9th century, and by the late Middle Ages much of Europe's woven cloth was made of a mixture of wool and linen. Most European monasteries cultivated flax, as linen was prescribed for ecclesiastical garments and shrouds.

Britain's naval strength developed with the aid of linen sails and ropes made from home-grown hemp. Both these raw materials became so important that Henry VIII had legislation passed requiring flax or hemp to be widely grown.

By the mid-18th century, with the emigration of Protestant linen workers from Roman Catholic Belgium and France, Ireland had

Flax

■ The flax plant (*Linum usitatissimum*) with its wiry stems and distinctive soft blue flowers is just one of some 180 species in the *Linum* genus native to temperate and sub-tropical regions. Used by people since prehistoric times, flax provides fibres from its stem for linen and also linseed oil, which is crushed from its seeds.

■ Different varieties have been developed to maximize production of these raw materials: those with long and relatively un-branched stems up to 1 m (39 in) high are grown for fibres, whilst the shorter-stemmed varieties with larger seeds are cultivated for oil.

Flax field, UK. The striking soft blue flowers help identify this distinguished source of fibre and oil.

73

become a major centre of linen weaving. Irish linen and fine hand-made lace came to acquire world fame, a reputation that still exists today. From Ireland, the skills of flax weaving were taken to North America with the early colonists, and a sturdy cloth, 'linsey-woolsey', made from flax and wool became the main clothing material in the colonies. It was not until the 19th century that cotton grown cheaply in America and woven into cloth in England overtook linen as the major plant fibre used for making clothes.

Whilst helping it compete with other fibres, which are now processed more cheaply, the special qualities of linen have singled it out as a modern luxury material. Two to three times as strong as those of cotton and, up to 1 m (39 in) long, flax fibres are naturally smooth and straight. Our English word 'line' is derived from the Latin *linum*,

74

Flax fibre production

The flax grown for use in textiles is harvested not by cutting, but by **pulling the stems**, to preserve the full length of the fibres, which run along them. The flax 'straw' as it now becomes is then dried and the seed pods removed. To extract the fibres the straw must be **retted**, a process that breaks down the plant material binding the fibres together. This is done either by leaving the straw on the ground for some weeks so that the action of the dew and bacteria gradually rots away the non-fibrous parts or by **soaking** in tanks of warm water, a much quicker process.

The next stage is known as **scutching** and it involves the retted straw being fed between cylinders which separate the fibres and remove any woody pieces or bark. In this way, both long and short fibres are obtained for textile use together with a certain amount of 'shive' – waste woody matter.

After scutching, the flax fibres are **combed** and slightly **twisted** before being spun into yarn. Traditionally, they were **bleached in the sun** to remove the natural yellow colour (to which which the expression 'flaxen-haired' refers), but most flax is now bleached chemically.

To produce very fine and supple yarns, flax fibres are drawn through a trough of warm water and **spun** while wet, a process that softens them. Heavier yarns are spun dry. Renowned for their hard-wearing properties, the yarns are finally knitted or **woven into cloth** alone or mixed with other fibres such as wool and cotton.

The special processing required by flax fibres, unlike cotton, is very time-consuming and has never been highly mechanized. In addition, linen made from **hand-processed flax** is of better quality than that prepared mechanically. The current high price of linen reflects both these factors.

The antiquity of cordage and weaving

Fibres from plants were probably helping our **ancestors** to perform a multitude of tasks a good deal longer ago than we have generally been encouraged to believe. They almost certainly formed the **cordage** used to attach thousands of beads to the clothing of an elderly man, buried

some 25,000 years ago, whose remains were found at Sungir, 200 km (125 miles) north of Moscow. According to fibre expert Damien Sanders, who identified the cords of twisted **honeysuckle stems** used to drag the enormous oak stump into place which formed the centre of the

ancient monument Seahenge (2050 BC), 'My feeling is that cordage, along with fire and stone tools, is one of the pre-human tech-nologies [pre-dating *Homo Sapiens sapiens*], and that simple weaving as a development of cordage will even-tually prove to be almost as old.'

meaning 'flax', and a similar ancient west Germanic term. Flax fibres have special water-absorbing and drying qualities, too, enabling fabric made from them to 'breathe' well, and a lustre that is produced by their capacity to reflect light from both inner and outer surfaces.

The individual country producing most flax fibre in the world today – around 220,000 tons – is China, but almost this amount is currently produced by France (the world's second biggest supplier), Spain, the UK and the Netherlands combined.

Hemp for 'green' fibres

That most famous item of modern clothing – the pair of jeans – has come to be associated entirely with cotton. Available in any number of ever-changing styles and textures, denim jeans have become an enduring icon of fashion that looks set to stay.

When Levi Strauss made the first pair of jeans, however, they were intended as tough protective working clothes. They were made not of cotton cloth but of a much hardier fabric – a lightweight canvas woven from the fibres of *Cannabis sativa* (from which the term 'canvas' is derived), a plant more commonly known as hemp (see right and overleaf). The word 'jeans' appears to have its origins at least as far back as the 16th century. Literary references of the time mention 'yerdes of geanes fustion', referring to a hemp fabric being exported from Genoa in Italy. The cloth Levi Strauss came to use in America to make jeans was imported from Nîmes in France and it was a corruption of the French name for this material serge de Nîmes that gave rise to the word 'denim'.

It was the special qualities of hemp's stem fibres – much longer than cotton fibres, with eight times their tensile strength and four times their durability – that made it so useful, not only for hard-wearing cloth, especially sail cloth, but also for the manufacture of ropes and cordage of many different sorts, stretching back into prehistory. But in America the development of cotton plantations was to provide a much cheaper raw material, and in Britain the importation of cheaper hemp from abroad as well as the influx of cotton and other tropical fibres undermined local production. Though hemp was still being grown in Britain during the Second World War, for making twine and rope, the increased use of synthetic fibres severely curtailed its use.

By the 1960s, hemp had become much more famous for the resinous chemicals produced by the plant (often referred to as cannabis or marijuana) than for its fibres. Believing that any cultivation of hemp posed a danger to the public, it became a

Hemp leaf on harvested hemp straw, Dorset, UK. Described as one of the most versatile and useful of all plants, comfortable and long-lasting clothing is being made once more in Europe from hemp's stem fibres.

75

Hemp

■ Native to central Asia and valued for its fibres, which were used for rope, textile and paper manufacture, as well as for its useful oil and resin, hemp (*Cannabis sativa*) has a long and complex history of human use. Both cloth and rope made from hemp were being used in China over 4,500 years ago and the plant is mentioned in an ancient Chinese pharmacopoeia said to date from 2737 BC. The nomadic Scythians, who occupied the Russian Steppes from about 700 BC–AD 300, also grew hemp and used it for cloth, ropes and as a narcotic, and the plant probably reached central Europe in the Iron Age by 400 BC.

■ In Britain, the Anglo-Saxons seem to have been cultivating hemp from about 800–1000. By the Middle Ages, however, it was being grown all over England – mainly for its fibres (for clothing, sails, ropes and fishing nets), but also for oil (from the seeds), which was burned in lamps, for thatching straw and for medicinal purposes. The names of villages such as Hemel Hempstead and Hemphill are evidence of the importance of the crop at this time. Vital for the sails and rigging of ships, the increased use of hemp across Europe was linked to developments in maritime trade, which made it a strategic crop. Henry VIII passed legislation requiring everyone with arable land to grow ¼ acre of hemp or flax for every 60 acres.

criminal offence to grow the crop in the UK without a special licence, a situation that remains unaltered today. The commercial use of hemp was brought to an end by the Misuse of Drugs Act, which outlawed its cultivation in 1971.

Over the years, however, plant breeders have been able to produce varieties of *Cannabis sativa* that contain very low levels of the chemical delta 9 tetrahydro-cannabinol (THC), which produces the famous psycho-active effects, but have a higher stem fibre content instead. The crop that is now poised to stage a triumphant comeback in Britain is a variety that contains less than 0.2% of THC. Already grown for fibre and seeds on a large scale in a number of countries worldwide, including China (the main producer), Romania, Spain, Chile, Ukraine, Hungary and France, only one company has held a licence to grow the plants for commercial use in the UK since 1993.

A new age for hemp?

Described as one of the most versatile and useful plants known to man, hemp has been heralded as the environmentally friendly raw material for the 21st century. The great revival of interest in hemp is focusing not only on its fibres (which are finding a range of new uses from glass-fibre substitution and panelling for vehicles to major new components in house construction), but on the nutritional and medicinal properties of its seeds and oil.

Needing no fertilizer, herbicides or pesticides to grow well (and therefore meeting major requirements of environmental sustainability), hemp plants – which are annuals and which flourish in temperate climates – grow extremely fast. Under suitably warm conditions they are able to reach 4 m (13 ft) in height in only 12 weeks. This means, of course, that hemp is a prolific source of fibre. Helping us to overcome our reliance on imported fibres such as cotton and on synthetics, hemp is ideally suited to the production of organic clothing.

Like flax, hemp's stem fibres are extremely strong, but are generally stiffer because of the amount of lignin they contain. Until relatively recently, this stiffness had been a factor in restricting the wider promotion of hemp for clothing. But hemp fabric does soften with use and recent advances in processing the fibres mean that a much softer fabric can be produced, making it much more comfortable to wear and giving it the look and feel of linen. As an added bonus, the hollow core of the hemp fibres allows hemp fabric to 'breathe'. Recent cultivation trials in southeast England have shown that hemp textiles can be produced in a sustainable way for the same cost as flax. Versatile, comfortable to wear and long-lasting, a variety of clothes made from hemp or hemp-mix fabric (for example, with cotton or silk) are now being marketed in the UK and other parts of Europe.

More plants for clothing

Many other plants supply us with the raw materials that are or can be used for clothing. They include nettles, the cork oak, kapok, hardwood and softwood trees, and (indirectly) species of mulberry.

Ramie

Another very tough fibre – claimed by some to be eight times stronger than cotton and three times stronger than hemp – is ramie. Looking rather like the stinging nettle to which it is related, but without the sting, two varieties of *Boehmeria* provide fibres that are known as ramie. Many native names exist for the plants, but *Boehmeria nivea* var. *tenacissima* (native to Malaysia) is generally called green ramie or rhea, whilst the better-known *B. nivea*, with soft white hairs on the underside of the leaves, is called white ramie or China grass.

White ramie, native to China and Japan, and the only variety that is cultivated on a commercial scale, is a hardy perennial shrub that can grow to a height of about 3 m (10 ft). About 70% of the plant consists of fibres that are very tough and water-resistant – in fact, they become much stronger when wet. To extract them from the bark and sticky outer tissues in which they are contained, these parts are first peeled from the stems, then scraped, washed and dried before being treated with caustic soda to dissolve the residual pectins and gums. Before spinning into yarn, the fibres are often chemically bleached and also softened to stop them from becoming too brittle.

After perfect processing, ramie – pure white in colour, with a silky lustre – is said to be one of the longest, strongest, silkiest and most durable of all fibres extracted for human use from plants. Although lacking the elasticity of wool or silk and the flexibility of cotton, it is able to keep its shape without shrinking and its smooth lustrous appearance improves with washing. Ramie fibres can be separated almost to the fineness of silk and have been made into lace and virtually transparent fabrics in China. It is recognized in the clothing industry as a premium, high-quality product, but its labour-intensive production, difficulties with harvesting (the plants tend to grow in uneven stands), plus variations in the quality of the fibres supplied, mean that ramie has not been promoted as a major world textile fibre.

Ramie plants have been introduced to the warmer parts of Europe and North and South America in relatively recent times. Highly productive, they can be harvested up to six times a year under good growing conditions. The main producer countries are currently reported to be China, Brazil, the Philippines, India, South Korea and Thailand.

Cork clothing

■ Cork has been used for centuries in shoe manufacture, but now it is available for clothes. After many years of research, a new and very versatile lightweight material has been made in Italy from very fine layers of cork, attached to backing materials such as cotton or viscose.

■ Soft and smooth, resistant to salt-water, scratches and stains, the new cork material can be printed or coloured, and is suitable for a variety of clothing applications, as well as accessories and footwear.

Giant nettles

■ Thriving in forested regions at altitudes above 1,500 m (4,900 ft), on land unsuitable for crops, the giant nettle *Girardinia diversifolia*, which can reach a height of over 3 m (10 ft), is still an important source of fibre for the people of the Koshi Hills district of Nepal. Clothing, bags, sacks, mats and nets are all made from the strong, smooth and lustrous fibres of this mighty relative of the common stinging nettle (commonly known as allo), which is covered with thorn-like stinging hairs.

The main importing countries are Japan, Germany, France and the UK, but only a small proportion of world production (thought not to exceed about 130,000 tons a year) enters international trade as most is used in the countries where it is grown.

Often blended with cotton, wool, silk or man-made fibres, ramie has been used commercially, chiefly in Asia, to manufacture a range of woven and knitted fabrics – from those resembling fine linen to coarse canvas – for all kinds of clothing, from shirts and shorts to sweaters. Garments including ramie have also become available in other parts of the world, including Europe from time to time. A current major end use of ramie in Britain is for sewing thread, ribbons and tapes and upholstery fabrics. In Asia, other uses of ramie include hats, fishing nets, cordage, sacking, carpet backing, sailcloth and table linen. Short ramie fibres from processing wastes are also used in the production of high-quality paper, such as for bank notes and cigarette papers.

Stinging nettles

Though much maligned and commonly regarded as an unpleasant weed today, the stinging nettle (*Urtica dioica*) offers a range of uses (including food, medicinal, dye, compost and traditional insect repellant) for which it deserves a much higher place in our esteem. Perhaps the most significant of these is the provision of strong and versatile fibres that have been used in Europe since prehistoric times – alongside those of flax and hemp – for making cordage and cloth. Because of the difficulty of removing the fibres from nettle stems and preparing them for use, it is thought that only small quantities were ever made, but stinging nettle fibre was still being woven into cloth into the early 19th century in Scotland and Scandinavia. In northern France and Germany, where it was known as *nesseltuch*, nettle fibres were reported to have been bleached to make a cloth as white as linen.

From the beginning of the 18th century, attempts were made to commercialize the production of nettle fibre, especially in Germany, but success was variable. During the Second World War, considerable research was done in Britain into the properties and uses of nettle fibre, among other things, for a very tough paper that could be used as a reinforcing agent for strengthening plastic panels in aeroplanes. Research into the applications of nettle fibre is still continuing, but methods of cultivation and processing have now improved and the future of large-scale nettle fibre production, for use in contemporary clothing, shows greater promise.

Wedding dress made from cork material, with lace overlay, Sardinia, Italy. In a new process developed in Italy, cork from native oaks is being used to make, amongst other things, a lightweight material suitable for clothing, shoes and handbags.

Mulberry trees

■ *Morus nigra* and *M. alba* are deciduous trees growing to a height of 6–9 m (20–30 ft). The leaves, which cover the densely spreading branches, contain traces of essential oils that are attractive to silkworms and a sticky, rubbery sap that is believed to influence the composition of the silk. Each strand of silk, produced by glands and secreted from the silkworm's mouth, is three times stronger than a strand of steel of the same density.

■ The use of mulberry leaves as food for silkworms stretches back thousands of years, but the long-lived trees have also been grown for their bark (in the case of *M. alba*, a traditional material for Chinese paper making) and for their blackberry-like fruits. In the Middle Ages they were eaten as a purée or used to add colour and flavour to wine.

Mulberry trees

Some of the most famous wedding dresses in history have been made in part by caterpillars. Though silk (the most expensive fibre in the world) is not produced directly by plants, it could not be made without them. Many different sorts of silk-producing caterpillars worldwide feed on many different plants, but the most famous are the mulberry trees, whose leaves provide the staple food for the common silkworm *Bombyx mori*. Hatched from the eggs of silk moths, these caterpillars eat the leaves until they start to spin their silk cocoons, which house them as they transform into moths. Most silkworms will never reach this adult stage, however, as their cocoons are carefully unravelled and made into fine silk thread.

Two species of the very large mulberry family, *Morus nigra* and *M. alba* (see left), have become the traditional food of *Bombyx mori*, which produces the finest white to yellow silk and also the largest proportion of the world's silk supply. Both are believed to be native to Asia; the white mulberry (*M. alba*) to the mountainous regions of central and eastern China, and the black or common mulberry (*M. nigra*) to the mountains of Nepal and the southern Caucasus.

Designed to help disperse the seeds to which they are attached, kapok floss or fibres develop inside cylindrical fruit pods, which split open when ripe. For commercial use the pods are gathered just before this and the floss, which falls easily from them once they are open, is spread out on the floor of open-sided sheds to dry. Frequent turning ensures the floss dries evenly, expanding as it does so before the seeds and other foreign matter are removed. Kapok fibres are so light and fluffy that workers wear fine protective masks and the drying sheds are enclosed in fine wire mesh to stop the fibres floating away.

While kapok fibres have been used in sound insulation for aircraft (it is one of the best sound absorbers per unit of weight), recent research into their excellent oil-absorbing properties has indicated their great potential for soaking up oil from seawater – a future remedy, perhaps, for oil spills at sea.

Fine fibres from plants

Some of the most unlikely plants have been used to make the most **delicate** of fabrics. In the mid-19th century, an extraordinary experiment was conducted in **County Carran**, in Ireland, with the aim of creating profitable work for poor families in the wake of the disastrous potato famine. **Lace** was made from the very fine fibres extracted from the stems of a range of plants, including sweet peas, honeysuckle, marsh mallow, Solomon's seal, stinging nettles and ox-eye daisies.

Pineapple clothing

If you thought silk and cotton made the finest shirts, have a look at one made of **pineapple fibres**. The Philippine Islands have a long history of making cloth as fine as gossamer from fibres extracted from the leaves of the pineapple (known there by its Spanish name *piña*) and this highly skilled practice continues today. Formerly more widespread, the industry is now centred in the province of Aklan on the **island of Panay**.

Making the fabric

Various pineapple varieties are grown in the Philippines, but the one most commonly cultivated, and with a high fibre content, is the **Spanish Red** or Philippine Red (*Ananas comosus*). Mostly grown in smallholdings, the leaves reach up to 2 m (6 ft 6 in) in length and are harvested at 18–24 months. To extract the two kinds of fibre found in the leaves, they are first **scraped** with a broken china plate or saucer. After folding over a few inches of the leaf base, the first set of fibres are pulled off and set aside. The leaf is then scraped again with a piece of **coconut shell** so the finer fibres, more fragile than human hair, can be extracted. Fibre bundles are then washed and further scraped – this time with a shell – under running water. After partial drying in the sun, the fibres are beaten with a **bamboo stick** to separate them and hung up to dry. Finally, the fibres are combed before being knotted together, strand by strand, and coiled ready for **weaving on looms** made of wood and bamboo. After weaving, the cloth, which is remarkably strong and durable, is washed in rice water or

citrus juice, before embroidering or making into articles.

Recently, the production of textiles, from *piña* **fibre** has staged an impressive comeback. In the 1850s, a single town in Panay boasted some 60,000 weavers but by 1986 fewer than 50 mostly elderly people were weaving *piña* fibre in the Philippines. The subsequent initiation of a development project and training centre, plus a huge increase in pineapple planting, has meant that there are now **several hundred weavers**, and many people involved in production – from planting to weaving.

It takes about a day to weave 25–50 cm (10–20 in) of *piña* cloth by hand. **Ten plants**, which will supply about ten leaves each after two years, produce enough fibre for about 1.8 m (6 ft) of fine and 2.7 m (9 ft) of coarse cloth.

Traditional items of clothing made from *piña* cloth include handkerchiefs, fine shawls and the 'Maria Clara' costume, as well as traditional *barong* shirts. Many wedding dresses, christening gowns and some of the garments that comprise the Philippine national costume are now made from *piña* fibre. The international market is taking notice of this **exquisite** hand-made fabric, still too expensive for most Philippinos to buy.

Kapok

■ Kapok comes from the fruits of one of the tallest trees in the tropical forest – the silk-cotton tree (*Ceiba pentandra*) – pollinated by bats and bees.

■ Before the arrival of the conquistadores, Mayan and Aztec peoples revered the tree and regarded it as sacred. Its size and stature led them to regard it symbolically as a link between the earth and the universe, its deep root system and magnificent spreading canopy seeming to support the sky whilst reflecting the social ordering on earth.

■ Kapok is thought to have spread naturally from South and Central America to various parts of Africa, but to have been introduced to Southeast Asia by humankind, where it is found in many countries today. Though the trade has diminished in recent years because of competition from synthetics, Thailand and Indonesia (and especially the island of Java) are important producers of kapok, exporting mostly to Japan, China, Europe and America.

Wood pulp

Some of our silkiest clothes, and some of the most hard wearing, will almost certainly be made from wood pulp. Everything from underwear and swimwear to flame-retardant military combat uniforms could all be made from eucalyptus logs.

Synthetic fibres now account for about half the total fibre consumption in the world. Whilst about 40% of global textile production, which includes fabrics such as acrylic and polyester, are derived from oil (itself the result of the compression of plants for millions of years), around 10% of all our textile fibres are made directly from the cellulose contained in living plants. Viscose (a 'rayon' fibre, the name given to the first artificial silk and now used to make a range of fabrics), lyocell (also sold under the brand name 'Tencel'), diacetate (used to produce fabrics such as 'Dicel') and triacetate are perhaps the best known of these.

It was the desire to produce a material with the look and texture of silk, but not the cost, that brought about the invention of the first synthetic fibres. As early as 1664, the English chemist Dr Robert Hooke had thought there might be a way to make 'an artificial glutinous composition' to rival the fibre produced by the silkworm. But it was not until the mid-19th century that the first discovery of the usefulness of plant cellulose was made – not just for fibres, but for materials that would later be used to make hard films and plastics.

Wood chips for pulp. Wood pulp is the main source of the cellulose used industrially for conversion to man-made textiles such as viscose, lyocell and triacetate. It is made into products as diverse as velvet dresses and military uniforms.

82

In 1846 the German chemist Christian Schönbein made the crucial discovery that cellulose in cotton could be made soluble. This happened when he used a cotton apron to mop up some sulphuric acid and saltpetre that he had spilled. As the substances mixed, the apron was inadvertently turned into cellulose nitrate and exploded as it dried. The scientific interest that the event initially aroused was directed, not surprisingly, at the potential of this new but unstable compound as an explosive. But more inventions followed. In 1885, another English chemist, Sir Joseph Swan, produced cellulose nitrate in the form of filaments and yarn, which he used as electric light bulb elements. Some seven years later, three British scientists discovered and patented another process for dissolving cellulose by converting it into sodium cellulose xanthate, or viscose. A little later, they discovered a way of making cellulose acetate – another soluble cellulose derivative that could be spun into fibres and made into solid plastic.

Schönbein's apron had reacted so spectacularly when it soaked up the acid because cotton fibres are almost pure cellulose. Cotton linters thus came to supply the raw material for the new fibre industry, but as demand outstripped supply, wood pulp was substituted and remains the form in which most cellulose is commercially available today.

The type and source of timber made into pulp has changed over the years. At one time the pulp used in Britain, for example, was made mostly from softwoods such as spruce and pine species from northern Europe and Scandinavia, with the addition of birch, beech and aspen and other common temperate hardwoods. But the demand for ever-increasing amounts of pulp brought about the need for faster-growing trees, and tropical and subtropical woods such as eucalyptus and – to a lesser extent – *Acacia* species now form a substantial part of the dissolving pulps produced. Much of the pulp now used in Britain comes from huge plantations in South Africa and Brazil. Both these countries have encouraged major investment in eucalyptus trees, which will produce usable timber in 8–12 years from seedlings grown in nurseries. Certain species will also produce shoots that can grow up to 5 m (16 ft) per year after the trees have been felled, and coppicing is practised in some areas to ensure a continuous supply of wood.

The world's biggest factory producing 'dissolving pulp' is near Durban, in South Africa. It supplies around 600,000 tons of wood pulp a year for use by the textile industry and by the producers of a huge range of cellulose derivatives from different sorts of paper, plastic and cellophane to thickeners used in the chemical and food industries. The timber is grown on nearly 202,000 ha (500,000 acres) of land in KwaZuluNatal and Mpumalanga. Nearly all this pulp is exported to markets in Europe, America and Asia. Much of it is derived from *Eucalyptus grandis* (or related hybrids), which is a very fast-growing

Making viscose

Wood fibres, the walls of which are mainly made of cellulose, are so strong and tough that they can only be separated with the help of powerful machinery and chemicals.

Once felled, logs are stored for **several weeks** to reach a uniform moisture level before being cut into small chips ready for pulping. They are then fed into computer-controlled digesters where **chemicals** liberate the fibres in the wood. This 'cooking' process reduces the logs to cellulose pulp, which is then washed and bleached and finally pressed into thick sheets ready for making into viscose or other **synthetic** materials.

To produce viscose fibres, the wood pulp sheets are steeped in a solution of **caustic soda** to convert the pulp to alkali cellulose. This is then pressed to remove excess liquid and the resulting material is ground into fine crumbs and added to **carbon disulphide**, which reacts with it to form sodium cellulose xanthate. More caustic soda is used to dissolve this compound and it is then filtered before being squeezed through the fine holes of a spinneret or jet, into a bath containing a mixture of salts and sulphuric acid. This **final stage** solidifies the new viscose fibres and they are either spun into yarns in the form of **continuous filaments** – principally for dress-making fabrics and furnishings – or cut into shorter lengths for spinning into other sorts of viscose cloth.

species introduced from Australia. Other eucalyptus species used include *E. smithii*, *E. nitens* and *E. dunnii*, while two *Acacia* species (*Acacia mearnsii* – see page 89 – and *A. decurrens*) account for 10–15% of the pulp or 'furnish' as it is also known.

Generally speaking, the tree species used to make pulp depends on where the mills are situated and the economics of growing or importing plantation or other trees. The USA uses a variety of mixed southern hardwoods as well as conifers such as western hemlock and various 'southern pines', spruce and Douglas fir. India, on the other hand, uses a mixture of tropical hardwoods, eucalyptus species and bamboo, amongst others, whilst Japan and Thailand have used mangrove trees, which grow abundantly in coastal regions of the tropics, to supplement yields of non-native eucalyptus.

Despite a preference for pulp made from 'controlled' plantation species, since it is generally of higher quality, non-European countries may use tropical hardwoods, if the price is right, cut from remote areas of rainforest.

Trees for babies

Wood pulp is keeping millions of babies comfortable and dry. Alongside small amounts of plastic and an absorbent gel, it is the basic raw material that makes up most disposable nappies **worldwide**. In Britain alone, around 9 million disposable nappies are used every day.

The trees involved vary according to the country in which the nappies are made and the source of the **wood pulp** used. Many of the nappies available in Europe are made from pulp obtained from plantation trees grown as monocultures in Scandinavia and North America. A percentage of the pulp used, however, still comes from clearing previously untouched forests, for example, ancient 'old growth' forests in Canada. About a third of the **pulp** used to make some nappies comes from sawmill residues. After

felling, the trees are reduced to pulp by chemical processing and this pulp is **purified** to increase absorbency, by bleaching – traditionally done using chlorine, but increasingly achieved by an oxidation process.

The production of disposable nappies, however, is now a major **cause of concern**, not only because of the raw materials consumed but because of the problems associated with their disposal. In Britain, the nappies currently thrown away account for around 4% of total household waste, and weigh about 1 million tons. About 25% of the **nappy waste is paper** together with plastic and chemical components, the rest is human sewage.

Some of the plastic used has been assessed to last **indefinitely** (or perhaps take up to 500 years to fully

decompose) and the rotting process – in combination with that of other household waste – releases **harmful gases** and acids, which can dissolve metals and leach into groundwater.

Reusable cloth nappies (though they are likely to be made of non-organic cotton) do not, of course, carry these problems. Disposable nappies have also been shown to use over eight times as many irreplaceable raw materials and to require up to **thirty times more land** for growing the natural materials than reusable nappies. Many washable nappies are now greatly improved in design and comfort, being **shaped and fitted**, and are increasingly promoted for use by local authorities and hospitals in the UK. A return to them is widely regarded as the best way forward for our environment, and for us all.

Dyeing and printing with plants

Most of our clothes are dyed or printed. For thousands of years plants have supplied most of the basic raw materials for these operations and in many traditional societies throughout the world they continue to do so. In most economically developed countries, however, the large-scale industrialized dyeing of textiles has come to rely on standardized synthetic dyes derived from petroleum and coal tar. Since the widespread takeover of synthetics, plant-derived dyes, which can be much richer and more subtle than their chemical equivalents, have tended to be used only by interested individuals or in small cottage industries. But a growing awareness of the unacceptable 'footprint' of synthetic dyeing in terms of damage to the environment (20% of textile dyestuffs, for example, currently enter the waste water during the finishing process, causing high levels of pollution) and to human health are prompting new legislation. In some places initiatives are under way involving the new use of traditional dye plants. One such plant is madder (*Rubia tinctorum*), formerly of great importance in Europe and the Middle East (see page 282). Its roots are the source of a range of red, pink, violet or brown dyes, depending on the mordant used. Madder is currently the object of renewed interest in terms of its commercial development as a clothing dye in Europe.

Following on from research previously undertaken in the UK, a project involving several European countries is now actively promoting the reintroduction of another very important plant-derived dye, natural indigo (see right) (still the focus of a significant industry in parts of India and China), as a commercial product. Several different plant species produce the natural blue dye known as indigo. In Europe, the best known of these is woad (*Isatis tinctoria*). In various tropical and warm temperate countries, other sources of natural indigo include several plants in the genus *Indigofera* (in particular *I. tinctoria* and *I. suffruticosa*), *Lonchocarpus cyanescens* (particularly important in West Africa), *Persicaria tinctoria* in Southeast Asia and *Strobilanthes cusia* in China. The EU-funded Spindigo project, coordinated from the University of Reading, is growing three indigo-producing plants, including woad, in different climatic environments in Finland, Germany, Italy, Spain and the UK in order to demonstrate the potential as commercial crops within Europe.

Around 80,000 tons of synthetic indigo are produced each year worldwide, much of it to dye denim, a large proportion of which is used to make the vast quantities of jeans that we buy each year. But the chemicals used in this process are highly polluting and hazardous to health. By providing a natural source of indigo – much in demand by

Indigo

■ Indigo has a long and distinguished history, dating back to prehistoric times. Famous for its rich and beautiful colour, it had become an important item of Indian trade in antiquity and was valued as a dye, a pigment for paints, a medicine and for cosmetic use. Whilst the legendary Bluebeard is said to have taken his name from the customs of the ancient Persians of dyeing their beards blue with indigo from *Indigofera* species, Europe's Celts used indigo from woad on their bodies. By Medieval times, a large industry of indigo production had developed in Europe based on the cultivation of woad and it became the source of great wealth for many farmers and merchants. However, this industry declined from the early 17th century because of competition from imported indigo produced from tropical *Indigofera* species.

■ Until the development of a synthetic substitute, all blue textiles were dyed with natural indigo. These included the blue serge uniforms worn by the British police force and hospital staff, as well as military and naval personnel, the Chinese 'Mao' suit and workman's clothes worn by millions of people and which inspired the term 'blue-collar worker'. Denim material was also originally dyed with indigo, and when a range of synthetic blues became available, synthetic indigo was used to give denim its distinctive look.

85

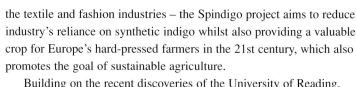

the textile and fashion industries – the Spindigo project aims to reduce industry's reliance on synthetic indigo whilst also providing a valuable crop for Europe's hard-pressed farmers in the 21st century, which also promotes the goal of sustainable agriculture.

Building on the recent discoveries of the University of Reading, the project aims to provide indigo producers with novel technology that they can use in situ, for the extraction and purification of the dye. This means that farmers will be involved in the entire production process and be able to sell direct to the European market.

Dyeing auxiliaries from plants

Though labelling may soon tell us that our jeans, for example, have been coloured with natural indigo, we are unlikely to be aware that many other plants are still used as vital auxiliaries to the dyeing and printing processes. Derivatives of commercial plant oils, for example coconut, palm, soya bean and linseed, play an important part in the manufacture of detergents used for cleaning crude wool before it can be spun. These oils also help prepare and soften the surfaces of many woven and knitted fabrics before printing or dyeing can take place. Pine oil from a number of pine species grown in Europe, America and India is widely used too as a wetting and dispersing agent, helping to ensure that dyes will 'take' properly and spread evenly when applied. Plant gums are particularly useful. Gum arabic, the reddish sap tapped

Wild plants for dyes

Almost all plants will produce a colour of some sort. Some fabrics can be very difficult to dye satisfactorily, but in most cases the **pigment** extracted from the plant is 'fixed' to the material to be dyed with the aid of a **mordant**. This is a term that includes a range of substances, chiefly the salts of metals such as tin, aluminium, iron or chromium, but also including urine, wood ash and animal dung. A mordant forms a chemical bridge between the dye and the fibre molecules, making the colour **fast**. The colour achieved from plants can be very variable and will be influenced by factors such as the time of day picked, soil type and weather conditions, the **method** of dyeing and the mordant used. Different parts of some plants will also give different colours.

A number of familiar temperate plants produce a range of subtle shades, varying according to the mordant used. **Various yellows**, for example (the most common colour produced by all parts of many plants), can be extracted from onion skins, ragwort flowers, privet leaves, bracken fronds and Spanish broom shoots. **Green** can be obtained from bracken fronds, ivy and stinging nettles, whilst **reds and browns** are produced, for example, from a number of lichen species, privet berries, blackberries, cherry wood and alder bark. Almost all the textiles made before the mid-19th century from **all over the world**, exhibited in museums today, are dyed with plants and natural minerals.

Dogon tribal elder in Mali wearing indigo-dyed clothes. The production of natural indigo is a complex process, which varies from place to place. Indigo-yielding plants do not contain indigo itself, but colourless compounds that must be extracted and combined.

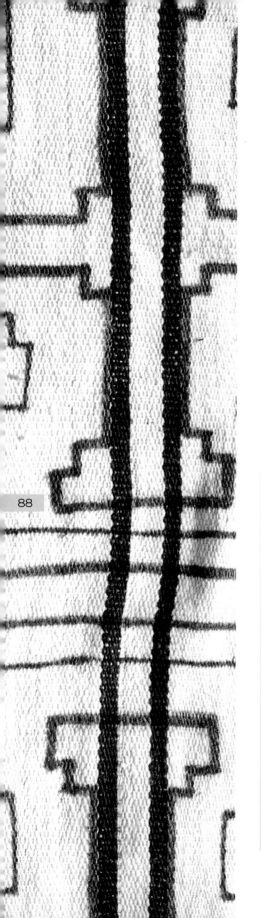

from spiny acacia trees that grow wild in Africa (see page 280), is used to give fabrics a soft finish and compact feel, as well as an even surface for printing. This gum was, in fact, vital for the invention of the lithographic printing process in the 18th century. Indispensable for this technique today, the application of gum to the parts of the plate required to print an image makes them receptive to the printing ink, while the areas left free will repel it. Carrageenan (see page 33), derived from seaweeds, is also used for these purposes.

An even pick-up and consistency of dye or ink, vital in large commercial dying or printing operations, has been greatly assisted by the use of natural gums (such as karaya, tragacanth, guar and locust bean), as well as alginates and natural starch derivatives (often made from potato or cereal starch). These are important too as thickening agents. They control the flow and elasticity of textile dyes and inks and allow multiple patterns to be printed that are sharp and bright, without penetrating the cloth too deeply. The solubility of these gums and their derivatives, plus their resistance to alteration by other chemicals, makes them highly valued in the textile trade.

Piro Indian skirt (detail): white and brown cotton with plant dye design.

Synthetic dyes: hidden dangers

'Many of the [8,000] chemicals used in fibre production and textile manu-facturing can cause serious damage to human health, not only for the workers directly involved, but also for consumers if these are retained as residues in the final product …. One estimate in Germany suggests that 30% of children suffer from textile-related allergies, most of which are triggered by dyes.' Robbins, N. and Humphrey, L. *Sustaining the Rag Trade* (IIED, 2000)

There are many different kinds of synthetic dye, but one group, the azo dyes, are known to be **particularly harmful**. About one-fifth of all tex-tile dyes fall into this category and many of these include compounds that are known to form carcinogenic substances. These can harm both the people making them and all of us who come into contact with them in our clothes. The use of azo dyes in clothing and bed linen has now been **banned in Germany**, but many other countries still use them.

In parts of the world some clothing companies are now using **low-impact dyes** and are phasing out dangerous chemicals used in textile processing. However, a common international standard for all aspects of textile production, which addresses **environmental, social and health concerns** and which includes codes of practice and labelling relating to dyeing, printing and finishing, is urgently needed.

Plants on the soles of your shoes

When it comes to shoes nothing quite matches the quality and feel of leather. Chinese warriors, Egyptian nobles and England's Anglo-Saxon men and women all used leather to protect their feet. Most of us own leather shoes or boots and take this strong and versatile material for granted. Without the grasses, grains and forage crops that feed our animals, there would of course be no leather, but it is the compounds in a range of other plants that have traditionally given it its special qualities. These compounds are known as tannins and until the end of the last century, when chemical alternatives were introduced, all hides were tanned with them to turn them into leather. Untanned hides exposed to moisture would very soon disintegrate and rot. Our museums contain several thousand examples of leather shoes and sandals (in Britain, many dating from Roman or Medieval times), excavated by archaeologists in a perfectly preserved state only because they were tanned with plant materials.

A wide range of plants contain tannin, but certain trees that have a high content have come to be most widely used. Traditionally, the sources that were plentiful and close to hand were those that were used. England, for example, depended very much on oak bark. But today those countries that are most industrialized tend to import tannin, often in spray-dried extract form, from the largest and cheapest commercial sources.

Because they must withstand great heat as part of the shoe making and moulding process without shrinking, and also be very flexible, the leather uppers of almost all the shoes produced in bulk are tanned with chromium. Most soft and pliant leather clothes and gloves are also made in this way. But leather shoe soles and accessories that need to be tougher, more resilient and almost stretch free – handbags and belts, for example – are tanned with plant extracts. Most leather that must stand up to heavy industrial use, as well as traditional items such as saddles, are also tanned with vegetable materials.

The current imported favourite in the USA, Japan and Britain is mimosa or wattle extract, which comes from the Australian tree *Acacia mearnsii*, or black wattle. In 2001 these countries imported around 5,500 tons of extract. This tree (see right) is grown in huge plantations in South Africa, Brazil, Kenya, Tanzania and Zimbabwe.

The next biggest tanning extract import is quebracho, the name given to a number of related trees of the genus *Schinopsis*, indigenous to the Chaco region of Paraguay and Argentina (see overleaf). The very tough quebracho wood has, along with wattle, supplied most of the world trade the last century. If the wild stands of these very slow-growing trees

Black wattle

■ The province of Natal in South Africa has been the main black wattle (*Acacia mearnsii*) producing area for over half a century. Black wattle was originally introduced from Australia as a shelter tree to plant around crops and homesteads, but plantation cultivation was encouraged to provide a replacement for depleted South American quebracho trees, which had come to supply most of the world with tannin extracts.

89

Quebracho

■ Several different Latin American trees with very hard wood have been called quebracho in the past, but only two *Schinopsis* species, *S. quebracho-colorado* and *S. balansae*, have been exploited on a commercial scale for the production of tannin extracts.

■ The slow-growing, very dense wood of these trees, which take 80 years to reach maturity, is so hard that it frequently broke the axes and the teeth of circular saws used to cut it down. The name 'quebracho' comes from the colloquial Spanish term *quiebra-hacha* or 'axe-breaker', but this did not stop the trees supplying more tannin for world use for many years than any other vegetable material. The ruthless over-exploitation of the trees, which grow singly or in small groups and whose natural distribution includes some 200,000 sq km (77,200 sq miles) of the Argentine and Paraguayan Chaco region, led to a serious depletion of numbers.

■ The heavy heartwood of quebracho trees is cut into chips and soaked in special vats to extract the tannin. This extract is particularly suitable for the sole leather of shoes, as it has weight-giving properties and produces a tough, firm feel.

■ Nearly 2,300 tons of quebracho tannin extract were imported into Japan, the UK and the USA in 2001.

become further depleted, however, plantations of fast-growing wattle established in South Africa may replace it.

The third favourite of the UK, USA and Japan is sweet chestnut (*Castanea sativa*), of which nearly 600 tons were imported in 2001. Chestnut is a traditional source of tannin for many European countries, and up to the 1950s another species, *C. dentata*, was grown extensively in the USA. Italy is the major supplier of chestnut today. With its traditional prowess in the fashion trade and in the face of a sharp decline in consumption elsewhere, Italy is now Europe's largest producer of vegetable-tanned leather for shoes.

Other traditional supplies of tannin have come from the acorn cups of the valonea oak, which is native to Turkey, and the sumac tree from the same area. Observing the practices of indigenous Indians, North America's colonizers used the bark of native hemlock trees for tanning leather. Australia, meanwhile, has always had a plentiful supply of wattle and different eucalypts, some of which have very high tannin contents, though none are currently commercialized in any quantity.

In tropical regions, mangrove bark is widely used and cutch from *Acacia catechu*, which is indigenous to India and Burma, is still popular. Another source from India is the fruit of the myrabolans tree. Tara extract – from the pods of wild *Caesalpinia spinosa* trees, exported from Peru – is becoming increasingly popular because its resistance to darkening, on exposure to light, is important in the manufacture of automotive upholstery leather.

More plants for our feet

The prevalence of synthetic compounds such as polyurethane and PVC for making shoe soles today has meant that the amount of leather tanned with plant extracts has decreased sharply. But other plant materials are still to be found in our footwear. Wooden sole units are still extensively used in clogs and similar footwear, while cork is sometimes used for the veneers on these units and as a core material for heels and soles (as well as for comfortable 'insoles' worn under the foot inside the shoe). Hard-wearing fabrics such as canvas made from natural fibres may also form the uppers, while synthetic fibres made from wood pulp as well as cotton may be tying up our shoes.

Natural rubber is in evidence too. Many wellingtons and agricultural boots are made from a base of natural latex and it is used as an adhesive, especially for linings in many modern shoes.

A shine from plants

Whether our shoes are made from leather or PVC, it is likely that we will polish them with waxes made from plants. Since it is harder than beeswax, carnauba wax from the Brazilian wax palm (see page 40) is

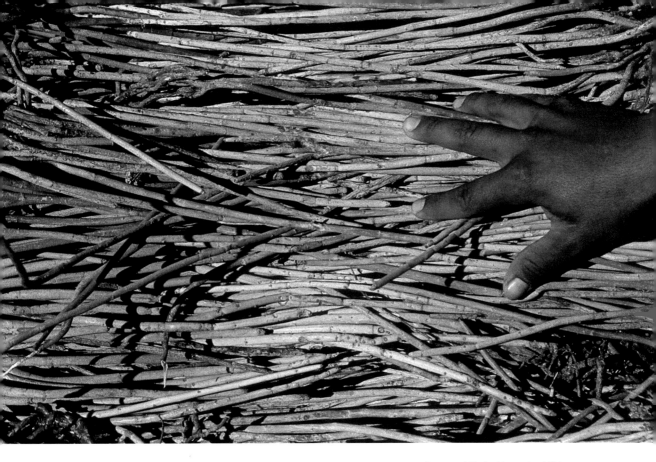

one of the main plant waxes used, forming 10–15% of the total volume of some products. Because of its hardness, only small quantities are needed to produce a shine.

Candelilla wax from Mexico is also found in some shoe polishes and wax from esparto grass and bamboo species has been used in different regions. Carrageenan, from the seaweed known as Irish moss, may be present in polishes, too, as it holds down and smoothes out the tiny rough projections on the surface of the leather.

Stems of *Euphorbia antisyphilitica* harvested in the Chihuahua Desert, Mexico, and ready for processing. After boiling in water with sulphuric acid, wax is skimmed from the surface and formed into blocks. Its strong adhesive properties make it useful for leather and various other polishes.

The tanning process

In industrialized countries, many tanning processes that were once very lengthy and laborious are now **highly mechanized** and very quick, lasting sometimes just a few days. This is true, for example, for the hides that are used to produce flexible, light leathers which, after chemical cleaning and the removal of hair, are placed in **revolving wooden drums** containing tanning solutions. Rigid,

heavy leathers, however, are still **tanned in pits**. Due to the lack of mechanical agitation, this is a much more lengthy process and can last many weeks. **Different grades** of 'tannages' are recognized, and since each extract has its own specific properties, different mixtures will be used for different ends. The tannins cause the collagen fibres in the hide to stabilize and become robust,

protecting them from microbiological decay. For some leathers, **vegetable and chromium extract** combinations are used. After tanning is completed the leather is usually bleached, rinsed and then dressed with special chemical solutions. Plant oils such as rape seed or gums like carrageenan are then used to smooth and polish the leather, helping to **waterproof** it and prevent it from cracking.

To cap it all: plants for hats

Some people say that no outfit is complete without a hat. While some traditional societies have maintained this age-old custom – often denoting status or the wearer's geographic origin – the generalized wearing of hats, common in Europe and America until the late 1950s, has now been abandoned. Most of Britain's Royal Ascot goers, however, and many a wedding guest would feel quite under-dressed without one.

In Britain, traditional 'boaters', sun hats, and many creative designs, especially for women, are still made from a base of different plant materials, but all of them are described by members of the hat trade as 'straw'. In botanical terms, straw refers to the stalk on which a grain has grown or to a quantity of the dried stalks left after harvesting, but other commonly used plant materials all share this same general term, as well as man-made viscose fibres.

Many of the hat-making 'straws' used in Europe are either grown in or exported from China. The names they are traded under, such as *xian* or 'paribuntal', *baku* or 'Philippine straw', have tended to obscure the identity of the plants involved. The range of materials currently used for straw hats, however, includes: abaca or Manila hemp fibres from the Philippines; hemp, jute and sisal; various palm fibres, including raffia and buri, a very fine, expensive straw from Sri Lanka and Malabar; the palm-like *Carludovica palmata*; rice and wheat straw; and Chinese seagrass. Hat straws referred to as 'toyo' or 'Chinese paper yarn' are made from twisted coated paper – sometimes rice paper – or recycled newspapers combined with materials such as cellophane.

Hats for keeping cool

Though many of our modern hats are largely decorative, others, like the boater, were originally designed to protect the wearer from the elements or rigours of the climate. Two very well-known natural fibre hats, made famous during the last century, were designed to combat the fierce heat of the tropical sun.

The pith helmet or topee, which protected the heads of British army officers in India from the 1860s onwards, was made from the pith of leguminous swamp plants native to India – *Aeschynomene aspera* and *A. indica*. Covered with a white cotton cloth on the outside and lined with a green material, the inner cork-like pith of these plants (also known colloquially as *sola* or *shola*) formed the famous lightweight helmet. This insulated the head against the heat of the sun, but was impervious to air and water, making it uncomfortably hot to wear. The panama is another hat designed to protect the wearer from the powerful tropical sun, but unlike the pith helmet it is still widely used today. Famous, but misnamed, most hats come today from Ecuador and gained their name because they had become popular with those constructing the Panama Canal, and were bought by the 'Forty-Niners' at the time of the Gold Rush. During the 1880s, Ecuador began to export the hats in large numbers through Panama to the United States and the hats were the country's biggest export item until the 1940s (see also below). Panama hats are also made in Mexico in the state of Campeche.

Straw hats on sale at Mitla market, Oaxaca, Mexico.

The panama hat

Panama hat making in Ecuador goes back some 300 years and is practised today in two regions of Ecuador: the province of **Manabi** and the highland region around the town of **Cuenca**. The fibre used comes from the leaves of the palm-like plant *Carludovica palmata* (known locally as toquilla), which thrives in the northern coastal region of Jipijapa and Montecristi, in Manabi. Since this is the region where the hats were first made and from which the fibres are exported to other parts of the country, they are also known as jipijapas or **sombreros de Montecristi**.

Various grades of panama hat are available and their origin is often distinguishable by their weave and style. Those from Cuenca, for example, often exhibit an **intricate herringbone design**. Generally, the cheapest panamas to buy, which may have taken only a few hours to make, have the coarsest weave and therefore the lowest 'grade'.

However, it is in the town of Montecristi that just a handful of master weavers remain, producing hats so finely and **skilfully woven** that they are considered the finest straw hats in the world. Some of these hats will even hold water.

To obtain the fibres, unopened leaf shoots of the **toquilla plant** are harvested with machetes in succession from the base of the plant. The hard exterior is stripped away to expose the **cream-coloured** 'fingers' inside. These are then split into thin filaments or strands with the thumb and fingernail or with a special comb and soaked in **boiling water** before removal, drying and grading. As they dry, the strands shrink and curl to form long cylindrical tubes about 90 cm (36 in) long. After treatment with sulphur, the fibres are washed and dried again. Most of this **processing** is done by very poor women in coastal and rural Andean communities.

The best weavers work with fibres so fine that it is hard to distinguish any weave pattern in the finished hat. The finest hats are said to be woven only at night, when it is cooler (to stop the straw from being stained by the weavers' sweat), and to require eight days of **hand weaving** while immersed in water to keep the fibres soft. Woven from the crown in concentric circles from the centre outward, when the end of one strand of fibre is reached it is tied to the next one with a **tiny knot**. The traditional weaving of the finest hats is a complex, uncomfortable business. With the chest resting on a cushion on top of a wooden block, and arms extending to the floor, each hat is woven at **arms' length**. While these very fine hats take at least two months to complete, the best and most expensive Montecristi hats take up to eight months to finish. Some are so soft and finely woven that they can be **rolled up easily** like napkins and are sold in narrow rectangular boxes made of local balsa wood.

Though very few master weavers are left, about 40,000 people are estimated to be currently involved in the processing and weaving of panama straw hats in Ecuador.

Plants that **feed** us

Wild plants were almost certainly our first foods. Before people had developed their skills as hunters, herders and agriculturists, and many thousands of years before the advent of the pre-wrapped, pre-sliced loaf of bread baked from engineered strains of high-yield wheat, native plants were our most immediate source of nourishment. Around the world, people must have experimented with and eaten different parts of the plants they found growing locally, selecting and then domesticating those they found most useful, both for their particular qualities, such as yield or flavour, and for their suitability to different growing conditions.

Of the estimated 422,000 flowering plant species, over 50,000 are reported to be edible. More than 80% of the world's food is derived directly from plants, but just a handful of species provides most of what we eat today. Just two plant families contain some of the most important species. The *Poaceae* (*Gramineae*) or grass family, which gives us wheat, rice and maize, supplies over half the protein eaten worldwide, while the *Fabaceae* (*Leguminosae*) provides peas, beans and all kinds of pulse grains.

Other families important for the food plants they contain include the *Rosaceae* for fruits such as apples, pears, plums, cherries, almonds and apricots; the *Brassicaceae* (*Cruciferae*) for many of our green vegetables; the *Arecaceae* (*Palmae*) for coconuts, dates, sago, oil and other palm products; the *Euphorbiaceae* for cassava; the *Musaceae* for bananas; and the *Solanaceae* for potatoes, red peppers, chillies and tomatoes. But the food plants we find on our supermarket shelves – whose features such as shape, size, colour and flavour are, in most cases, the result of complex breeding and development – represent only a tiny fraction of the diversity that has been created over thousands of years by ordinary people. By observing the natural variation in plants of the same species and selecting individuals with particular traits, many different forms (also known as varieties, cultivars or landraces) of a single crop species have been developed. In the case of our most important and valued food crops, thousands of such types exist. Such horticultural variety is an important reflection of the world's cultural diversity.

A huge range of potato varieties, for example, has been developed by Andean farmers. They have traditionally known which will do best in their high altitude fields, or which yield varying tastes, textures and colours, and which are good for immediate eating or freezing for long-term store. Such cultural knowledge – handed down from one generation to the next – continues to play a critical role in maintaining the world's enormous diversity of crop types.

Most of the major crops that currently dominate global food markets are the result of improvements achieved with the help of these little-known or wild relatives. Their genetic material has played a major role in boosting them with characteristics that have enabled them, for example, to increase productivity and resistance to pests. Or they have become better able to cope with difficult growing conditions, or the make-up and appearance of the crop itself has been altered, at the same time increasing its commercial value to processors.

Previous page: Rice paddy in Indonesia.
Opposite: Tomatoes grown organically.

98

The origin of our foods

Unlike the people of many non-industrialized societies who can still grow or gather much of what they consume, the **precise origin** of what we eat and drink is often unknown to us. Plant materials are present in most of our food, from fish fingers to champagne. But as more and more is pre-processed, the task of decoding and unravelling the list of ingredients on the packaging often obscures any connection with plants at all.

Even though a 'country of origin' may be stated, the fact that the size, colour and flavour of many fruits and vegetables may be due to the genes of some **little-known relative**, nurtured by people we are largely unaware of, on the other side of the world, is never apparent and certainly not advertised.

Whether eating a tomato or a bar of chocolate, we tend to worry more about its **price** than where it came from or how, by whom and under what conditions it was produced. Our 'cheap' foods – including many of the 'basics' obtained from plants, such as tea, coffee, bananas and vegetable oils – are mostly the result of production on a massive, industrial scale involving high **chemical inputs** and the serious exploitation of millions of people, everyday. So, as you unpack your carrier bag, **spare a thought** for all the plants you've bought and the people whose lives we hold in our hands.

Changing times

It has been estimated that the introduction of genetic material held within crop relatives, by crossing them with mainstream varieties, is worth around $115 billion a year in crop increases alone. Genetic diversity, then, is generally regarded by scientists as an 'insurance policy', enabling mankind to respond to future agricultural challenges.

Until relatively recently, plant breeding relied only on the crossing or sexual hybridization of related species or varieties followed by selection, to produce plants with desired characteristics. This 'conventional' plant breeding has been expanded to include a variety of techniques such as tissue culture, involving hybridization at cellular level, and the use of the plant chemical colchicine to increase genetic variation.

The so-called 'Green Revolution' of the 1960s and 70s used conventional plant breeding techniques to create high-yielding varieties of major crops such as rice, wheat and maize. It also initiated the trend towards the monocultures (the cultivation of just one crop) that have become standard in Western agriculture. Although such techniques have resulted in massive increases in yield of up to two to four times, depending on the crop, and generally made harvesting easier, they have also facilitated the spread of virulent pests and diseases to which wild or less manipulated populations often have some natural resistance. In the late 1970s, vast areas of Indonesia (which had formerly relied on hundreds of rice varieties) were planted with just a few varieties that were closely related. Excessive use of pesticides on these crops destroyed the natural predators of a particular pest – the brown planthopper – without reducing the hopper itself, which subsequently caused widescale destruction.

This agricultural model is in sharp contrast to the traditional farming systems in which a single field is likely to be planted not only with many different crops, but with many different varieties of these, each with differing nutritional benefits and times of harvesting. As 'poor' countries devote more land to 'cash crops', often for export to repay foreign debts, this traditional system is being rapidly replaced.

The latest method of crop 'improvement' is genetic modification (GM), which involves the transfer of individual genes from other species (indeed any other organism, from bacteria to fish or animals) directly to the cells of the crop plant, in which they will 'express' the desired characteristics, thus bypassing sexual breeding. Already accounting for 46% of the world's soya and nearly 11% of its maize, genetic engineering is discussed on page 150. Fiercely opposed by an ever-greater body of farmers, scientists and concerned individuals worldwide, such developments are widely regarded as a serious threat to global food security, raising new questions about safety and the sustainability of intensive agriculture. Those in favour, however, argue that genetic engineering offers great potential in improving crop yields and reducing agrochemical use.

Plants for breakfast

The chances are that, whatever we decide to eat for breakfast, most of it will have come directly or indirectly from plants. Without plants, of course – and particularly the grasses – there would be no milk, butter, cheese or yoghurt, and no meat. About half the world's grain harvest is used to feed livestock – American animals, chiefly cattle, consuming about a third of the world's grain, including over 70% of that grown in the USA.

It would be impossible to describe here the thousands of plant foods that are available to us all around the world. But by looking first at some of the ingredients of a typical Western breakfast, we can consider the plants involved in the production of a broad range of foods and drinks that millions of people are likely to consume in meals or snacks throughout the day.

Wheat field, UK. One of the world's most important food plants, and ranking first among cereal crops in quantity produced, nearly 600 million tons of wheat are grown each year on more than 200 million ha (494 million acres) of land.

99

Bread and butter

Wheat

■ Along with barley, the main grain upon which Neolithic agriculture was founded and flourished, today wheat (*Triticum* spp.) is still the preferred staple food throughout the Old World and is one of the world's most important food plants.

■ An indication of the great antiquity of wheat as a cultivated plant is the 20 cultivated species and over 30,000 varieties that are recognized today, presenting great variation in shape, size and colour of wheat grains and ears. The two most important species are bread wheat (*T. aestivum*) and durum or macaroni wheat (*T. durum*). The wheats can be broadly grouped according to hardness of their kernels:

■ The 'hard' bread wheats: these are richer in gluten and produce a flour particularly useful for bread making.

■ The 'soft' bread wheats: these make better cakes, pastries and biscuits.

■ Durum wheat: though 'hard' and higher in protein than the bread wheats, this wheat is less used for baking since its dough does not retain carbon dioxide to the same extent and is often ground into semolina, the basis of pasta, including macaroni, spaghetti and noodles. Couscous and bulgar are also products of durum wheat.

■ Wheat cultivars are very versatile and are grown throughout temperate, Mediterranean and subtropical regions of the northern and southern hemispheres. As a crop, wheat is often widely classified according to the time

A vailable in an endless variety of shapes and forms, bread is indispensable to most European breakfasts. The British certainly love it: nine out of ten of them eat bread every day. The Germans, however, come top of the league, eating over 80 kg (176 lb) per person per year.

Bread is one of the oldest foods known to man. Fine loaves were eaten by the Egyptians around 4,000 years ago (made from highly proteinaceous emmer wheat, *Triticum dicoccum*) and its use has been linked with some of the most important events in history. The need for wheat – to make bread – was one of the principal reasons underlying the expansion of Rome, the conquest of Egypt providing access to the wheat growing regions of the Nile Valley and northern Africa.

Bread has been used as a powerful status symbol for different peoples at various times: while the ancient Egyptians paid their officials with it, by the Middle Ages fine white loaves were considered food fit only for the nobility in most parts of Europe. This 'staff of life' came to be used in only its finest, whitest form for celebrating Roman Catholic Mass, and one of the world's major religious disputes has centred on whether bread is to be swallowed as the body and blood of Christ or as its representation.

Whatever type of bread we eat today, whether Jewish bagel, Irish soda bread or Asian naan, none of it could be made without the fruit of different grasses; cereal grains. One of the most prominent is rye (see page 102), which grows further north than any other cereal – up to the Arctic Circle – and on land too poor for most.

The grain that supplies most of the world's bread, however, is wheat (see left). Eaten not only in this form, but as semolina, bulgar and couscous, in every sort of pasta and a huge number of farinaceous foods, including cakes, biscuits, pastries, pies, sauces, coatings, toppings and fillings, wheat currently provides 20% of the world's calories and constitutes the staple diet of one-third of the world's population, coming second only to rice.

Ranking first amongst the world's cereal crops, in terms of production, nearly 600 million tons of wheat are grown each year on more than 200 million ha (494 million acres) of land. There is an enormous number of different kinds of wheat. The one we are most likely to eat in our toast is *Triticum aestivum*, or bread wheat, which itself comprises a variable group of many thousands of varieties.

Modern mass-produced bread wheats are quite different plants from those first gathered from the wild by our ancestors. These wheats evolved about 10,000 years ago from wild species of *Triticum* and the related genus *Aegilops*, in the 'Fertile Crescent' of Southwest Asia and the eastern Mediterranean, alongside barley and pulses. Wild wheat ears with their long bristles or 'awns' (which help the grains to find a foothold in the earth) are very brittle and shatter on touch, but this and other 'undesirable' characteristics have been selectively bred out of the commercial wheat crops grown today. In 2001, the world's major wheat growers were China, the EU, India, the USA and the Russian Federation, whilst the top five exporters were Argentina, Australia, Canada, the EU and the USA.

Wheat production is now so successful that in both Europe and America farmers are being paid not to produce the crop. Yet scientists are constantly trying to produce higher-yielding varieties of wheat, with resistance to attack from diseases such as rusts (fungi with very complex lifecycles), which attack the fruiting stalks and leaves.

it was sown, e.g. spring wheat and winter wheat (sown in the autumn and needing to experience a chill before producing flowers).
■ Over 587 million tons of wheat were grown in 2001.

Bread

Although nutritionists emphasize the benefits to be gained from eating bread in most forms, particularly that made from **wholemeal flour**, many believe that the quality of British bread has deteriorated over the years. Various continental breads have now become fashionable, but about 86% of Britain's bread (some 10 million loaves each day) is industrially produced, bearing little resemblance to the kind of bread baked using more **traditional methods** and ingredients in small local factories or at home. Bread diversity once reflected cultural diversity, with regional varieties representing the variety of grains as well as style.

Ironically, perhaps, a typical **British flourmill** can produce as many as 60 different types of flour by blending different varieties of wheat (from the 25 that are now grown in the UK and a small number that are imported) and **extracting flours** at different stages in the milling process. Once reliant on the traditional

English cone or rivet wheat (*Triticum turgidum*), which produces a 'soft' flour (which does not rise well), and large imports of 'hard' North American wheat, plant breeders have now developed UK wheat varieties that, in combination with modern **baking** technology, are suitable for bread making.

Processed by **high-speed mixing**, which agitates the dough to reduce the fermentation period, ingredients such as glucose syrup, hydrogenated vegetable oils, soya flour, emulsifiers and ascorbic acid or vitamin C are now common in British bread, especially that made with a proportion of soft flour. **Regulations** governing bread and flour production require that both white and brown flours, other than wholemeal flour, have synthetic B vitamins added to them, such as niacin and thiamin, and minerals such as calcium and iron, to **replace nutrients** lost in the milling process. While wholemeal flour is milled from the entire wheat

grain, **white flour** has had most of the wheat germ (the embryo of the young plant, containing protein, oil and vitamins), bran (the outer seed coat, which is a natural source of roughage) and an intermediate layer called the aleurone (rich in protein, minerals and vitamins) removed.

The part ground into flour is the starchy endosperm (only 80% of the grain), the **food reserve** of the young plant. Until 1997, this was generally bleached with chlorine or benzoyl peroxide in the UK. Brown flour consists of about 85% of the grain, with some of the wheat germ and bran removed.

Although 'craft bakers' are still responsible for about half of western Europe's bread production (which totals around 25 million tons of bread each year), this situation is changing rapidly as **industrial bakers** increasingly take a bigger share of the market. Many believe that we should insist on wholesome bread – especially for our children.

101

Rye

■ Rye (*Secale cereale*) is now produced in the largest quantities by the Russian Federation, Germany and Poland, where it is milled into flour or used in its whole, 'cracked' or flaked forms as the main ingredient of a number of traditional recipes.

■ Pumpernickel from Germany (which itself boasts over 300 different kinds of bread), white and dark loaves, sourdough and Scandinavian crisp breads are all made from a base of rye flour.

102

In 2000, the Consultative Group on International Agricultural Research (CGIAR) spent about $27.1 million on wheat research. Alongside national research institutions, it has been developing and distributing 'improved varieties' of wheat with greater resistance to diseases and tolerance of environmental stress.

While most wheat is grown for food – especially bread – it is also the source of a wide range of products used in very diverse industries – from starch, malt, dextrose and alcohol to building materials.

Butter

In many economically developed countries, particularly in the USA, dairy farming is dominated by economies of scale and the buying power of vast corporations. It is estimated that over half of America's cows now live on intensive farms, spending their entire lives indoors, on concrete floors, tethered to milking machines. They are now bred to produce unnaturally large quantities of milk (double the quantity of 30 years ago and up to 100 times more than they would in their natural state), none of which their calves will taste. Furthermore, their shortened lives (now around five years, as opposed to the 20 or so of 50 years ago) are supported by highly processed feeds and a battery of drugs and chemicals, including the notorious genetically modified Bovine Growth Hormone, designed to increase productivity.

Many intensively reared cows are given food that is based on the waste products recovered from large-scale oil-seed processing, such as the residue left after the extraction of oil from sunflower seeds, shea nuts, African oil palm kernels, copra (dried coconut) and rice bran. Citrus pulp, wheat, molasses and vegetable oils are also often added. Other plants that may end up in processed cattle food include oats, lucerne or alfalfa and sugar beet, in the form of pulp.

Where dairy cows are lucky enough to be able to graze outside, a quick look at a typical field in Britain reveals that the grass the cows are eating is very uniform. Of the 10,000 or so known species of grass, only a dozen have been regularly sown at different times as pasture or forage grasses in conventional systems. Thousands of hectares (acres) of what would otherwise be much more heterogeneous pastureland is now sown with just one or two plant species, strains of which have been developed to maximize the production of milk. The principal plant involved is ryegrass. Two species are widely sown: perennial ryegrass (*Lolium perenne*), native to Britain, mainland Europe and parts of Asia, and Italian ryegrass (*L. multiflorum*), a species introduced to Britain, which grows prolifically and has a high nutritive value.

Meadow with clover, UK. An important pasture plant.

In today's sown pastures, especially where Italian or perennial rye-grass forms the dominant cover, it is usual for farmers to include one or more clover species in their seed mix. Clovers are rich in protein over a longer period than grasses and increase the food value of the herbage. As legumes, they greatly assist the plants they grow with since nitrogen compounds (necessary as fertilizers) are produced by bacteria housed in special nodules on their roots, and are released directly into the soil as the roots decay.

Hay was once the normal winter food for dairy cattle, but because wet weather makes it an uncertain crop, silage is now very widely used. Packed into large silos or individual plastic sacks, the action of bacteria ferments the grass and effectively pickles it, keeping it in a usable condition for many months.

Meadowlands

Left to its own devices, all **grassland** is a community of widely differing species and varieties of plants – an early stage in the succession of plant colonizers of a bare patch of earth that, given the right conditions, would eventually succeed to shrubby vegetation and, in time, **forest**. Almost all meadows and pastures are 'artificial', in the sense that they are continually cut or grazed, and thus maintained as grasslands by man or animals. But over **long periods** of time, certain plant species have become adapted to these regimes and, in association with one another, can be identified as distinct vegetation types. Some of the most common of European pasture grasses include the bromes, fescues, meadow tussock and oat grasses, sweet vernal grass and Yorkshire fog. They are often accompanied by **wild plants** such as yarrow, daisies, buttercups, vetch, dandelion, chicory, cow parsley, hogweed and hawkbit, which together help create **idyllic scenes** that have come to symbolize a much-loved aspect of the countryside.

A victim of decades of intensive farming, including drainage schemes and the **indiscriminate use** of pesticides and herbicides, it is now estimated that England has lost 95% of its ancient, 'unimproved' grassland over the last 50 years. Only 97,000 ha (240,000 acres) of wildflower meadow remain, scattered in fragments across the country – **precious reserves** of plant and animal diversity, containing many rare, though once common, species. Despite government-sponsored schemes and other **initiatives** encouraging farmers to maintain their 'unimproved' grasslands, they are still being replaced. **Organic farming**, however, seeks to work in harmony with systems of land use that allow biodiversity to thrive. Hay made from meadows containing a large number of plants that are collectively rich in minerals and other nutrients is highly valued since it is relished by **grazing animals**, especially sick ones, which may refuse otherwise good-quality hay.

Margarine

There is hardly a processed food eaten today that does not have some edible vegetable oil in it. Around 30 million tons are traded around the world each year, much of it used in cakes, biscuits, pies, confectionery, salad dressings, snacks and for frying. But the largest use of edible oils in Britain is in margarine and other 'table spreads', which accounts for some 285,000 tons per annum.

With over 95 million tons of vegetable oils and fats produced worldwide in 2001, alternatives to butter are available today in a wider range of products than ever before. Processes for the thickening of plant oils by hydrogenation (the bubbling of hydrogen gas under pressure through the oils in the presence of a catalyst such as nickel) have made it possible to incorporate many which would otherwise be unsuitable for use in margarine and other foods. Other basic procedures involving solvents (which wash the oil out of the plant tissue), and steam, for bleaching, refining and deodorizing, make quite different oils indistinguishable and therefore interchangeable, since their taste and other characteristics can be entirely removed.

Major vegetable oils used for margarine-making include palm and palm kernel (extracted from different parts of the African oil palm fruit), soya, sunflower (see right), rapeseed, coconut, peanut, cottonseed, corn and safflower. With the exception of palm oil, which is pressed from the fleshy fruit layer surrounding the palm kernel (which yields oil of a different composition), all these oils are crushed from seeds or fruits. For large-scale production, mechanical

Sunflower

■ With an oil content of 25–48% (though able to reach 65%), the seeds of sunflowers (*Helianthus annuus*) are important sources of oil for soft margarines, domestic cooking oils and salad dressings. Seeds are also used for confectionery, in snack and bakery products, and in bird and pet foods.

■ Argentina produces the most sunflower seeds in the world – over 3 million tons in 2001, roughly equalling the annual production of the whole of the EU, where most are grown by France and Spain.

■ Throughout the 20th century, it was Russia that dominated world supplies, becoming famous for breeding plants with giant flower heads, each containing up to 1,000 seeds. The Russian Federation and the Ukraine are still amongst the world's top producing nations, with China and the USA also major growers. Recent research has shown that the sunflower was possibly first domesticated in both Mexico and eastern North America over 4,000 years ago.

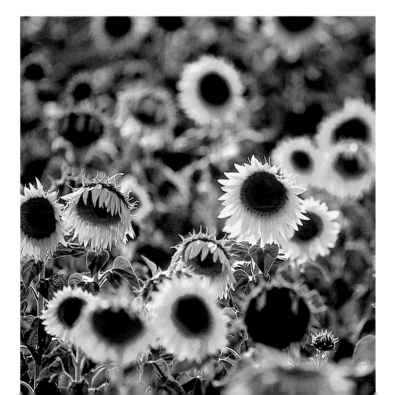

Sunflowers in Spain. An important source of oil and seeds.

African oil palm

■ African oil palm (*Elaeis guineensis*) is native to tropical West Africa, and in some areas may be found growing in large natural groves. Since the trees will thrive equally well in other areas of high rainfall within the world's humid equatorial belt, many countries now grow them in vast plantations, and their oil has become a major revenue earner (see page 16).

■ Large fruit bunches, weighing up to 10 kg (22 lb) and containing up to 1,000 individual plum-sized fruits, are borne by the female trees after about five years. These turn from green to orange-red (or black in some varieties) when ripe. Between two and six bunches are produced by the trees each year and are harvested either by climbers using ropes, or by men on the ground who cut the fruit bunches from the crown with a sharp blade on the end of a long pole. The fruits produce two different oils: 'palm oil' – pressed from the oil-rich mesocarp layer – and palm kernel oil, pressed from the seed. These oils have many commercial uses, including soap, candles, margarine, ice cream, bakery goods and snack foods.

presses or expellers mostly perform this job, as well as oil solvents, such as light petroleum. Attempts to process the oil obtained from cottonseeds (available in huge quantities as a by-product of cotton ginning) led to major advances in the vegetable oil industry as a whole. Until the beginning of this century, cold-pressed cottonseed oil was considered inedible because it contains a bitter pigment, gossypol. But experiments involving fuller's earth and steam eventually succeeded in purifying the dark-coloured, semi-toxic oil into a colourless, tasteless and harmless one, and techniques for hydrogenation and fractionation (fat-splitting) followed. China currently produces nearly a third of the world's cottonseed oil – over 1 million tons in 2001.

Saturated oils: The various seeds processed contain varying proportions of oil. Some of the highest concentrations are to be found in those of two tropical palms: the coconut (desiccated coconut containing 70% oil) and African oil palm (both palm and palm kernel oils comprising about 50% oil) (see left). The huge quantities of oil extracted from these two palms have made them important suppliers to the food industry, but one aspect of their chemical composition has led to the adoption of other plant oils for some culinary uses, including non-dairy spreads. Both coconut and palm kernel oil are very high in saturated fats – 92% and 51% respectively. The move towards healthier eating patterns has drawn attention to the suggested merits of unsaturated rather than saturated oils or fats. Although saturated fats are preferred by manufacturers as they are less likely to go rancid, they have been linked with cardiovascular disease and various cancers. It is somewhat ironic, therefore, that milk substitutes, such as 'non-dairy creamers' and artificial whipped toppings, available for use by those concerned about their fat intake, have been widely made from coconut or palm oils. At least in the USA the term 'tropical oils', listed on food labels, is equated by the public with high levels of saturated fat. Elsewhere in the world, the extensive use of these oils in a large number of processed foods is not usually readily apparent.

Recent research is now showing that animal or other saturated fat is not necessarily bad for us. It is, rather, in the processing and in particular hydrogenation (hardening) of commercial oils that a major problem lies. Hydrogenation causes an increase in the degree of saturation and produces trans-fatty acids from unsaturated oils, which at high intakes are believed to adversely affect blood cholesterol levels.

Unsaturated oils: At the other end of the scale, the oil crushed from safflower seeds (see opposite) is highly unsaturated – comprising 91% unsaturated fatty acids. Along with sunflower seed oil, also

African oil palm nursery, amid plantation, southwest Cameroon.

containing a naturally high level of polyunsaturated linoleic acid, it is a constituent of many soft margarines and is widely used for cooking.

In recent years, spreads that include an oil traditionally relished for other culinary purposes – especially for salad dressings – have been developed. Held in the highest esteem by peoples of its native Mediterranean region, olive oil – obtained from the fruits of the evergreen tree *Olea europaea* – is also unsaturated and is regarded by some as a virtual 'elixir of life', promoting a long and healthy existence. Extensive research has shown that olive oil contains a number of compounds that are very beneficial to most functions of the human body: its fatty acid structure, vitamin E and other antioxidant content in particular making it, it is claimed, the oil best suited to human consumption.

Available in various grades, 'extra virgin' olive oil – considered the best – is the result of the first cold pressing of the fruits, without the heat used subsequently to extract oils of lower grades. Unlike most commercial oils today, good olive oil is rarely refined to tastelessness and, with different regions producing oils of different flavours and characteristics, remains much sought after by gourmets.

The olive oil used in spreads, however, is often used only in small quantities (for example, around 20% of the total product), supplementing ingredients such as skimmed milk, whey and other vegetable oils (such as rapeseed) that are highly unsaturated.

Safflower

■ The safflower (*Carthamus tinctorius*) is a member of the thistle family and is thought to have originated in the Near East or Central Asia.

■ It was first cultivated for the deep red and yellow dyes produced by its flowers, for which it was once widely used for dyeing silk and making rouge. It is the seeds, however, that have attracted much commercial attention, as the source of a neutral-tasting edible oil – traditionally much used for culinary purposes in India. The discovery that it has the highest proportion of polyunsaturated oils and the highest linoleic acid content of any known seed oil (important in the human diet, and one of the few fatty acids that cannot be synthesized by humans), has made it an increasingly popular oil for margarines, salad dressings and for cooking with.

■ Used clinically to treat coronary disease and thrombotic disorders in China, safflower extracts have been shown to slow down blood clotting. Major producers of safflower oil are India, Mexico and the USA.

107

Colour for butter and margarine

The **natural pigmentation** in vegetable oils is generally removed, but the pale yellow appearance of many of our margarines and some butter is often due to the addition of other colorants from plants. One of the most common is the red-orange dye, described on product labels as **annatto E160(b)**, processed from the seeds of *Bixa orellana*, a shrub from tropical America. Other sources of the yellow pigments used in dairy products include **carotenes** (often obtained from carrots, maize and alfalfa), **turmeric** (from *Curcuma longa*) (labelled as E100), **paprika**

extracts (from *Capsicum annuum*) (E160c), and **lutein** (E161b), extracted from marigold petals and alfalfa. Yellow or orange pigments are found in all higher plants, but they are especially prominent in carrots, green leafy vegetables, tomatoes, apricots, rosehips and oranges.

The **thick red paste** made from the waxy outer coating of the seeds of *Bixa orellana* is widely used by many Amerindian groups for painting the body, and sometimes the hair, and for colouring threads, ceramics, implements and weapons.

Citrus fruits: a fresh start to the day

Citrus fruits have become an established part of many Western breakfasts. Most commonly they are consumed in the form of orange or other citrus juices but they are also used in the form of fresh fruit or in preserves, such as marmalade. The production and processing of citrus fruits is certainly very big business and it is hard to imagine that only 100 years ago oranges were considered an expensive luxury in North America and northern Europe.

The genus *Citrus* comprises not just oranges but numerous relatives, including lemons (*C. limon*), limes (*C. aurantiifolia*), grapefruit (*C. paradisi*) and three ancestral species from which recent DNA evidence suggests all the commercial citrus fruits derive. These are the lemon-like citron (*C. medica*), much used today for crystallized or candied peel; the pumelo or shaddock (*C. maxima*), the biggest of all the citrus fruits; and the mandarin or tangerine (*C. reticulata*). Still the subject of experimentation and manipulation by plant breeders, a large number of hybrids also exist, the origin of which is not easy to unravel.

Best known and most widely marketed today are the orange, lemon and grapefuit. The sweet orange (*C. sinensis*), which thrives in climates with abundant sun and reasonably dry air, is thought to have originated in southern China, the citrus group having probably evolved in this region of Southeast Asia and perhaps India. The orange is now the most widely grown citrus fruit in the world and, forming vast plantations in California and Florida, is the USA's largest perennial fruit crop. Most oranges for world markets are grown in Brazil and the USA, with China, Mexico, India, Spain, Italy and Iran also major producers. The high proportion of vitamin C contained in oranges as well as limes

Citrus fruits

The fruit of all *Citrus* species, known botanically as a *hesperidium*, is a berry with a leathery skin, formed from the inner pith and outer layer. The outer layer is covered with tiny glands containing **aromatic oils**. Present in the leaves and other vegetative parts of the plant, they are responsible for the **distinctive aroma** that is released when peel is crushed or grated. The bergamot orange (*C. bergamia*) is grown almost exclusively for the oil it yields, which is much used by perfumers and for the flavouring of alcoholic drinks.

In some countries where temperatures are never cool, oranges **remain green** even when mature. Coolness promotes the release of orange pigments and, if temperatures fluctuate, the fruits may **alternate** from one colour to another. To overcome this, some batches of oranges are treated with ethylene, which promotes the **development** of a uniformly 'orange' appearance.

The part of the citrus fruit we eat is actually formed from fleshy hairs that act as **juice sacks**. They contain solutions of sugars and acids, as well as vitamin C and the vitamin B complex, carbohydrates, minerals and other nutrients.

gave effective protection against scurvy, and this promoted their spread around the world. They were so prized in northern Europe that special greenhouses or 'orangeries' were built by the wealthy in the hope of a fruit supply of their own.

Three main hybrid types of sweet orange are distinguished: the navel, the common and the blood orange. Of the common types, 'Valencia' is the most widely grown and its richly flavoured juice has traditionally set the standard by which other orange juices are judged. Though 'freshly squeezed' juice is now widely available in Europe, much of the packaged orange juice sold is made from filtered, freeze-dried concentrate, reconstituted with water and controlled amounts of pulp and peel. The benefits of drinking these juices are not as certain as those to be gained from drinking freshly squeezed juice.

Marmalade and other orange preserves (as well as some liqueurs) are made exclusively from Seville oranges, a form of sour orange which is too acidic to eat raw. Many of these oranges are grown in southern Spain, though cultivation has spread to tropical and subtropical countries. With the increasing destruction of forests in southern and Southeast Asia, many of the wild relatives of oranges, tangerines, grapefruit, lemons and limes are now endangered. Pectin, the essential setting agent needed to make many preserves and jams, is itself extracted for commercial use from the peel of lemons, oranges, limes and grapefruit where it is naturally present in large quantities. The peel and pulp of apples is another major commercial source of pectin.

Orange trees in a plantation, São Paulo state, Brazil. Along with the USA, Brazil is a major exporter of oranges – the most widely grown citrus fruit in the world. Now cultivated in many countries, oranges prefer a warm, dry climate.

109

Plants that make life sweet

Sugar maple

■ Before the general availability of sugar in Europe, the only practicable sweetener was honey. Sugar was first prescribed as an expensive medicine, generally to help disguise the taste of bitter herbs. In North America, however, Indian groups had long used the sugar maple (*Acer saccharum*) as a source of sweetness. After cutting the bark of the trees in late winter or early spring, the sap collected was boiled down to make a thick syrup.

■ The practice is continued on a commercial basis today (networks of plastic tubing transport the sap to processing factories) and maple syrup has become a famous Canadian export and tourist attraction.

■ The sugar content of sugar maple sap (about 2%) is much lower than that of sugar cane; it takes about 182 litres (40 gallons) of sap to make 4.5 litres (1 gallon) of syrup.

Sugar is, of course, essential to marmalade and jam manufacture, and it is likely to be present on the breakfast table not just in a bowl of its own, but in much of the other food and drink we consume during our first meal and throughout the day. Between us we add vast amounts of sugar to our tea and coffee and sprinkle it again on many breakfast cereals that already contain it.

All green plants make sugars and these can be stored in their fruits, roots, bulbs, stems or flowers. For thousands of years, people throughout the world have been using the sap or juice extracted from plants, and especially their stems, to obtain sweet syrups. The most notable examples include the wild date palm, the sugar palm, sorghum and the sugar maple. Today, however, just two plant species provide the household sugar we consume: sugar cane (*Saccharum officinarum*) (see below and box, opposite) and sugar beet (*Beta vulgaris*), the refined sugar (sucrose) produced from both these sources being identical.

Sugar as a whole is much maligned and generally regarded as bad for us, but it is really only the quantity that many Westerners consume (estimates of some 40 kg/88 lb per person per year in Europe, Australia and North America have been made) that has health experts worried. We need sugars to maintain body temperature and provide energy, but most of our essential supply is metabolized by the body itself from carbohydrates that we consume, such as starch, much of which is contained in quantity in cereal grains and edible roots. Sugars are also present naturally in small amounts in many of the fruits and vegetables we eat, in the form of fructose and as glucose – the latter found especially in grapes and onions. It is these two sugars that form the sucrose we use as table sugar.

Sugar cane

Chewed for its sweet sap since ancient times, sugar cane is believed to have evolved from a wild relative in the South Pacific region, probably in New Guinea, spreading throughout the Pacific, Southeast Asia and the Middle East by human migration many centuries ago.

Although sugar from cane grown by Arabic peoples was in demand in Europe by the 12th century as an expensive luxury, it was not until the establishment of plantations in Brazil and the West Indies early in the 16th century (following the introduction of the plants to the New World) that Europe had its own secure supply. Made possible only by the inhuman exploitation of generations of slaves shipped from Africa, entire countries were given over to sugar cane cultivation and huge areas of tropical forest destroyed to make way for this new

Sugar cane plantation in the Sonora Desert, Mexico. Fertilizers are being added directly to the water that will irrigate this crop.

cash crop. Though processing methods are now greatly updated and scientifically controlled, harvesting is still very labour intensive, the cutting of sugar cane by hand being still common.

After harvesting, the stalks are taken to factories where they are crushed and the juice extracted. This water and sugar solution is then purified, concentrated by evaporation, and the resulting brown sugar crystallized. The sugar is then refined to various degrees of whiteness in several stages.

Despite the confusion over the properties of brown and white sugar, there is nutritionally very little difference between them. Molasses, the excess sticky brown liquid that is separated from the newly formed raw sugar crystals by means of a centrifuge, is sometimes re-added to white sugar to make it brown. Used also for the manufacture of rum, it can, of course, be bought separately as molasses or black treacle. Refined white sugar is virtually 100% sucrose.

In 2001, world sugar production reached 130 million tons, of which sugar cane accounted for 75% and sugar beet the remainder. Major producers of sugar cane were Central and South America and India. Because of the unfair rules of international trade and rich-country protectionism, however, sugar exports from tropical countries have been seriously undermined by sales of sugar derived from sugar beet, from heavily subsidized European exporters.

Sugar beet

The Romans grew sugar beet as a vegetable but its use as a commercial source of sugar dates only from the beginning of the 19th century, following the investigations of a German chemist who extracted 6.2% of the sugar from the roots of a white variety. Today's improved varieties and modern methods of extraction have boosted the figure to around 20% and large quantities of sugar beet (a biennial root crop) are now grown in France, Germany, eastern Europe and the USA. In

Sugar cane

■ Amongst the largest of the tropical grasses and resembling bamboo is sugar cane (*Saccharum officinarum*), which take 12–18 months to mature and may grow to around 5 m (16 ft) in height. There is great variation in stem colour, width of cane and hardness of its outer skin.

■ Today's modern cultivars have been bred to give optimum yields under different cultivation conditions, and breeding programmes continue to focus on increasing sugar content, improving resistance to pests and disease, and environmental suitability. Formed as a result of photosynthesis, sugary cell sap accumulates slowly in the pithy centre of the canes and by harvesting time the pith cells may contain around 10% sucrose, depending on the variety.

■ Bagasse, the name given to the dried cane stalks after the sugary juice has been extracted, is used for a variety of purposes, including the manufacture of paper, plastics and building materials such as fibreboard, fuel for generating electricity and cattle food.

111

Molecular manipulation of sweeteners

Limited supplies of the raw materials and the limited stability of the compounds themselves have helped prevent the widespread use of **taste-modifying proteins** in our food to date. However, much research and development has focused on their manipulation at molecular level and production in transgenic organisms. This has enabled 'tailor-made' molecules to be produced on an industrial

scale. An example of this is the protein '**brazzein**', first identified in the West African plant *Pentadiplandra brazzeana*. Said to be 2,000 times sweeter than sucrose, it is now being produced experimentally in GM maize. It is reported that genetically modified, tooth-friendly and non-calorific **sweeteners** have not yet taken over from sugar, as expected by some, partly because of the protectionist

nature of the sucrose market, and reticence on the part of the public to accept such products.

Much attention has also been focused on the genetic manipulation of **sugar beet** in tandem with the insistence by US regulatory authorities that refined sugar is an 'inert chemical', containing **no genetic material,** and so does not need to be labelled if it comes from GM plants.

Sugar cane plantation cut from Atlantic rainforest, near Pernambuco, Brazil. Such plantations are responsible for the destruction of large areas of this unique and highly diverse, but now threatened, forest type in Brazil.

all, over 50 countries – almost all in temperate regions of the northern hemisphere – grow sugar beet, on some 6 million ha (15 million acres).

The sugar is contained in the whitish conical roots of the plants, which are harvested with an average weight of about 1 kg (2.2 lb). To extract the sugar, the roots are first shredded and then heated in running water. After the removal of impurities, the clear liquid obtained is sconcentrated and crystallized to give a sugar indistinguishable from that of sugar cane. Molasses is also produced from sugar beet. The pulp that remains after the beets have been processed (along with the rosette of leaves from each plant) is widely used as cattle food.

Alternative sweeteners

Although its markets are still highly protected, sucrose, whether from sugar cane or sugar beet, has faced much competition in recent years because of the introduction of a variety of sweeteners made from other sources. Three broad types of sweetener have emerged:

● Bulk calorific sweeteners, such as corn syrup and inulin, a naturally occurring carbohydrate found in many plant families but mostly extracted from chicory.

● Synthetic, very-low-calorie super-sweeteners, such as saccharine (made from petroleum), cyclamates (banned in the USA in 1969, but still legal in many other countries), aspartame (200 times sweeter than sugar), formed by the bonding of two amino acids – aspartic acid and phenylalanine – and Acesulfame K (less sweet than aspartame, but more stable).

● Super-sweeteners derived from plants.

Biotechnology has transformed the way in which many of these sweeteners can be made.

Chief among sugar's rivals have been corn syrups, made from hydrolyzed maize starch, and invert sugar syrups, made by treating this same starch with enzymes or acids that convert it into glucose molecules, which can then be converted into high-fructose syrups by further enzyme action. Commercial development of various types of corn syrups in the 1970s brought about significant changes in the food industry, as they became widely adopted in place of sugar in many processed foods, especially soft drinks, and in alcoholic drinks and baked goods, where quick fermentation is required. Corn syrups are used not only for their sweetness but for various other properties they confer. These include improving the texture of some processed foods and providing the suspension which helps to keep ingredients evenly mixed, enhancing colours and, in ice cream and frozen desserts, depressing freezing to prevent crystal formation. Three main classes of sweetener are produced from maize: corn syrup, glucose and high-fructose syrups (also known as high-fructose corn syrup), which is 170% sweeter and 30% cheaper than sucrose and now accounts for 42% of the sweetener market in the USA. A sweet syrup is also produced commercially from brown rice. This is done by culturing the rice with enzymes, which break down the starches, and then heating the resulting liquid until the right consistency is reached.

Some interesting alternatives to sweet syrups and sugar, however, exist as natural components of other plants. Three plants in particular, all from tropical West Africa, have aroused much scientific and commercial interest because of the sweetening and flavour-enhancing properties of chemicals known as TMPs – taste-modifying proteins – contained in their fruit or seeds, and which are active at extremely low concentrations.

The first of these, a herbaceous plant that grows up to 3 m (10 ft) high, *Thaumatococcus daniellii* (see right), has produced one of the sweetest substances known to man – thaumatin – a protein 2,500

Thaumato-coccus daniellii

■ In the humid rainforests of its native West Africa, *Thaumatococcus daniellii* is well known to local people. They use its broad, flexible leaves as disposable plates or for wrapping food prior to cooking, and the long thin leaf stalks are harvested for mat making. Children suck the sweet arils of the fruits' seeds and these are also used as a source of sweetness in food preparation.

■ The taste-modifying protein thaumatin, extracted for commercial use from the seeds by initial immersion in a solution containing aluminium sulphate and sodium metabisulphate, is transformed into a powder after filtration and freeze-drying. Effectively non-calorific, it is therefore suitable for dieters and diabetics.

■ Such extraction, however, is expensive, and so most research has focused on isolating the sweetness genes and engineering them into bacteria and yeast strains. One US biotech company has claimed that it can produce thaumatin protein using recombinant DNA technology at a price that competes with that of extraction from the plant itself. It is also reported that the gene-encoding thaumatin can be engineered directly into fruit and vegetables to improve sweetness.

113

Other plant-derived sweeteners

Lippia dulcis from Mexico, known to the Aztecs. Modern scientists have isolated from the leaves and flowers and further synthesized hernandulcin, which is 1,000 times sweeter than sucrose.

Sorbitol, which takes its name from the trees and shrubs belonging to the genus *Sorbus*. It is one of the most common low-calorie sweeteners. Today, the compound is made by the chemical conversion of glucose.

Neohesperidine, derived from citrus rind, is the source of a substance (NHDC) 500 times as sweet as sucrose.

Glycyrrhizin, extracted from liquorice roots and widely used in the food industry. It is 50–100 times sweeter than sucrose.

114

times sweeter than sucrose. Crimson fruits, which develop just above the soil surface, contain the substance in the soft jelly-like aril that surrounds the black seed. Although extremely sweet, thaumatin leaves a liquorice-like aftertaste and is therefore not viewed as a conventional sweetener.

In 1993, a Korean company, Lucky Biotech, and the University of California were granted international patents for the genetic modification of thaumatin marketed (by Tate & Lyle) under the brand name Talin. GM thaumatin is now nearing market production.

In Japan, it appears widely in chewing gum, soft drinks, canned coffee, flavoured milks and even cigarettes. It is also sold as a flavour intensifier, since it will enhance a variety of sweet and savoury tastes (including peppermint and coffee), while suppressing any bitterness.

The bright red fruits of the serendipity berry (*Dioscoreophyllum cumminsii*), only about 1.5 cm (¾ in) long, grow in grape-like clusters of 50–100 fruits. While the seeds contain very bitter substances, the fruit pulp that surrounds them contains a protein called monellin that is intensely sweet – around 3,000 times sweeter than sucrose. Although natural, monellin has not been mass-marketed to the food industry, largely because of the problems inherent in large-scale purification of the protein, which is sensitive to heat or acid. However, genetically engineered monellin has now been produced.

The fruits of a third West African plant, *Synsepalum dulcificum*, aptly named the miracle berry, have the unusual power – in the words of the *Kew Bulletin*, which published details as long ago as 1906 – 'to change the flavour of the most acid substance into a delicious sweetness'. The taste of a lemon or some other sour substance, it noted, becomes wonderfully sweet after chewing just one of the plum-like fruits – a sensation said to last for up to an hour. While West Africans have traditionally used these fruits to sweeten palm wine and beer, the glycoprotein responsible (miraculin) has undergone much investigation for use as a food supplement, either as an additive, to alter the taste of food, or taken separately before eating.

The Japanese are the commercial promoters of another plant-derived sweetener stevioside, a white crystalline powder 250–300 times as sweet as sucrose. First isolated in France in the 1930s from the leaves of *Stevia rebaudiana*, this plant has long been used by the Guaraní Indians of Paraguay (who call it *caa-ehe*, 'sweet herb') for sweetening their tea-like yerba maté. *Stevia* is banned from use in human food in Europe and the USA (though it can be found in some 'dietary supplements'). It has, though, been widely accepted in other countries, including Japan, as a tabletop sweetener, and in food products such as ice cream, bread, pickles, seafood, processed vegetables, soft drinks and sugarless chewing gum.

Not so sweet: the world sugar trade today

Described by the historian Toby Musgrave as 'one of the world's most indulgent and least necessary crops', sugar production caused unimaginable **suffering** and the **death** of hundreds of thousands of people during the colonial era. The world's sugar trade today is still the cause of misery for countless people in the developing world. Physical conditions for those involved in the cutting of sugar cane continue to be **extremely tough** in the tropical countries where most sugar is still produced. But it is the unfair rules enforced by Europe – now the biggest exporter of white sugar to the rest of the world – that are destroying prospects for some of the least developed countries.

From amongst the 121 sugar-producing nations worldwide, sugar cane is grown very efficiently and provides **thousands of jobs** for poor people in countries such as South Africa, Brazil, Colombia, Malawi, Guatemala, Zambia and Thailand. To produce sugar from sugar beet, conversely, is much less cost effective – costing over 50% more. The EU's sugar production costs, in fact, are over double internationally competitive rates. Yet, in 2001, western Europe accounted for a **massive** 40% of white sugar exports, more than any country in the world and totalling nearly 7 million tons. Fixed high prices and huge subsidies from the EU allowed the less efficient **European producers** to sell their sugar – derived from sugar beet – very cheaply, undercutting other producers. Such practices consistently depress world prices.

Large British and other European companies have lobbied fiercely to prevent the least developed countries being given the **market access** they enjoy, causing great social and economic hardship. It has been estimated by Oxfam that Mozambique will have lost the possibility of earning £108 million (almost 75% of the EU's annual development aid to that country) by 2004, through exclusion from access to Europe's sugar markets. In

'Tell those people that our children are dying of hunger and disease. If our salaries could increase, life would be different.' Sugar cane harvesters, Mozambique, 2002

Mozambique the sugar sector is the single largest source of formal employment, but nearly three-quarters of the population live in poverty. The **complexities** and contradictions of the EU sugar regime directly affect not only the take-home pay of some of the poorest workers in the poorest countries, but the cost of many items purchased across the world.

Big winners are the beneficiaries of the European sugar processing industry, which has been able to maintain **very high** profit margins, and, of course, the large-scale farmers in areas such as East Anglia, Lower Saxony and the Paris Basin, who receive huge subsidies to grow sugar beet.

Research has shown that in Britain the intensive farming of

sugar beet involves heavier pesticide use than for most other crops, and causes **soil erosion** that is far from beneficial to the environment.

The rich world, supported by the World Bank and the IMF, insists that developing countries must trade their way out of poverty, but the unfair rules and **double standards** of global trade do not allow this to happen. As Oxfam point out, changing these rules and practices is essential to make trade fair. In their briefing paper *The Great EU Sugar Scam* (which has been used as a major reference here), Oxfam concluded that no agricultural sector is in need of more **urgent reform** than that of sugar, yet the subject is not even included in the latest EU reform proposals.

A bowl of cereal 'for everyone'

Cereal facts

■ In 2001, some 2,086 million tons of cereals were grown on over 670 million ha (1,655 million acres) around the world: wheat, rice and maize account for at least 80%, while barley, oats rye and pearl millet made up about 19% of production. The crops of 19 different grass species provided the remaining 1%.

■ Despite our current dependence on grains and the expected increases in global population, the world's croplands have been decreasing in size in recent years (due to urbanization, land degradation and the loss of land to non-food crops such as cotton), but yields have been increasing.

116

The world's cereal bowl

Ceres, the Greek goddess of grain, has given her name to the plants that provide most of the world's staple foods – feeding not only people but most of our domesticated animals too. **Cereal grains** are generally considered to have been a prerequisite for the development of the world's major civilizations: **maize** for those of South and Central America; **wheat and barley** for the Near East and Mediterranean basin; and **rice and millet** for the populations of the Far East.

To most of us, a bowl of cereal means cornflakes, muesli or some other product that arrives on our breakfast table in an attractive box, needing only the addition of milk, and perhaps sugar. The main ingredients are, as the names tell us, derived from cereal grains such as maize, wheat, rice and oats, perhaps in the company of nuts and soft dried fruits. However, these grains are often so highly processed that it is difficult to remember what it is we are really eating. To many rural peoples a bowl of cereal would be more likely to mean: in Southeast Asia, a portion of plain boiled rice; in South and Central America, a dish of cooked maize kernels; and in Africa, a kind of porridge made from pounded sorghum or millet.

Various other cereal grains, however, that are relatively little known outside their traditional production areas make highly nutritious foods. Two of these are teff (*Eragrostis tef*), the most important crop in highland Ethiopia, noted for its resilience to drought and which produces tiny but highly proteinaceous grains, and Adlay or Job's Tears (*Coix lacryma-jobi*), which is of local importance as a cereal in south Asia.

Several plants that do not belong to the grass family, but which produce cereal-like grains (known as pseudocereals), are also valuable sources of protein. The amaranths (about 60 species belong to the genus *Amaranthus*) are a good example of these. Native to South and Central America, three species of *Amaranthus* have been cultivated since pre-Colombian times. The Aztec ruler Montezuma received tributes of thousands of pounds of amaranth seeds every year, and Andean peoples also held the crop in high esteem, both for dietary and medicinal uses. In Peru, where cultivation of kiwicha or Inca wheat (*A. caudatus*) has been boosted in recent years, the beautiful colours of the stems, leaves and flowers create a fiery patchwork across the mountainsides. Over 100,000 tiny seeds with a higher protein, iron and calcium content than wheat may be held on one seed head, weighing up to 2.7 kg (6 lb). The amaranths are also important sources of leaf vegetables in the lowland tropics of Africa, Asia and the Caribbean.

Quinoa (*Chenopodium quinoa*), native to Peru, was once the cherished crop of the Incas, a vital and sacred food. Although it was banished by the Spanish when they conquered the country, in favour of wheat and barley, it was never forsaken and is still cultivated throughout the Andean region as a staple at high altitudes. A major and highly nutritious source of protein for millions of people, its white, yellow, pink or black seeds (about 18% protein) are traditionally toasted or ground into flour. They are also boiled, made into breakfast foods, added to soups or stews, and fermented to make chicha beer. Bolivia's

annual quinoa export market is estimated to be worth some $1 million.
In 1994, two American researchers were granted a patent on plants of
the country's traditional 'Apelawa' variety, causing outrage in Bolivia.
Following protests, the patent was allowed to lapse in 1998.

Buckwheat (*Fagopyrum esculentum*), thought to be native to northeastern Asia, was introduced to Europe in the Middle Ages and is now
widely cultivated, notably in China and the Russian Federation. Its
three-cornered seeds, which resemble beechmast, led to its German
name *Buchweizen* ('beech wheat'), later corrupted to 'buckwheat'.
Ground into flour, it is used to prepare a number of foods, such as
bread, pasta, dumplings and pancakes.

Rice

With over half the world's population dependent on it as their staple
food, rice, alongside wheat, is one of our two most important crops.
Unlike wheat, however, most rice is grown in less industrially
developed countries, mostly in Asia, where it is the primary source of
protein and carbohydrate. Over 595 million tons of paddy rice were
grown in 2001, in 112 countries on some 10% of the earth's arable
land – the three greatest producers being China, India and Indonesia.

Twenty-one wild species of rice are scattered in Asia, Africa,
Australia and the Americas, but only two of these species, Asian rice

**Winnowing millet, Tirelli village,
Mali.** The Dogon people of Mali depend
on millet – a crop adapted to
produce grain in extremely hot and dry
conditions. By contrast, rice is mostly
grown in the wet tropics.

117

Asian rice

■ Because of its long history of cultivation in many countries, thousands of
local rice varieties have arisen from
Asian rice, out of which two major
groups or 'races' are recognized:
■ 'Japonica' types: originally from
Japan and Korea.
■ 'Indica' types from the tropics and
subtropics of India and China.
Some also recognize a third type, the
'javanica', from Indonesia.
Grains are generally divided into 'long',
'medium' or 'short', and classified
according to the texture of their
endosperm. But, in addition, four main
methods of cultivation also distinguish
the many cultivars, determined by
degree of access to water (from various
combinations of irrigation, flooding or
rainfall).

118

**Rice growing in terraced fields in
Madagascar.** Indonesian people
migrated to Madagascar long
ago, taking with them the valuable
knowledge and techniques of rice
production. Today, rice is a major
staple on the island.

(see left) and, to a much lesser degree, African rice (*Oryza glaberrima*), are cultivated. Like its wild relatives, rice is a swamp plant. A
specialized stem anatomy, which allows oxygen to reach the roots, has
enabled it to be grown in paddy fields, which are flooded for much of
the growing season. Some deep water or floating rices can cope with
water up to 5 m (16 ft) deep. 'Upland' rice varieties, however, upon
which nearly 100 million people depend as their daily staple food, are
grown as rainfed crops like other cereals.

Recent years have brought a range of rice varieties for different
culinary purposes to our supermarket shelves. So-called 'wild rice'
belongs to the distinct genus *Zizania*. Grains of *Z. palustris*, an aquatic,
fresh-water grass native to the Great Lakes region of North America,
were an important source of food for North American Indians and are
now grown commercially. Most rice-based breakfast cereals are
produced from polished white grains (of *Oryza* cultivars) from which
the nutritious outer layers have been removed. These 'puffed' grains
are synthetically enriched with vitamins lost in the milling process.

Millet

Millet is a collective term that refers to several small-seeded grass
species native to Africa and Asia. The most important of these are pearl
millet (*Pennisetum glaucum*), finger millet (*Eleusine coracana*), proso
or common millet (*Panicum miliaceum*), and foxtail millet (*Setaria
italica*). Better adapted to dry and nutrient-poor soils than most other
crops, the millets can survive and produce grain in extremely harsh
conditions, in some of the world's hottest, driest regions, which may
receive as little as 300 mm (12 in) of rainfall a year. Nearly all millet is
grown by small-scale farmers, for home consumption as food or for
local trade. It is also cultivated for grazing, green fodder and silage.

Pearl millet: Also known as bulrush millet and with many other
local names, pearl millet accounts for nearly half of global millet
production and has become the staple food for millions of people in
Asia and Africa who live on desert fringes or on very arid lands. In
India – the world's largest producer – it makes up about two-thirds of
the millet crop. Critically important for food security, it will give good
yields in some of the most inhospitable regions, and is the second most
important native African cereal after sorghum. A species in which there
is a great deal of variation, the cereal grains are generally ground into
flour for making into flat breads or eaten whole. Pearl millet has also
been introduced as a grain crop to the Americas and Australia.

Finger millet: Bearing several spikes or 'fingers' at the top of each
stem and requiring slightly more water than other millets, finger millet
is widely produced in the cooler, higher-altitude regions of Africa and
Asia. It is often eaten as porridge, ground into flour for making into

119

Rice variety: who reaps the benefit?

'Today, all those who care about the future of Asia must recognize the importance of rice in the formation of [India's rural] societies and in their future. Rice must be … controlled by the small farmers – the people – and not by foreign corporations with no interest in or understanding of it. Thousands of farmers are already showing that growing rice organically can produce more food and nutrition, not just for humans, but for all species who share our planet with us. It can be done.'
Vandana Shiva, Director, Research Foundation for Science and Technology, 2000

Rice was first grown thousands of years ago in India and China and is an **integral part** of most Asian cultures. The thousands of ancient rice varieties are the result of careful selection and cultivation by local people over long periods, to suit different environments and needs. Rice is much more than just a source of calories, however, and Indian scientist, Vandana Shiva reminds us that it has come to represent **many things** to Asian people: culture, history, landscape, religious and social ideas – indeed, to many people, life itself.

Many traditional rice varieties were lost forever or became seriously threatened during the 'Green Revolution' of the 1960s and 70s when farmers were encouraged to abandon them in favour of a small number of genetically uniform, high-yielding varieties. New insect pests and diseases were also introduced. The serious **genetic erosion** of ancient rice varieties that ensued (India alone had some 200,000) led to efforts to collect and conserve genetic

material of those that remain. The International Rice Research Institute (IRRI) now holds over **86,000** accessions in its gene bank of *Oryza sativa*, *O. glaberrima* and wild species, from more than 110 countries. Such material is described as being 'held in trust for the rice-producing and rice-consuming nations of the world'. The major **rationale** underlying such collections is to 'ensure the conservation and continued availability of genetic resources for rice improvement'.

Private rice?

Plant breeding programmes continues then, to 'improve' rice varieties and develop **new cultivars** with higher yields or adaptation to particular stresses or environments. Such research and development is often justified by the claim, as made by Future Harvest, that its aim is to 'achieve sustainable food security' and so 'help the poor'. While this may, in certain cases, be true, it implies from the outset that most farming as carried out by people in developing countries, is **unsustainable**. It also implies that it is unproductive and that the way forward is for the production of surpluses 'to meet the demands of expanding local, regional and national markets'. Numerous examples – sugar, coffee and bananas, for example – show that the operation of such **markets** at national level is intended to benefit large commercial producers. The development of new rice varieties accounted for a **doubling in yields** between 1960 and 2000. But the commercial imperatives of the biotechnology industry have obscured the root causes of the problems facing small-scale producers.

An example of this is the development of '**golden rice**', hailed as a 'miracle cure' for blindness. Over $100 million have been spent to produce this transgenic rice at the Institute of Plant Sciences in Zurich. Genes taken from daffodils and bacteria have been introduced into a rice strain to produce a yellow-coloured rice with high levels of betacarotene, which is converted to **vitamin A** within the body. Plans were made to transfer this 'golden rice' to India – but what for? It has been pointed out that vitamin A rice is likely to fail in preventing blindness as it will meet only a **small amount** of the required daily intake. Most vitamin A will still have to be provided by alternatives that already exist, such as green leafy vegetables and fruits.

The corporate control of agriculture is evidenced by the continuing attempts of biotechnology companies to patent food crops such as rice. A **notable example** of this occurred in 1997 when the US-based corporation RiceTec Inc. was granted a patent on novel 'lines', seeds, grains and their progeny of India's famous basmati rice, grown anywhere in the world. Some 20 different claims were made, 15 of which were, however, rejected by the US Patent and Trademark Office in August 2001. Basmati rice is grown on nearly 2 million ha (4.9 million acres) across India and Pakistan and the attempt to **patent** varieties, which could threaten the livelihoods of millions of Punjabi farmers, was widely condemned as being both ethically and morally unacceptable. By the end of 2001 some 250 patents had been granted on rice varieties, over 60% of them corporately owned.

large flat breads or fermented to produce traditional beer. India and China are the chief producers for international trade. It can be stored for many years and is largely resistant to insect attack.

Proso or common millet: This species, which will tolerate a large range of temperatures, is widely grown for food in eastern Asia, India, parts of the Russian Federation, the Ukraine, Kazakhstan, the USA, Argentina and Australia. Traded internationally (along with foxtail millet) for use as bird seed, the compact, branching seedheads, which resemble an old-fashioned broom, gave rise to the name 'broomcorn'.

Foxtail millet: Also suited to moderate climates, most foxtail millet is grown in China, where it was considered a sacred plant as long ago as 2700 BC. It is also cultivated in other Asian countries, Africa and some parts of southern Europe.

Sorghum

Since its domestication in Africa over 5,000 years ago, sorghum (*Sorghum bicolor*), the world's fifth most important cereal after wheat, rice, maize and barley, has been subject to intensive selection by different peoples, and comprises a great array of wild, weedy and cultivated annual forms. The cultivated forms are generally classified into five basic 'races', or cultivar groups, plus ten hybrid combinations of these. Although principally a plant of tropical, semi-arid conditions, particularly adapted to drought, sorghum can survive within a wide range of environmental conditions, and will tolerate water-logging as well as high rainfall.

An important human food, sorghum grains may be boiled or roasted but are often ground into flour to make flat breads and porridge or fermented for beer. Small quantities of sorghum are used by commercial food industries for flour, malted drinks and beer. In China, sorghum is widely distilled to produce vinegar and a popular spirit. Other uses include syrup from sweet-stemmed cultivars, stems for fuel and thatching and brooms from 'broomcorn' varieties.

Maize

Millions of Europeans and Americans start the day with a bowl of maize (see overleaf) (or corn, as it is also known), not in the form of whole grains, but as cornflakes. Stripped of the fibre and nutrients present in the outer layers and the embryo of the grain, the pre-cooked starchy endosperm is rolled and toasted with the addition of a range of synthetic vitamins, malt flavouring, salt and sugar. For many people of the tropics, however, maize in its unadulterated form is a major part of their diet. It has about the same number of calories as wheat or rice (after which it ranks third in importance worldwide as a cereal crop) but, though lower in protein and deficient in the amino acid lysine, it is higher in fat and richer in thiamine. It is often grown in the company

Sorghum production

Sorghum production can be broadly divided into two groups:

■ **Local varieties** grown in extensive systems mostly for food and chiefly in developing countries (in Africa, Southeast Asia, India, Central America and China).

■ Modern, **high-yielding hybrids** grown as an intensive, commercialized crop in the developed world (chiefly the USA and Australia) and parts of Latin America and the Caribbean, mainly for livestock feed.

121

Maize

■ Maize (*Zea mays*) has been tradi-tionally classified into the following main groups according to the appear-ance of the grains and the variation in their starch: sweet, pod, dent, flint, pop and floury.

■ Sweetcorn, the variety eaten widely in the USA and Europe as a veg-etable, is so named because some of the sugars in the endosperm are not converted into starch and so are 'sweet' to taste.

■ Dent corn (which accounts for 95% of all the maize and is mostly fed to animals) dries with a small depression or dent in the surface of grains. Like flint corn, which also has a hard endosperm, dent corn was one of the traditional crops of North American Indians, who showed European settlers how to grow the plants successfully.

■ Pop corn, developed as a crop by Andean peoples, contains very dense endosperm inside a tough coat. Heating causes the endosperm to explode and the grain to swell and burst, creating the familiar snack.

■ Floury corn, conversely, has a soft and mealy starch.

■ Over 609 million tons of maize were grown around the world in 2001, of which nearly half was produced in North America. The country's most important crop, the majority was fed to farm animals; silage maize has now become a major crop in many industri-alized countries. Other major producers include China, Brazil and Mexico, but only a small proportion of the total output enters world trade.

of beans as they supply two essential amino acids in which maize is particularly deficient.

Testimony to its enormous importance as a food plant, the scientific name for maize, *Zea mays*, was established by Linnaeus: *zea* meaning 'cause of life' and *mays*, from the name *mahiz*, 'our mother', given to the plants by the Amerindian peoples of Cuba and Haiti. A highly adaptable plant, maize is cultivated today in almost every continent.

A great number of different myths, still told by native peoples of South and Central America for whom maize is a staple food, give to our ears implausible accounts of the origin of maize. In some of these tales, a fox or a small bird steals the grains from heaven; in others, the moon in the form of a man brings them to earth as a precious gift. But the ancestry of corn has long puzzled Western scientists because it differs so much in appearance from teosinte, the wild plant most closely related to maize. However, studies of DNA have shown that teosinte is indeed the wild ancestor, and archaeological fieldwork points to the Mexican highlands as the place of domestication, 6,000 or more years ago.

By the time Europeans first reached the Americas, some 300 major maize varieties were being grown, from Canada to Chile. Many thousands of cultivars have now been described, amongst which there is enormous variation: plants, for example, may range from 60 cm to 6 m (2 to 20 ft) in height and 'ears', containing the cobs (whose colours can also vary greatly), from just a few centimetres (inches) to 60 cm (2 ft) in length. Most of these cultivars are now represented in gene banks around the world.

Modern breeding has resulted in cultivars with increased yields and resistance to disease, as well as those with attributes such as higher lysine (an amino acid important in nutrition in which it is naturally deficient), 'waxy' corn (comprising starch composed of amylopectin) and 'high-sugar' corn.

In 2001, the alarming discovery was made that transgenic DNA was present in ancient cultivated varieties of maize growing in Mexico's Oaxaca province, a region included in the crop's centre of origin and diversification, and elsewhere in the country. Mexico banned the planting of GM maize in 1998 but, in 2002, the grain was still being imported from the USA. One of the main arguments against the acceptance of GM crops is that escaped transgenes may become incorporated, in just this way, in wild crop relatives by crossing, posing a threat to their integrity and survival. In 2000, the genetically modified 'Starlink corn' caused widespread contamination of the crop growing in America's western corn belt, resulting in the recall of thousands of tons destined for food use.

The myriad uses of maize

While 'white' maize is the type widely used for human consumption around the world, 'yellow' maize (otherwise chiefly used for animal food) is preferred in central Europe and the Mediterranean. In the Americas, indigenous peoples grow an **array of varieties** which produce cobs of very beautiful colours or combinations of colours. For local people, such colour differences indicate the various characteristics of the grain for cooking purposes as well as the cultivation requirements of the plants.

Central and South American peoples have developed a number of delicious and nutritious ways of **eating maize**.

■ In Mexico and Central America, grains are treated with lime (in a soaking process, thus making their niacin content available) before grinding into a meal for use in **tortillas** (flat pancakes traditionally accompanied by beans).

123

■ Portions of cooked, mashed maize, wrapped in maize or banana leaves, filled with sweet or savoury mixtures (of beans, peppers or meat), known as **tamales**, are eaten in many parts of Latin America.

■ Chicha, a fermented **maize beer**, is widely drunk in Andean regions.

■ In Latin America, Africa and Asia, a **maize porridge** may also be found, whilst Italian **polenta** and North American **hominy** are likewise made from wheat.

Maize flour and starch have come to play a major if unsuspected role in many of the items and **industrial processes** we take for granted. Cornflour was used by the Aztecs to thicken their famous chocolate drink **xocoatl**, and today it is used, like maize starch, to thicken a huge range of **processed foods** from sauces and instant soups to confectionery. Maize starch is also used to make adhesives, pharmaceuticals, cosmetics, cloth and paper size, and the glucose and fructose syrups with which a large number of processed foods and drinks are sweetened.

The fermentation of dextrose (a form of glucose) has created a new group of bioproducts:

■ **Organic acids** (namely citric and lactic acids), which have numerous food and industrial applications.

■ **Amino acids**, such as lysine, used as a supplement in animal feeds.

■ **Additives**, such as monosodium glutamate.

■ **Vitamins**, such as vitamins C and E.

■ **Food gums**, such as xanthan gum.

Other, more traditional maize products include the **food oil** – highly polyunsaturated – extracted from the germ of the grain, and alcoholic drinks such as whisky, gin and beer.

A vast array of products, from paints and biodegradable plastics to synthetic rubber, can all now be made, at least in part, from processed maize. The conversion of maize starch to ethanol for use as an **alternative fuel** is a major new development, as is the biodegradable food packaging, carpet and clothing material made from polylactide (PLA) polymer, also formed from maize-derived dextrose (see page 261).

Coffee and tea: plants that stimulate

Coffee beans drying in the sun, near Lago Agrio, Ecuadorian Amazon.

124

Coffee (*Coffea* spp.) is an immensely valuable crop. There is hardly a country in the world that does not drink it and, as the primary export of many developing countries, 20–25 million people (mostly small-scale farmers) rely on its production for their livelihoods. Without coffee, countless breakfasts and dinner parties would be incomplete and social gatherings the world over would lack an important symbol of hospitality. The Japanese, who bathe in coffee grounds believing them to have health-giving properties, and the Turks, who scan the dregs of their coffee cups for omens of the future, would also have to look elsewhere.

The drink that revives hundreds of millions of us is made from the roasted seeds of a tropical evergreen shrub, now grown in some 80 different countries. Three different species of coffee bush are grown commercially, though only two are cultivated on a major scale:

● *C. arabica*: the source of arabica coffee, and the most widely grown. It produces about 70% of the world's coffee, chiefly from South and Central America, and to a lesser extent from its native Ethiopia.

● *C. canephora*: gives robusta coffee and accounts for around 30% of the world's coffee. It is grown in Southeast Asia (where Vietnam is now the world's largest robusta-producing country), central and West Africa and Brazil.

The discovery of coffee is attributed in legend to an Ethiopian goatherd who noticed that his goats had become unusually frisky after eating the ripe red berries of wild coffee bushes that grew in the forested mountains of the country. The goats' activity was, no doubt, due to their ingestion of caffeine, a natural stimulant contained in the coffee plant. By the 2nd century, local tribesmen were making small cakes from the pulverized fruit mixed with grain and animal fat to sustain them on long journeys and to relieve fatigue. The berries were also fermented and mixed with water to make a stimulating drink.

Arabic peoples were the first to brew coffee as we know it today, and by the 15th century, with the opening of coffee houses, the drink had become extremely popular across the Islamic world. Coffee drinking spread quickly throughout Arabia and into Turkey and the surrounding areas. To offer any visitor a cup of the strong black brew became an important gesture of hospitality.

The Arabian drink did not reach Europe until the early 17th century, but it did not take long for coffee houses to become all the rage. Dubbed 'penny universities' since it was possible to exchange gossip and learning there for the price of a cup of coffee, and as forums for political debate, they were regarded by anxious politicians as centres of sedition. Attempts were made to close down the houses, but such an uproar ensued that the idea had to be abandoned. Many of the City of London's financial institutions had their beginnings in coffee houses.

Making a cup of coffee

In Britain, 90% of all the coffee drunk is 'instant', made from soluble granules of a spray- or freeze-dried coffee brew. Real coffee fans, however, stick to the drink brewed directly from the freshly ground and roasted beans.

Much of the ground coffee available pre-packaged is made not from one single type or batch of beans, but from blends of a number of different coffees that are carefully selected to give a desired taste. The variety of coffee bean, the region it is grown plus the method of its preparation and subsequent roastings combine to produce a distinctive appearance and aroma. Unblended coffee from one particular area or plantation is referred to as 'pure', 'original' or 'single-estate' coffee, and attracts the attention of coffee connoisseurs. After picking, by hand or machine, the sweet pulp that surrounds the coffee beans inside each fruit has traditionally been removed in one of two ways.

● The dry process: involves first drying the whole fruits in the sun or artificially and then removing the dried pulp, the fine protective endo-carp or 'parchment' and the silvery skin surrounding the seeds.

Coffee

■ All cultivated coffees are small trees, which may grow to heights of 10 m (33 ft) if not pruned, with glossy, evergreen leaves and white, sweet-smelling flowers. Over 7–9 months, following fertilization of the flowers, the berries mature. Inside the sweet pulpy outer layers are two coffee seeds (or 'beans') surrounded by a delicate silvery seed coat.

■ Since coffee bushes cannot tolerate frost, they are grown primarily in tropical and subtropical countries with an average rainfall of 1.5–3 m (5–10 ft) per year.

■ Robusta is a hardier plant than arabica; it is able to tolerate higher temperatures and drier conditions, and has a greater resistance to pests and diseases, making it easier and cheaper to grow. However, it produces an inferior-tasting coffee, with a higher caffeine content than arabica, which grows best at mild temperatures (15–24°C/59–75°F), often on high ground.

125

Wild coffee in peril

The continuing destruction of Ethiopia's once extensive montane forests, for agriculture, fuel wood and timber, is threatening the survival of its **wild coffees**. These plants form the dominant understorey shrubs, from which cultivated *Coffea arabica* developed. Also home to the endangered Ethiopian wolf, less than 2,000 sq km (772 sq miles) of forest now remains. The natural **immunity** to 'rust' held within wild Ethiopian coffee has already proved invaluable in helping our cultivated coffees fight this dreaded disease, and in future the genetic diversity of wild coffees could provide breeding programmes with essential characteristics for

resistance to other pests, improved quality and higher yields. Although some wild coffees are represented in gene banks – where, since the seeds will not remain viable for long periods of time, they have traditionally been conserved by growing plants – their **safety**, due to plants dying, cannot be assured. Attempts are being made to conserve **Ethiopian montane forest** by establishing networks of coffee gene reserves, but much remains to be done.

About 50 wild coffee varieties also exist in **Madagascar** (where most, too, are threatened). A distinguishing feature is that they have low to negligible amounts of caffeine.

Shade-grown coffee beans
ripening on the bush in Ecuador.

● The wet process: which produces 'mild' coffees with a superior flavour. The ripe fruits are first de-pulped by machine, exposing the thin parchment, then washed and left to ferment for 12–24 hours. During this process a chemical alteration takes place in the beans, producing substances that eventually develop into the characteristic coffee aroma and taste. The beans are then dried in the sun for about a week, whilst still encased in the parchment and semi-transparent skins. These layers are not removed until just before shipping.

A third method, the pulped natural method, has emerged in countries with low humidity. This involves pulping but not fermenting the beans, resulting in a high-quality coffee with the characteristics of both wet- and-dry processed coffees.

Roasting, mostly with hot air – a precision job, performed with very sophisticated machinery – reduces the moisture in the dried green beans and brings out their aromatic oils. This is the all-important final stage, which develops the optimum flavour, aroma and appearance of the beans, and the taste of our cup of coffee.

Coffee substitutes from other plants

Many substitutes for coffee have been drunk over the years. The best known is **chicory** (*Cichorium intybus*), which became popular during the 18th century and is still favoured by the French. The bitter, fleshy taproots – brown externally, but yellowish-white inside – are chopped, roasted and ground for use either alone or as a coffee additive. **Other substitutes** include figs, dates, acorns, malt, barley, peas, maize, oats, chick peas and lupin seeds, which, after roasting and grinding, may be flavoured with steam that has passed through roasted coffee. **Dandelion roots** (*Taraxacum officinale*) are also made into a coffee-like drink. They possess tonic and stimulant properties but lack the caffeine of coffee beans.

Instant coffee: what's in the jar?

Unlike **Beethoven**, who, it is said, measured out exactly 60 coffee beans for every cup of coffee he brewed, most British and Australians prefer a quicker, simpler drink – largely prepared by someone else. To make a jar of instant coffee, coarsely ground beans are brewed under **pressure** to make a concentrate. The water from this brew (recycled at very high temp-eratures) is then often removed by **spray drying**. As the concentrate dries, it may be tumbled with steam to induce the formation of granules.

Another method is to freeze-dry the coffee brew into **thin sheets**, which are then cut into granules. The temp-erature of these sheets is allowed to rise under vacuum, resulting in the water 'boiling off' at a very low temp-erature, without the coffee solids becoming wet. This leaves small dry **particles** with no heat damage and minimum loss of volatile aromatics. Since coffee essences and subtler flavours may be lost during both pro-cedures, however, they are captured and returned to the processed coffee.

Unless extracted (often with the help of solvents, steam or water), this coffee contains the drug caffeine. **Caffeine** is a cellular stimulant. It increases the metabolic rate, stimulates the heart and mimics the feelings produced when adrenalin is released. Though the inspirational effects of pure coffee are not to be denied (**Bach**, for example, wrote his *Coffee Cantata* in praise of the drink), an excess of caffeine can cause anxiety, dizziness, heart palpitations and even mild delirium.

THE REAL PRICE of a cup of coffee

'The coffee farmers in Latin America are suffering from the worst crisis in a hundred years.' *Mugged: Poverty in your Coffee Cup*, Oxfam International campaign report, 2002

Most of the coffee so many of us enjoy each day is grown at the expense of millions of poor farmers, who have now been pushed to crisis point. The coffee we buy makes a large profit for the coffee companies but, as Oxfam report, growers in developing countries are having to sell their coffee beans for much less than they cost to produce, and suffer huge hardship as a result.

The cultivation of coffee has traditionally been carried out on small family farms, often as part of a diversified system involving other crops. Bushes are planted at fairly low densities under shade trees (often fruit trees and other hardwood species), giving a range of benefits to plants, people and the environment. About 70% of the coffee consumed worldwide is still produced in this way.

Encouraged by the World Bank and the IMF to increase export earnings, governments of some of the world's poorest countries have, in turn, encouraged coffee growing as a way of reducing their huge national debts. With this impetus, over the last 20 years the cultivation of coffee has become increasingly intensive, abandoning the 'shade-grown' system in favour of the industrial cultivation of high-yielding coffee varieties, with heavy use of chemical fertilizers and pesticides. Around 40% of the coffee cultivated in Colombia, Central America, Mexico and the Caribbean is the so-called 'sun-coffee', grown in the open. In the newer, industrialized systems, the berries are harvested by machine, irrespective of whether all of them are ripe, and this results in coffee of lower quality. Using these methods there have been big increases in production in Brazil, and a dramatic increase by Vietnam, now the world's second biggest producer.

A threat to livelihoods

A new threat to small-scale coffee farmers and the environment lies in wait in the form of GM coffee, for which the University of Hawaii was granted a US patent in 1999. Being developed by a private company, research has focused on engineering coffee fruits so that they do not ripen until a chemical is applied. Estimated to be able to increase yields and decrease labour by as much as 50%, the further industrialization of coffee would be inevitable.

Some 8% more coffee is already being produced than consumed. The result of this is that there has been a dramatic fall in the prices paid to farmers for their beans. According to Oxfam, farmers receive only a quarter of the price they were paid in the 1960s, the price of coffee in 2002 being lower than it was 30 years ago. With a flooded market, the major coffee roasters have been paying the lowest possible price for beans, making very considerable profits. But it is a disaster for growers, many of whom cannot afford to buy food or other basic necessities, and have been forced to leave their homes. As coffee traders go out of business, too, so banks and entire national economies suffer – the value of coffee exports having dropped by $4 billion in the last five years. The adoption of the free-market system means that some of the poorest people least equipped to do so are obliged to negotiate in an open market with some of the richest and most powerful.

127

Good coffee, fair prices

■ 'Shade-grown' coffee, produced under conditions that are better for both the environment and small farmers, has much to recommend it. In El Salvador, these plantations account for 60% of the country's forest area, and elsewhere in Central America such plots have ensured the survival of the remaining forest.

■ Organic coffee is often grown under shade in a natural forest or farm environment, and is therefore beneficial to producers, consumers and the animals and plants with which it coexists.

■ As its name suggests, Fairtrade coffee bypasses conventional coffee dealers and pays fair prices directly to farmers or farm co-operatives. As well as ensuring that coffee workers receive a guaranteed minimum price, which covers production costs and a living wage, Fairtrade criteria encompass the right to join trade unions and work in conditions that follow health and safety and environmental standards, also encouraging sustainable methods of crop cultivation.

128

129

Tea

■ The drink that is now enjoyed by over half the world's population is made from the leaves of a small evergreen shrub related to the garden camellia, *Camellia sinensis*. Indigenous to the lower montane forest of mainland Asia from southwestern China to Assam in northeastern India, two main varieties of *C. sinensis* are formally recognized: *C. s.* var. *sinensis* (known as China tea), and *C. s.* var. *assamica*, which produces Assam tea.

■ Tea plants require even temperatures with moderate to high rainfall (at least 1.3 m/4 ft 3 in per annum) and high humidity throughout most of the year. Both varieties are grown mainly in the subtropics and mountainous regions of the tropics, generally between sea level and just over 2,100 m (6,900 ft). Near the Equator, tea may be grown at heights of 1,200–1,800 m (3,900–5,900 ft) and at elevations of some 2,100 m (6,900 ft) in Kenya and the Darjeeling district of northern India.

Tea

Autumnal, pungent, brisk, weedy and chesty are just some of the terms used regularly by professional tea tasters as they sample teas produced on plantations all around the world. Along with the packers and blenders, their job is to decide which of the 3,000 or so different types or grades of tea will enter our homes in tea bags or in 'loose' form in more expensive packets.

In 2001, just over 3 million tons of tea (see left) were produced globally from over 25 countries. Accounting for 80–90% of world production, by far the largest producer is Asia, chiefly the countries of India, China, Sri Lanka and Indonesia. Of these, India is the largest individual tea-producing country, growing nearly 30% of the world's tea, but Sri Lanka is the world's biggest exporter. East Africa is also now a major producer, with Kenya exporting 400,000 tons. Indonesia, China and Vietnam are also important producing countries.

While about half of all tea grown is consumed domestically, North Americans drink about 45 billion cups of tea each year – most of it 'iced'. But the British are the world's biggest tea drinkers, getting through some 70 billion cups per year, followed by Ireland.

The Chinese have been drinking tea for over 4,000 years and its earliest uses were almost certainly medicinal. For some 200 years, teas exported from China were the only ones known to the Western world. Since the Chinese had long resisted attempts by traders to gain any established rights there, the British in particular were keen to grow their own supply in India. Several expeditions to China were launched to try to bring back seeds or plants, and this eventually led to the first experimental plantings in India in around 1818.

In 1823, however, an important discovery was made: *Camellia sinensis* var. *assamica* (Assam tea) was found growing wild in the hills of Assam in northern India by two employees of the East India Company. Over time, this led to large-scale commercial planting in Sri Lanka, now the world's biggest exporter (where 1 in every 20 workers is employed in the tea industry), and India. Introduced to East Africa at the beginning of the 20th century, it has now become a very important crop there. As its name suggests, most China tea is still grown in China, in the southern and eastern parts of the country, and in Taiwan.

In Britain during the early 19th century, tea was still very much a drink for the rich, and they objected very strongly to the desire of poor people to drink it too. One unpleasant substitute sold to the unwary was apparently made from ash leaves and sheep's dung. By Victorian times, however, the taking of 'afternoon tea' had become a British institution. The custom is said to have originated in 1750 with Anna, wife of the 7th Duke of Bedford, who complained of having a 'sinking feeling' in the late afternoon.

The tea we drink

Grown on plantations, usually as a single crop, tea bushes are pruned severely to keep them at 0.5–1.5 m (1 ft 7 in–5 ft) in height and to encourage the formation of twigs. Every 7–10 days the bushes produce a 'flush' of tip growth, which is picked by hand in developing countries, and thrown into baskets carried on the pickers' backs. Experienced pickers are reported to gather up to 30 or 35 kg (66 or 77 lb) of leaves and buds each day. Very arduous work, most employees are still very poorly paid and live in conditions of extreme poverty. Teas traded as part of the Fairtrade initiative, however, guarantee the workers a price for their crop that will cover their costs, allow for investment and provide a living wage. Good housing, healthcare, safe working conditions and children's education are also important issues addressed by the scheme.

While the best-tasting teas are 'loose leaf', produced on single estates, and consistently good enough not to need blending, the popular brands of tea, whether loose or in bags, are blends of many different tea crops. Combined by experts who grade for strength, flavour and colour to produce the same blend continuously, there are often around 20 different teas in one blend. Plant variety, the size, age and part of the leaf picked, region of origin and processing method all determine the final appearance and flavour of the tea we drink. Good teas will tend to have a bright appearance whilst the cheaper the tea, the muddier the colour.

Previous page: A tea plantation on the edge of rainforest, in Queensland, Australia, which is harvested mechanically.

Tea processing

In general, **China tea** produces a drink with a flowery flavour that is lighter and finer than that made from Assam leaves. The major distinction that is made between the teas we buy, however, refers to their processing method, giving us either 'black' or 'green' teas. **Black tea** makes up about 78% of the commercial tea drunk throughout the world, including the famous British 'cuppa' and that most often enjoyed in the rest of Europe and North America. Green tea remains more popular in the Far East.

To make black tea, the leaves are first '**withered**' by placing in special 'troughs' where air is blown through them. This reduces their water content and begins the enzymatic processes that will release the tea's **aroma**. The leaves are then 'rolled', passing through machinery which ruptures the cells and releases enzymes, causing chemical oxidization of the phenolic compounds present. This process, known as **fermentation**, takes place in a warm, damp atmosphere until the leaves, having absorbed oxygen, become a bright, shiny copper colour. Finally, the leaves are **dried** with hot air to stop the fermentation and remove excess water. It is during the drying that the tea leaves acquire the characteristic colour of black tea.

Well-known grades of black tea include the **Pekoes**, such as Orange Pekoe, which gives a light, pale tea, and the **Souchongs**, such as Lapsang Souchong. This tea originally acquired its distinctive flavour from the smoke from burning rope that was placed beneath trays of leaves to quicken the drying process.

Green tea is unfermented: the freshly picked leaves are steamed and dried without the withering stage and thus retain a faint but distinctive 'grassy' taste and their green colour. Examples of green teas include Imperial, grown in China, Gunpowder from China and Taiwan, and Yamashiro from Japan. **Oolong** tea is half-way between black and green. It is a very fine tea, produced by a short withering process, and is lightly fermented to give a delicate flavour of peaches.

131

The soft, hairy tips of the tea leaves are said to produce the best and most expensive teas, but the time of picking is crucial.

The stimulating effect of tea is due to the caffeine contained in the leaves, whilst the characteristic brown colour and 'bite' come from the tannins present. The flavour is a product of the fermentation process and the tea's essential oils. Some teas, such as Earl Grey, have added flavouring in the form of oil from the bergamot orange.

Other 'teas'

Countless other plants are widely used throughout the world to make stimulating or refreshing drinks, often widely termed 'teas'. These include the maté made from the holly species *Ilex paraguariensis*, widely drunk in Paraguay, redbush 'tea' from the South African *Aspalathus linearis*, chamomile, and numerous other herbal 'teas' widely available in Europe and America.

Taking tea

In Britain today 93% of all tea drunk is made from **tea bags** (also made from plant material, mostly wood-pulp, but also manila hemp). In 1970, this figure was only 2%.

Many Russians, conversely, still make tea in a very distinctive way, using a special tea-urn – the **samovar**. The custom arose after a regular trade in tea – finely powdered and pressed into large blocks for transportation – had been established between China and Russia in 1696. Broken off in pieces for brewing in the samovar, the tea is drunk either with lemon or jam or by sucking through a sugar lump.

In Tibet, tea is taken with the addition of rancid **yak's butter** and salt, whilst in Burma it is not only drunk, but eaten as a pickled food *lepet*. Japan's ancient tea ceremony has its roots in China, where, about 1,000 years ago, the preparation and drinking of tea had become ritualized as a means of contemplation for the Ch'an sect of Chinese Buddhists. Moving to Japan in the 13th century, the '**Ceremony of the Hot Water tea**' was embraced as a means of reaching the ideal state of perfect harmony and tranquillity at the heart of **Zen ideology**.

132

Plant foods from different peoples

Everywhere in the world plants sustain human societies. Most indigenous peoples who are still able to support themselves by various combinations of horticulture, hunting, fishing or gathering from the wild have traditionally benefited from a diet far more varied than is often acknowledged. Generally well balanced and without the highly processed and excessively sweet foods that exacerbate or cause physical and psychological ill health, knowledge of local plants and their properties has sustained cultures for millennia. Tragically, this knowledge and the self-sufficiency amongst communities that it

A Nenets woman picking mountain cranberries in Siberia.

A Weyewa boy eating rice and pork at a ceremonial village gathering on the island of Sumba, Indonesia.

supports is being everywhere undermined by global consumerism. Whilst this may also bring increased access to resources, for most these remain out of reach.

The famines that still occur with distressing frequency in tropical regions are caused not by a lack of agricultural skill, but by a combination of factors, often beyond the control of those affected. The huge pressures imposed by international financial institutions, including trade policies that oblige countries to devote land and resources to the production of export crops, are often a cause of poverty and hunger. They often promote migrations of people from their traditional homes, causing feuds, acute suffering and social strife. Complex international battles over land and resources, failure of local governments, and the effects of climate change, caused or worsened by environmental mismanagement on a global scale, are only making things worse.

Plants from cooler climates

E ven the frozen Arctic offers plant foods for people. The Inuit have traditionally included some plant food in their diets, mostly from lichens, berries and seaweeds, as well as flower blossoms and grasses, supplementing foods such as seal blubber and oil, fish eggs and whale skin (rich in vitamin C). Over 30 different plant species contribute to the diet of the Inupiat of Northern Alaska, which also includes moose, caribou, duck, fish and whale meat.

With the melting of the winter snows, various berries, including black crowberries, bilberries or whortleberries, cranberries and cloudberries, widespread across Inuit-occupied areas of the Arctic, are picked and eaten. Today, the Inupiat may eat berries mixed with oil and sugar, or with fish livers, or mixed with fluffy fat, like ice cream. A salad made from the leaves of roseroot is made by the Chukchi people of the Chukotka Pensinsula, while in the western Yukon-Kuskokwim delta region, medicinal teas are prepared from sourdock, fireweed and the plant known commonly as Labrador tea.

Plants from arid lands

In some of the driest regions of the world, where they are still able to do so, nomadic peoples such as the San Bushmen of the Kalahari and the Hadza of Tanzania eat a wide selection of fruit, seeds, berries, roots and important water-storing bulbs, as well as animal and bird meat when available. In the Kalahari, the sweet fruit and nutritious seeds of the manketti or mongongo nut tree (*Schinziophyton rautanenii*) has been an important source of protein, fats and sugars for some Bushmen groups, especially during 'dry' years. They also eat the tsamma or karkoer (*Citrullus lanatus*), a water-storing melon, which provides a source of precious moisture. The fruit contains up to 95% water for many months after cutting, and can also be made into a stew. The seeds, meanwhile, are ground to make an oily, nutritious coarse meal. Richer in vitamin C than citrus fruits, marula (*Sclerocarya birrea*) fruits and their nutritious kernels are also highly valued.

In Australia, Aborigine peoples also traditionally ate a varied range of plant and animal foods, which formed a healthy and well-balanced diet. The Bunaba people of northern Australia, for example, ate around 50 different 'bush' foods regularly, from a total of about 200 locally available. Today, the majority of Aborigines, resettled around towns, have diets high in processed foods, and as a result previously unknown disorders such as diabetes, heart and circulatory diseases and obesity have become a chronic problem. In central Australia, however, at least 140 plant species are still used for food. Of the many fruits

135

Rights for the San Bushmen

Masters of their environment, with a supreme understanding of the **uses and properties** of the plants around them, the San peoples have used the prickly, cucumber-like fruit of *Hoodia gordonii* for thousands of years to stave off hunger and thirst during long hunting trips. Without the San's permission, however, an appetite-suppressing compound within the plant was isolated and subsequently **patented** by the Pretoria-based Council for Scientific and Industrial Research (CSIR) in 1997, and licensed to a British botanical

pharmaceuticals company. The American pharmaceuticals giant Pfizer then paid a reported $32 million for the rights to develop the drug.

The case was taken up by the San and human rights advocates and a benefit-sharing agreement belatedly made. **Vast profits** are expected to be generated by the obesity drug (an appetite suppressant that may be available by 2007), since some 100 million people are estimated to be affected by obesity in the West. A percentage of the **royalties** are to be shared between all the San peoples

scattered across southern Africa – with the focus reported to be on education, the training of leaders and land purchase where the San are tenants.

After centuries of genocide, slavery and **dispossession** of their lands, and despite the recent return of some ancestral land to the Xhomeni people, most of the estimated 90,000 Bushmen scattered in and around the **Kalahari basin** are landless and impoverished, living as squatters on cattle ranches or banished to grim settlement camps.

eaten, two of the most important are the desert raisin, which is dried and ground into a paste, and the bush tomato. Other plants once commonly eaten by various Aborigine groups include:

● Red or purple pigface fruits, said to taste like salty strawberries.
● The sweet, sausage-shaped fruits of the common apple-berry.
● *Grevillea* flowers, which are sucked or used to make a sweet drink.
● Bush 'potatoes' such as *Ipomoea costata*.
● The fruits, shoots and roots of the boab or gourd-tree.

Just under half (47.2%) of the earth's land surface (6.1 billion ha/ 15 billion acres) is classed as dry land, of which about 1 billion ha (2.5 billion acres) are hyper-arid desert. The remaining 5.1 billion ha (12.5 billion acres), supporting nearly one-fifth of the world's population, is classed as arid, semi-arid or dry subhumid.

According to the UN, the increasing degradation or desertification of arid lands due to human activity, poverty, political instability and climate change is now directly affecting 250 million people and putting a further one billion at risk.

This alarming transformation of land that once supported vegetation is focusing much botanical attention on research into and the conservation of plants adapted to survive in arid lands. The yeheb (*Cordeauxia edulis*), an evergreen shrub from the desert regions of Ethiopia and Somalia, is one such plant, since it will produce nutritious, chestnut-flavoured nuts when conditions are too arid for other crops to grow. The carob tree (*Ceratonia siliqua*) is well known for the commercial use of its fruit as a chocolate substitute, but it cannot survive in extremely arid lands. In Somalia and Oman, however, an endangered cousin of the carob (*C. oreothauma*) has the all-important capacity to grow in drought-ridden areas, and has been the subject of a breeding programme aimed at introducing drought resistance into its better-known relative.

Foods from the forest

The extraordinary biological diversity that the world's rainforests represent has provided a huge array of useful food plants. Although our rainforests are now seriously depleted, a number of food plants that have developed or been cultivated over very long periods of time within them, provide millions of people with their staple foods, and many others around the world with additions to their diet.

Root crops such as manioc or cassava (see page 138), yams, taro or 'old cocoyams', 'new cocoyams' and sweet potatoes are particularly significant.

A much less well-known plant, though one that has great potential for cultivation in the wet tropics and which has been enthusiastically described as a 'supermarket on a stalk', is the winged bean (*Psophocarpus tetragonolobus*). Also known as a goa bean or asparagus pea, this plant is a legume traditionally grown in Papua New Guinea, Myanmar and Indonesia. Every part of the plant, which grows as a vine with climbing stems, can be eaten, from the immature seed pods to the dried seeds, shoots, leaves and tubers, which contain a high-quality protein, as well as the flowers. Now grown all over West Africa, Asia and the Pacific, some 1,500–3,000 varieties of *Psophocarpus* are recognized.

Rainforests have produced an enormous range of other foods (including many obtained from the great variety of palm tree species that grow there), which are important to local people in the tropics. Rainforests have also produced hugely important commodities, such as bananas, the world's most popular fruit, which we take for granted every day. From nuts and spices, fruit and vegetables, to chewing gum and chocolate, many foods that originated in rainforests liven up our diets.

Cassava flour is toasted on a griddle before being made into flat bread in Amazonia, Brazil.

Cassava

■ For over 500 million people, cassava or manioc (*Manihot esculenta*) is the most important food, eaten at almost every meal. The tubers that form from the plant's adventitious roots are high in starch with low levels of protein. Processed in a number of different ways, however, according to variety, they are a major source of calories in tropical regions, and are cultivated today in almost all subtropical and tropical countries.

■ A large number of cassava varieties are grown by the indigenous peoples of Amazonia, to which region the plant is native. Though many intermediates occur, these are broadly divided into 'bitter' and 'sweet'. Sweet manioc may be eaten after peeling and boiling, but bitter varieties need more complex processing and are generally grated and pressed, before being made into flat breads or toasted to make a coarse flour, farinha.

■ In Brazil, farinha has become an essential accompaniment to various dishes in urban and rural homes alike. Cassava is also used to make fermented drinks and sauces in the Caribbean and northwestern Amazon. In temperate countries, dried cassava starch may be used in the manufacture of processed foods such as soups, biscuits and confectionery and is available as tapioca, for use in puddings.

The Machiguenga – a rainforest people

For the Machiguenga Indians who live in the rainforest of southeastern Peru, **sekatsi**, as their manioc tubers are known, have a special importance. They are believed to be presents from the moon, which came down to earth in human form bearing both tubers and manioc plants. These were the **first crop plants** to be given to the people, with the instructions that they should be grown with care in their gardens and treated well.

Manioc, served with small amounts of meat or fish, is the Machiguenga's **staple food**, and they recognize many different varieties. Diversity, as reflected by the other vegetables and fruits they eat, is indeed a horticultural principle. Including plants useful for household purposes, medicinal and fibre plants, the Machiguenga typically grow around 80 different cultigens in their gardens. About 30 of these may be found in a house garden at any one time – the most commonly planted of the food crops including maize, pineapple, sugar cane, yams and cocoyams. Of the rest, crops such as pigeon peas, peppers, taro, beans, avocado, squashes, soursop, peanuts, guava, plantains, bananas and mangos, with their different tastes and textures, make interesting **additions** to their diet. Like other traditional rainforest people, the Machiguenga are **knowledgeable horticulturists** who have met their needs without irrevocably damaging their environment. Repeated incursions into their territory by missionaries, loggers, oil prospectors and colonists have **obliged** many to change their traditional way of life, and become involved in the extraction of timber and cultivation of cash crops such as cocoa and coffee.

Chewing gum

The Mayan peoples of Guatemala were chewing gum from the chicle tree (*Manilkara zapota*) long before its milky latex was processed and sweetened for the first commercial use in the late 19th century. From the height of the USA's chewing gum craze in the mid-1940s until the 1960s, however, booming demand for latex meant that trees were over-harvested and, at about the same time, synthetics began to appear. Today, most of the gum base used for chewing gum is produced synthetically but *chicleros* still tap the wild trees (which can reach heights of over 45 m/148 ft) in the tropical rainforests of Central America for their latex, which is moulded into blocks before being exported for processing. In recent years, the Japanese have been major buyers of chicle gum, but this market has now substantially collapsed. A growing interest in sustainably produced gum, however, has sparked enthusiasm for a rebirth of the chicle industry among producers. Latex from other trees native to Central America has also been used in chewing gum, as well as jelutong tapped from the Southeast Asian rainforest trees *Dyera costulata* and *D. lowii*.

Many different peoples have chewed substances exuded from the bark of trees. For centuries, mastic gum – the resin obtained from the small, evergreen, shrubby tree *Pistacia lentiscus* – has been chewed as a breath freshener, also helping to preserve the teeth and gums. Relatives of the pistachio nut, mastic trees are native to coastal Mediterranean regions, but are cultivated only on the Greek island of Chios, from which gum is exported to some 50 countries, of which Saudi Arabia is the biggest importer. Greeks use mastic as an 'all-natural' chewing gum in their cooking and to flavour retsina wine and ouzo liqueur. Highly prized in ancient times as a cure for every sort of ill, mastic is being investigated afresh for its possible use in dental care and for activity against diabetes, high cholesterol, ulcer disease and cancer, amongst other disorders.

Chocolate

Chocolate comes to us courtesy both of the South American rainforest – where wild cocoa trees (*Theobroma cacao*) are still to be found – and the peoples of Central America, whose use of cocoa beans is extremely ancient. Considered to be of divine origin, the tree and its fruit played an important part in the material and spiritual lives of the Maya and their predecessors as well as other societies of Central America, including the Aztecs. Cocoa trees were widely cultivated in the tropical lowlands of Central America and the beans were important items of trade and tribute long before the European invasions. From ancient times cocoa beans were widely used as currency, and, paid as tribute to rulers such as Montezuma, were stored in huge warehouses in the Aztec capital Tenochtitlan. Long after the Spanish conquest tributes and taxes were paid to private individuals and to the crown in cocoa beans. Their most famous use, however, was as the essential ingredient of invigorating chocolate drinks known as *xocoatl* – many recipes for which were developed.

In ancient Mexico, ground-roasted cocoa beans were mixed with hot or cold water, to which maize, ground chillies, annatto and vanilla were added, plus other vegetable additions of various kinds. Often consumed in the form of soups and sometimes made into a kind of sauce (perhaps the predecessor of the modern day mole sauces in Mexico), a preference existed for chocolate drinks beaten to a froth, large quantities of which were consumed at festivals and banquets. Such preparations were believed by the Toltecs and later by the Aztecs to form part of the diet of their plumed serpent god Quetzalcoatl, and it was this association that gave rise to the genus name *Theobroma* (established by Linnaeus), meaning literally 'food of the gods'.

The cocoa tree

■ In its natural rainforest environment, cocoa (*Cocoa* spp.) is an understorey plant, usually shaded by trees with dense foliage. While most are cultivated under shade trees in this way, some cultivars, induced to produce a dense upper foliage, are grown in direct sunlight. Cocoa trees can reach 20 m (66 ft), but are usually pruned to no more than 4–6 m (13–20 ft).

■ A vast number of different cultivated and wild varieties of cocoa exist today. Of these, two or three main types are recognized by experts:

■ The *criollo*, which seems to have been grown only in Central America before the arrival of Europeans, but which is presumed to have originated in the Amazon.

■ The *forastero*, a hardier tree native to the upper and lower Amazon.

■ The (debated) *trinitaria*, a cross between the former two.

139

Most of the traditional chocolate drinks encountered by the Spanish were bitter and unfamiliar to their tastes, but the introduction of sugar to the Americas did much to boost their acceptance, and by the end of the 16th century Europeans in the Americas were enthusiastically drinking sweetened chocolate drinks. In Europe itself, chocolate was at first drunk only by nobility and the Spanish court, but by the mid-17th century, chocolate houses had spread to Paris and London, and soon competed with coffee houses. In the 18th century, some began to add milk (or wine) to their chocolate drinks (the physician Sir Hans Sloane recommended chocolate with hot milk and honey as an aid to digestion), and the eating of solid chocolate now began to gain acceptance.

Although factory conditions were created for the mass production of chocolate in England in 1728, it was not until almost a century later that the first of a series of Dutch, British and Swiss inventors and entrepreneurs began to modernize chocolate production, developing new ways of processing cocoa beans, improving texture and flavour, leading eventually to the products that we are familiar with today. Curiously, whilst coffee tended to be regarded as a drink for men, chocolate came to be considered more suitable for women in parts of western Europe and by 1800 had fallen behind tea and coffee in popularity. In its solid form – and having gained an ancient reputation as an aphrodisiac – chocolate became a token of romantic love. Casanova is said to have preferred chocolate to champagne as an inducement to romance.

Harvesting cocoa beans

Roughly the shape of an American football, pods develop from flowers on the **bark** of the trunk or on large branches. Inside, the cocoa beans are held in a sticky white pulp.

After harvesting, using long poles for the higher fruit, the pods are split open and beans and pulp removed for **fermentation**, during which time the all-important precursors to the chocolate flavours are developed. The beans are then separated from the pulp by hand or machine, **dried and graded** before processing or shipping to consuming countries.

Throughout the tropical world,

cocoa plantations have relied for their vigour and disease resistance on genes from their wild and semi-wild **Amazonian relatives**. At their most diverse along the eastern jungle-covered fringes of the Andes, the lower Amazon basin and along the Orinoco river in Venezuela and the Guyanas, the natural home of cocoa is threatened now as never before by the rapid pace of forest destruction.

Between 1995 and 1998 about 2.6 million tons of cocoa was produced **worldwide** each year from some 40 countries, about 85% of which was exported. While only 20% of the total

was produced in tropical America, and 17% in Asia (where production increased 100-fold between 1976 and 1997), 63% was produced by West African countries, now the location of cocoa production on a massive scale. Just eight countries currently account for most of **world production**, of which the Ivory Coast is by far the biggest (producing over 100 million tons per year), followed in 2002 by Ghana, Indonesia, Nigeria and Brazil.

Cocoa pods develop from flowers on the main stem and branches.

141

Harvested cocoa pods. About 20–60 seeds (beans) are contained in each one.

Debates have raged over the nutritional and medicinal benefits of chocolate, but it is considered by many a well-rounded food, containing glucose, lipids and proteins as well as significant quantities of minerals such as potassium and calcium. Cocoa beans contain several stimulants, including caffeine, but most notably the alkaloid theobromine. They also contain phenylethylamine, a stimulant related to amphetamines found in the human brain, and the alkaloid salsolinol, which together have an antidepressant effect.

The level of phenylethylamine, it has been claimed, decreases when one is lovesick – explaining the craving to eat chocolate as a solace at such times. A more recent discovery by neurologists is that chocolate contains anandamide, a neurotransmitter which, like cannabis, heightens sensations and induces a feeling of harmless euphoria. In Japan, meanwhile, researchers have concluded that chocolate has antibacterial properties, though sugar will continue to cause tooth decay. The flavonoids found in chocolate, others claim, are good for the heart. The British Heart Foundation suggests, however, that fruit and vegetables are a much more suitable source of heart-preserving chemicals. The Swiss, followed by the British, are the world's biggest eaters of chocolate today, but the USA, followed by western Europe – where consumption has doubled since 1945 – are the leading consumers of cocoa.

Spices

Spices from the tropics earn around $2 billion each year in world trade. The lure of cloves, cinnamon, cardamom, allspice and, most importantly, pepper – a spice that originated in the western Ghats of India – drew Columbus and later merchants to 'discover' for themselves the rainforest regions of the earth. For Europeans, North America was a 'byproduct' of the maritime search for pepper, whilst at home Venice had

Making chocolate

Depending on their various qualities, two main types of cocoa are distinguished in trade: **'fine'** (used for the best quality chocolate, butaccounting for less than 5% of production) and **'bulk'**.

Chocolate is made by first roasting **fermented and dried** cocoa beans. In a process that allows the fat or cocoa butter to run off, the beans are then **crushed** into small pieces or

'nibs' before heating and **pressing** to form 'cocoa mass' (or 'cocoa liquor'). Raw cocoa powder is the pulverized form of this cocoa mass.

In the process known as alkanization or **'dutching'**, calcium carbonate is added to the cocoa mass to increase its pH and improve its flavour. In its most basic form, chocolate is made by mixing this cocoa mass with **cocoa butter** with

the addition of sugar and often milk. Various flavourings and additives, such as lecithin (as a stabilizer), may also be added. Since March 2000, the EU has allowed the addition of up to 5% vegetable fats of tropical origin other than cocoa butter in recipes for chocolate. Over **60 aromatic compounds** present in cocoa beans have been found to contribute to cocoa's unique flavour.

THE REAL PRICE of cheap chocolate

Recent investigations into working conditions in the Ivory Coast – the source of much of the world's cocoa – resulted in reports of appalling conditions endured by hundreds of thousands of people today. Working excessively long hours, and barely fed, 'cocoa slaves', including children, were reported to be prevented from leaving the plantations and severely beaten if they tried to escape. The Child Labour Coalition, however, has concluded that 'there is little documentation beyond anecdotal evidence' that child slavery is occurring.

Despite this conclusion, in 2001, the International Cocoa Council adopted a resolution to 'encourage member governments of the International Cocoa Organization concerned to investigate and eradicate any criminal child labour activity that might exist in their territory'.

Although large plantations are to be found in countries such as Brazil, Malaysia and Indonesia, almost 90% of cocoa is grown on smallholdings of under 5 ha (12 acres), where it provides a form of income for millions of people in the developing world. The fluctuation in market prices, however, plus serious losses through pests and disease, means that cocoa growing is a precarious business for small producers. By contrast, the average annual value of cured cocoa beans in global trade is estimated at around $3.1 billion: third in the world league of food commodities, after sugar and coffee.

The Fairtrade initiative offers products guaranteed to have been produced without slave labour or the unacceptable exploitation that appears to be the hallmark of much mass-produced chocolate.

---

become rich by the 16th century from the profits of the pepper trade. When parts of the warship *Mary Rose* were lifted from the seabed in Portsmouth harbour, where they had lain since 1545, peppercorns and peppermills were among the finds.

Who's going bananas?

Bananas, the fruit of the world's largest herb, are also the world's most popular fruit. Having overtaken the apple in popularity in 1998, bananas are Britain's favourite, too, bought by 95% of households. Worth £5 billion per year worldwide and £750 million in Britain, bananas are the most valuable food sold in British supermarkets. With more different kinds of banana available than ever – from red or 'apple' to 'fun-sized' – we can't get enough of them.

But the factors that ensure their cheapness and the uniformity of their taste and appearance for the consumer are causing acute suffering for millions of growers. Large transnational corporations control the plantations in Latin America where the cheapest bananas are grown and which currently account for some 64% of European banana imports. Despite making huge profits, in order to remain 'competitive' and with increasing global production and over supply, these companies pay their workers next to nothing and use large quantities of chemicals on the developing crop.

The Fairtrade Foundation reports that Costa Rican banana plantations are sprayed with agrochemicals by aeroplane 40 times during

The hottest chilli?

A quite different kind of 'pepper' made the headlines in 2000, claiming (*Plant Talk*, issue 22/23), to be the hottest chilli pepper **yet recorded**. Though native to the Americas, chilli peppers (*Capsicum* spp.) were taken to India five centuries ago, and have been used there ever since. Scientists in Assam were said to have found a variety of *C. annuum*, regularly eaten by local people, that measured 800,000 '**Scoville units**', a way of measuring 'hotness'. The previous hottest chilli had been the 'habañero' (measuring only 300,000 units), with the well-known 'jalapeño' only 2,500–5,000 units.

Other varieties of *C. annuum* provide us with **sweet or bell peppers** and paprika (or pimiento) as well as the hotter chillies.

143

Bananas in Ver O Peso market, Belem, Brazil. Different kinds of banana vary greatly in length and colour and may be red, green or yellow.

each cultivation cycle, amounting to around 11 million litres (2.4 million gallons) of fungicide, water and oil emulsion every year. An estimated 90% of pesticides is absorbed directly by the surrounding environment, including the food plants and water sources of local people. The average consumption of pesticides in Costa Rica per capita is 4 kg (9 lb) per year.

The same report reveals that according to one Costa Rican study around 20% of male banana workers there have become sterile through handling toxic chemicals, and that women working in packing plants have double the national rate of leukaemia and birth defects. One banana worker gave birth to a baby whose head was four times bigger than his body.

Meanwhile, the situation has become desperate for the small-scale farmers of the Windward Islands and other Caribbean producers, who cannot compete with the price and volume of the bananas produced by the huge transnationals. Encouraged to depend entirely on trade with the UK for the last 50 years, the recent intervention of the Single European Market and rules imposed by the World Trade Organization have made matters worse. 'Freetrade' rules insist that Europe must cease preferential access for Caribbean bananas despite the fact that the banana is of key importance to the economies of many Caribbean countries. Thousands of farmers have already been driven off the land, sinking into desperate poverty. Fairtrade bananas guarantee suppliers in Costa Rica, Ghana, the Dominican Republic and Ecuador a fair price for their crop, a 'social premium' allowing investment in health and education, safer working conditions and a reduction in the use of hazardous chemicals. Such standards, however, should not be the exception but become the rule.

144

Foods from the mountains

Potatoes

■ There are nearly 2,000 species of *Solanum* – between 200 and 235 of them tuber-bearing – distributed from the southwestern USA and Mexico, through Central America into South America, with a strong concentration in the Andes. There, local farmers distinguish several thousand wild and cultivated varieties. Particularly at higher altitudes, Andean farmers will often grow 10–20 potato varieties in a

In some of the highest areas on earth, as in some of the wettest, tubers play a central role in indigenous diets. In South America's Andean region, several tuber crops are important, including:
● Brilliantly coloured *oca* (*Oxalis tuberosa*), some varieties of which are sweet enough to be eaten like fruit.
● *Olluco* or *papa lisa* (*Ullucus tuberosus*), like *oca*, also striking, in shades of yellow, pink, purple and red.
● *Mashua* or *aiñu* (*Tropaeolum tuberosum*), closely related to the garden nasturtium, and one of the most resistant of tubers to cold.

The most commonly cultivated species of another, much more famous tuber, however, has risen to become the fourth most important crop in the world. The famous staple food of the Andes, revered by the

Colombian potatoes. An enormous variety of cultivated potatoes exist in the Andean region.

Incas and grown today at altitudes too high for most other crops (up to 4,200 m/13,800 ft), the potato is the world's most important non-cereal food plant. It is cultivated on around 18 million ha (44.5 million acres) of land, with an annual production of some 300 million tons. Each time we eat a packet of crisps, tuck into French fries or simply peel a potato, we have Andean farmers to thank. In southern Chile, remains of wild potatoes have been dated to around 11,000 BC, while evidence of cultivation – on the high plateau between Cusco and Lake Titicaca, on the Peruvian/Bolivian border – dates from around 5000 BC.

Thanks to the Andean tradition of exchanging potato seeds and tubers at regular local fairs and at family gatherings (chiefly by women), the great range of potato varieties has long been maintained. As families have come under increasing pressure to plant modern higher-yielding varieties or have been obliged to abandon their farms altogether, the future of many varieties has become threatened. The International Potato Centre in Peru holds the world's largest bank of potato germplasm, which includes some 1,500 samples of about 100 wild species, and 3,800 traditional Andean cultivated potatoes. But while various species have been used in potato breeding, most of the potatoes grown in the West stem from just a few varieties of the only species cultivated outside South America, *Solanum tuberosum*.

single field, while as many as 54 may be cultivated by one family in their tiny plots scattered across the mountainsides. The shape and colour of these tubers varies enormously, as does the texture and configuration of the 'eyes'. These traits are reflected in the descriptive names for potatoes in the Quechua language, such as 'cat's nose' for a long, flat potato, or 'potato that makes young brides weep' for one that is knobbly and hard to peel.

■ In the Andean region, potatoes are often consumed in the form of:

■ *Chuño* and *moraya*, made by exposing bitter potato varieties to nights of frost and then washing and trampling on them to expel the water (containing toxic glycoalkaloids) released by freezing. After drying in the sun, this 'freeze-dried' potato will keep for several years and can be added to soups or stews along with other tubers.

■ Flour made from potatoes that have been boiled, peeled and dried.

■ The alcoholic drink chicha.

In the USA, just three processing companies currently control the market for and the production of almost the entire potato crop. Farmers are contracted by these processors to grow particular varieties under particular conditions, for a pre-determined price. Breeders in Europe and North America have created varieties that can be used in intensive farming practices (meeting the requirements for optimal shape, size, starch and sugar content), but this has promoted genetic uniformity, producing crops that are highly susceptible to disease and reliant on heavy pesticide use. With the latest developments of genetic engineering, large agrochemical and seed companies now hold many patents on transgenic potatoes. Developed to overcome susceptibility to infection, pests and diseases (rather than improve quality), these have been field-tested in numerous countries, and by 2001, three varieties had been approved for commercial production in the USA.

Each time we eat a packet of crisps, tuck into French fries or simply peel a potato, we have Andean farmers to thank.

Potatoes are currently grown in 150 countries, with developing countries now producing over a third of the world supply. In 2001, the largest producer was China, followed by the Russian Federation, Poland, the USA, India and the Ukraine. One-fifth of the total harvest is fed to livestock, but potatoes still feed millions of us every day. In industrialized countries, however, the form in which they do so has changed dramatically in recent years due to the massive rise in fast foods. In the USA, only 26% of the potato crop is now consumed fresh, while in the Netherlands – where one-fifth of all arable land is given over to potato production – the proportion has fallen to 18.5%, as chips, crisps and other processed products increase in popularity. Potatoes are also a major source of starch – the raw material for an increasing number of non-food, industrial applications, as well as food uses, such as the production of glucose and dextrin – and alcohol.

Few single foods have as much nutritional value as the potato; one that is medium-sized, for example, contains about half the daily vitamin C required by an adult, is virtually fat free, and has more protein and calcium than maize. Europeans have come to hold it in special regard and, for the Irish, the potato occupies an unforgettable place in history. By the mid-19th century, millions of Irish Catholics had come to depend on potatoes – a crop they could conceal from marauding British troops. But in 1845 and 1846, fungus blight caused practically the entire Irish potato crop to fail. Unassisted until it was too late, the horrific famine that ensued is estimated to have killed about 1.5 million people and caused a million others to emigrate to the New World.

146

Food of the future

The modern supermarket shelf is deceptive in the variety of foods it tempts us with. When the wrappers are removed, we find that most of these foods are derived from just a handful of plant species. Nearly three-quarters of the world's food supply is now based on just seven major crops: wheat, rice, maize, potatoes, barley, cassava and sorghum. In fact, we cultivate fewer species in the West than the farmers of Neolithic times.

Some people believe that the price we have paid for huge increases in the productivity of our major crops over the past 50 years is an increasing uniformity and the under-utilization of additional or alternative food plants. This, it has been argued, is part and parcel of a movement to control and privatize our most important food sources – a trend amplified in the 1990s with developments in biotechnology (in particular genetic modification) and patenting. In 1998 Patrick Holden, Director of Britain's Soil Association, warned: 'The five major Agrochemical Companies envisage a future where only a handful of varieties of wheat, maize, rice and other food crops are grown commercially. They are working flat out now to ensure that, within a decade, most of the world's staple crops will be from genetically modified seeds which they have engineered.'

Soya beans

Soya beans (see right) are one of the most important agricultural commodities in the world. In the 1970s, soya overtook wheat and maize as the most important of America's cash crops – and became the world's primary protein plant.

Following the first commercial plantings of soya in the USA in the 1920s, and promotion by such entrepreneurs as Henry Ford, the plant experienced a huge rise in popularity. Containing 30–50% protein, soya's seeds were discovered to be very versatile and easily processed in many ways to make different foods. Textured soya protein can be flavoured to make anything from dog food to hamburgers, whilst its oil – used in the 1930s for the production of paints and plastics – now has many food and industrial uses. The source of flour, protein concentrates and isolates widely used in bakery and meat products, soya beans are also the most common source of lecithin, used in the food industry. It is estimated that soya is now present in around 60% of all our processed food. Soya milk from the crushed dried beans is widely available as an alternative to cow's milk, especially in baby formulas. But it is not humans for whom most of the world's soya crop is grown.

Soya beans

■ The soya bean or soy bean plant (*Glycine max*), an annual legume with white or lilac flowers, is believed to have originated in northern China and was first cultivated there in at least 3000 BC.

■ Rather than forming a large part of traditional diets, however, throughout Asia, soya is mainly used in a fermented form, as a seasoning. Many traditional Far Eastern uses have now been adopted elsewhere in the world. The beans are widely processed to give the cheese-like bean curd tofu, and fermented to produce miso and tempeh, as well as the unbiquitous soy sauce.

Soya beans. As well as being processed in many ways, they can be eaten whole, split or germinated to give bean sprouts.

147

Lupins

■ Some British farmers and food processors have become interested in lupins as a substitute for imported soya. White lupins (*Lupinus albus*) are now being grown on an increasing scale in the UK, for both their meal and their oil.

■ In Peru, Bolivia and Ecuador, meanwhile, the Andean lupin (*L. mutabilis*), known in Peru by its Quechua name tarwi, has been grown for thousands of years for its highly proteinaceous seeds. Containing more than 40% protein (as much as the world's major protein crops) and 20% oil, these outstandingly nutritious seeds have, together with maize, potatoes and quinoa, formed the basis of the traditional diet of many highland Indians.

148

As meat consumption soared over the last few decades in the world's industrialized countries, European farmers built up large cattle herds and boosted the mass production of poultry through battery farms. Faced by a shortage of fodder, however, soy meal was seen as the ideal solution and, with a particular focus on South America, soya cultivation was promoted in developing countries as a new export crop. Today, in the wake of the European BSE crisis, and the ban on the feeding of animal remains to farm animals, soya provides the most commonly used protein for the animals that supply the world's mass-produced industrial meat. Cheap to buy, the high protein content of soya meal promotes weight gain in cattle and egg laying in poultry, and produces good-quality fodder.

The USA is currently the world's leading soy producer and it accounts for almost half of global output. Technological advances in Brazil, however, including the development of new seed varieties and new fertilization techniques plus advances in pest and weed control, have made this the world's second biggest producer. Until 30 years ago soya beans could only be grown in temperate or subtropical climates at latitudes of 30° or higher, but according to the president of São Paulo's Research Support Foundation, 'As much technology has been incorporated into Brazilian soybeans as in today's new computers.'

The brave new world of the 'gene bean'

Large-scale soya cultivation is highly mechanized and depends on huge quantities of agrochemicals. The massive environmental damage involved (including chemicals leaching into the Amazon's river system) and the social problems exacerbated by its production in Brazil have been cause enough for alarm. But the issue of genetic modification in Brazil, whose yield of soya beans has outperformed that of the US without the use of GM varieties, has recently come to the fore.

In 2001, soya beans genetically engineered by the biotechnology company Monsanto, and developed to be used alongside and resistant to its best-selling Roundup herbicide, accounted for about two-thirds of the global area under cultivation of GM crops. Brazil is currently sandwiched between the USA as the leading soya producer and Argentina in third place, both having planted GM soya (which accounts for about 70% of US soy production). But for some years it has been under intense pressure to grow GM soya.

In 2002, despite resistance from many soya growers, Brazil's commission on GM foods recommended the immediate authorization of GM crops, in spite of an adverse ruling against the use of Monsanto's Roundup beans in 2000. According to some reports, up to half the soya planted in Brazil's most southerly state, Rio Grande do Sul, is in any case now transgenic, having been smuggled over the border from Argentina. Reports published in 2002 indicated that this transgenic soya may be costing Brazil a great deal in lost profits, since the yield was reported to be one-third lower than that from normal varieties. Meanwhile, Argentina's crop was reported to have brought disappointing yields with a large increase in the use of pesticides. As long ago as 1998, genetically engineered soya beans grown by American farmers were released into the general soy supply used in some 30,000 food products (including bread, margarine, cereals, sausages and baby food) in Britain. It was declared impractical for the farmers involved to segregate the beans. Segregation and labelling appears to have been strongly opposed since it would allow consumers, who have voted overwhelmingly in favour of such labelling, not to buy them. GM soy meal is estimated to be forming a large proportion of the feed consumed by animals that end up on our supermarket shelves.

Burning rainforest in Brazil for soya production.

Soya destroying the Amazon

First cultivated in Brazil in the south, by the 1980s soya had **expanded** into the savannah lands and rainforest of the Mato Grosso region. Sustained by large state subsidies, soya farmers are now, however, moving into the Amazon basin, replacing its rain-forests with an industrial monoculture – a threat seen by some as its **most serious ever**. Although authorities promise that no new forest will be cut

down, soya farmers are promoting the cycle of forest destruction.

Much of the land already used has been **expropriated** from millions of poor rural labourers or bought from cattle ranchers who have already cleared areas of forest and who then move on to open up regions that are even more remote. **Soya plantations** employ only one or two people per ha in comparison with the 30 or so for a

family farm and have thus added hugely to **rural** unemployment. Whilst 60% of its population suffers from malnutrition, soya beans, meal and oil are currently Brazil's chief exports, earning substantial foreign income. Some 31 million tons were exported in 2000, but a massive **expansion** is planned. To achieve this, it has been estimated that the land required can only be found in the Amazon basin.

149

Soya in our food: a hidden menace?

In the mid-1990s, New Zealand scientists warned of the possible dangers of the '**anti-nutrient**' properties of soya, which are never-theless lessened by fermentation, heat treatment or sprouting – and the isoflavones it contains, which have effects similar to the female hormone **oestrogen**. In 2002, soya was identified as being a possible

source of some birth defects.

Scientists had already reported that the **isoflavone compounds** known to cause thyroid and sexual development disorders in animals could increase the risk of breast cancer in women, brain damage in men and abnormalities in children. Now used more widely than ever in **convenience foods**, babies may

be particularly at risk. According to one expert, the amount of isoflavones in a day's worth of **soy infant milk** is equivalent to five birth control pills.

While isoflavones appear to encourage premature sexual develop-ment in girls, they inhibit an enzyme that is critical in the development of testosterone in boys.

Engineered plants: problem or panacea?

As has already been noted, our major food plants are the result of a very long and continuous process of selective breeding, involving many techniques. Genetic modification or engineering is different in that it involves the artificial insertion of individual genes from one organism into another (often from unrelated species) to confer specific traits, thus achieving huge changes that could never occur in nature and violating well-established natural boundaries. This means, for example, that plants can be given genes from fish, or any other organism, to make them behave in a particular way, or to become resistant to various stresses, such as extremes of temperature or particular insect pests. Toxin genes taken from the soil bacterium *Bacillus thuringiensis* (BT), for example, are the source of the most commonly engineered insecticides in many staple crops including maize, soya, oilseed rape, potatoes, rice and tomatoes. Naturally produced chemical defences are an important strategy for many plants but, unlike GM crops, the interaction between plants and insects has developed during a lengthy process of co-evolution, which ultimately allows individuals on both sides to survive, fulfilling complex functions within a given ecosystem. Present in every cell of the newly engineered plant, GM toxins are designed to exterminate insects wherever they may attack.

Widespread concerns about GM crops have included:
● Unpredictable side-effects caused by the interaction of new genes.
● Health dangers such as the creation of new toxins and allergens and the transferral of antibiotic resistance from 'marker' genes.
● Loss of biodiversity, and poor agronomic performance.

Another major concern is the increased use of agrochemicals. Glyphosate is the world's best-selling 'total' herbicide – it kills many

Feeding the hungry or lining pockets?

A major rationale of **biotechnology** is that it will help scientists to create the food needed to feed the world's hungry people. A report written for the World Bank and the Consultative Group on International Agricultural Research (CGIAR) in 1997 claimed that **transgenic crops** could improve food yields by up to 25% in developing countries and could help

feed an estimated additional 3 billion people over the next 30 years. In 2000, an FAO report stated, however, that although new techniques of molecular analysis could give a welcome boost to such **productivity**, 'agricultural production could probably meet expected demand over the period to 2030 even without major advances in modern biotechnology.'

In 2002, several southern African countries reacted with outrage to the suggestion that they should accept **GM food aid** from America to help overcome the worst food shortage in 50 years, in which 13 million people may die of starvation. Zambia's President Mwanawasa said that he would rather let his people die than feed them hazardous food.

types of vegetation, including crops. But once seeds have been genetically modified to tolerate it, farmers can blanket their fields with the
herbicide, which will kill everything else. Many fear that as plants,
especially at field edges, develop a resistance to such herbicides, they
will become 'super weeds'. Various serious human health risks are
also associated with the use of such herbicides or the ingestion of
foods treated with them.

Proponents claim that GM crops reduce farmers' reliance on
pesticides (biotech maize, for example, contains the bacterial BT gene
that makes its protein lethal to caterpillars). Evidence shows, however,
that many pests have developed immunity to these toxins (therefore
requiring ever larger doses of pesticide) whilst beneficial insects that
weren't targeted have been killed instead. Toxic side-effects on
mammals, birds and frogs are also feared. Research at the University
of Hawaii has shown that insects that survive BT bacteria transmit
genetic resistance to their immediate offspring.

Transgenic plants are also known to have an effect on neighbouring
plants, transferring genes to the wider environment. Concern has been
voiced over what is known as 'gene stacking', where more than one
GM trait is found in the same plant because of cross-pollination in the
field – the consequences of which are unknown.

A genetically engineered plant
under study at Cambridge University's
Centre for Plant Sciences (UK). Around
the world many research institutions
are involved in studies to understand
the genetic make-up of our major
economic plants.

PLANTS THAT FEED US

Harvested tomatoes, Mexico.
Subjects of genetic manipulation, the rate of softening of many commercially grown tomatoes has been reduced and their texture altered to make their shelf-life longer.

152

Seeds of Doubt, a report produced by Britain's Soil Association in 2002, found that most of the benefits claimed for GM crops did not stand up to examination. North American farmers reported lower yields, continued dependency on chemical sprays and widespread GM contamination of non-GM and organic crops, and few, if any, of the promised economic benefits. The report estimated that GM soya, maize and oilseed rape may have cost the US economy $12 billion in sub-sidies, lower crop prices, loss of exports and product recalls.

In the same year, Friends of the Earth urged EU agriculture and environment ministers to scrap proposed new seed regulations that would allow unacceptably high levels of GM contamination of 'normal' crops to be permissible, without warranting labelling. Recommending that the UK aims for 'levels of transgenic impurities that are near zero', conservationists warned that farmers could otherwise unknowingly grow millions of crop plants with transgenic characteristics. These could then cross-pollinate with other conventional and organic varieties, spreading contamination throughout Europe and into the food chain, thus removing consumer choice. In the USA, where multimillion dollar campaigns have resisted the labelling of products, it is estimated that at least 70% of processed food contains engineered ingredients.

According to Professor Richard Lacey, microbiologist, medical doctor and Professor of Food Safety at Leeds University: 'The introduction of genetically engineered foods is essentially creating unlimited health risks. The fact is, it is virtually impossible to even conceive of a testing procedure which could assess the human health effects of genetically engineered foods when introduced into the food chain, nor is there any valid, nutritional reason for their introduction. Given the huge complexity of genetic coding, even in very simple organisms such as bacteria, no one can possibly predict the effects of introducing new genes into any food. There is therefore no way of guaranteeing the safety of GM and the overall long-term effect.'

Notwithstanding these concerns, aggressive marketing by agro-chemical corporations has ensured that GM crops have been widely grown. From 1.7 million ha (4.2 million acres) in 1996, the area under GM cultivation had reached 52 million ha (128 million acres) by 2001, involving some 5.5 million farmers in 13 different countries. About three-quarters of the world's total GM crops are grown in the USA and Canada – mostly soya beans, maize and cotton.

A move to genomics?

A more recent development is a move towards applying gene mapping to conventional plant breeding, using the genetic sequence of plants in natural plant breeding programmes known as 'marker-assisted breeding'

(MAB) or 'genomics'. Unlike genetic engineering, which artificially transfers genetic material between or within species using recombinant DNA, genomics enables the genetic sequencing of plants to be used in sexual breeding processes for the development of new high-output varieties. Without bypassing the natural process of gene regulation and placement, this therefore avoids the unpredictable health and environmental risks associated with genetic engineering. Regarded by many scientists as the way forward (and accepted by Britain's Soil Association), it employs a more appropriate and sophisticated form of genetics. The European agriculture industry may adopt this methodology for the future. There is a risk, however, that this technology may be used to further reduce the genetic diversity of commercially available crop varieties, narrowing the gene pool.

Patenting our crops: whose food is it?

Hand in hand with developments in biotechnology are those of patenting – giving individuals or corporations exclusive rights to what, henceforward, becomes their private property. While this system was essentially designed for industrial inventions, such as electrical appliances, which are novel, useful and non-obvious, it has come to be used not only for biological material, including plant genes, but for plant varieties themselves.

Why should anyone be allowed to transfer a resource from the public domain to the private? Such activity, especially where it involves privatizing genetic material without the prior consent of or providing a fair return to the people whose culture it may form part of (as stipulated by the Convention on Biological Diversity, under which biological materials are the property of the country in which they originate – see page 236), is widely condemned as 'bio-piracy'. Genetic modification and patenting are issues perhaps more of control than of changes in technology. The USA has become notorioius for its refusal to ratify the CBD.

Agrochemical companies are now in a position of unprecedented legal control over the food chain. By 2001, six major agrochemical corporations owned nearly three-quarters of the 918 patents on rice, maize, wheat, soya beans and sorghum – crops that are major world staples. The many other food plants on which patents now exist include broccoli, black pepper and many traditionally-used plants of the Andean region, such as the *nuña* popping bean, quinoa, *yacon* and the root crop *maca* (*Lepidum meyenii*). Indigenous peoples' and farmers' organizations from the Andes and Amazon have denounced the recently granted US patents on *maca*, a valuable plant with medicinal attributes, able to withstand extreme cold, whose sweet, spicy, dried roots have long been valued as a nutritious delicacy.

Bio-piracy examples

Many examples of what many consider to be bio-piracy are well documented.

Yellow beans: After bringing home some yellow beans (*Phaseolus vulgaris*) following a holiday in Mexico, ActionAid report that the president of a US seed company managed to acquire a **patent** to them in 1999, naming them after his wife, 'Enola'. Despite the fact that these beans have been grown and eaten in Mexico for centuries, the patent provided **protection** for any yellow dried beans, making it illegal for them to be grown in the USA or imported without royalty payments to the company concerned. By causing an immediate drop (of over 90%) in export sales, the patent subsequently caused serious economic **hardship** for thousands of farmers in northern Mexico. A report prepared for the United Nations Development Programme in the mid-1990s estimated that bio-piracy was costing developing countries some $4.5 billion per annum.

Broccoli sprouts: Following the discovery by American scientists that they contain 20–50 times the amount of a **cancer preventative compound** found in mature broccoli, three-day-old broccoli sprouts – which have been neither invented nor genetically altered – were patented in 1998.

153

According to ActionAid, 'Control over crops, plants and their DNA is one of the defining issues of the early 21st century. It will determine who wields power over farming and the global food system.'

In 1998, the US Department of Agriculture and a private seed company were granted a patent on the so-called 'terminator technology' – the engineering of plants so that they produce sterile seeds, which will not germinate if planted. They have, however, now been forced to abandon this patent. Numerous other cases have shown that the developers of GM seeds aggressively prosecute their patents related to crops and their genes.

It has been estimated that several billion farmers around the world depend on seeds either saved from previous years or exchanged with farm neighbours. Agrochemical companies may require those who use their seeds to sign contracts agreeing not to save seed. Such contracts may jeopardize systems of food security that have evolved and been developed by people over millennia as a reflection of their culture and environmental knowledge.

Miracle plants or sustainable agriculture?

Hundreds of species with a long history of use by local peoples have provided a diverse range of nourishing foods. As they have attracted scientific attention, new examples have regularly emerged of 'miracle' plants (like soya) considered good candidates to help feed the 1 billion people estimated to be affected by 'food insecurity' and provide a better diet for many others. But an urgent priority must surely be the restructuring of the global economy, which has largely caused the crisis faced by so many, to reduce consumption by wealthy countries and promote sustainable, local agriculture for everyone.

A report by the US Surgeon General in 2001 warned that obesity-related problems were costing America $117 billion a year. Some fast-food companies in America are now said to be warning consumers that eating too many of their products may make them fat. Bombarded by advertising, and for many obliged to lead lives that are increasingly out of touch with the source of the foods we consume, this is a predictable trend. Where people are able to grow at least a proportion of their own food, such problems are much less likely to occur – and it's not even necessary to have a garden (see box, opposite).

Local food for local people

In the depths of a European winter, it would be perverse not to be grateful for the array of exotic, largely tropical fruits that liven up our supermarket shelves. Those on display, however, whether the soursop from tropical America, the spiny rambutan or the mangosteen from Malaysia, which demand our admiration, represent only a fraction of the different fruits grown in the world's tropical regions.

But what about the varieties of well-known temperate fruit and vegetables that beg to be revived? According to the Henry Doubleday Research Association, Britain has lost about 2,000 varieties of vegetables since the 1970s, when legislation to protect commercial growers against copying required seeds to be registered. Many rare varieties are now being grown by a network of 'guardians' across the country.

Seasonal food for local people

In parallel with the availability of tropical fruit, Europeans who live in temperate zones can enjoy many seasonal vegetables all year round – from tomatoes grown in Spain, to the pea variety 'mange tout' grown in African countries. But rather than take up valuable agricultural land to produce 'luxury' crops abroad, and thereby also contribute to air pollution (addressed by the concept of 'food miles'), why not make the most of the traditional food plants that can be grown at home sustainably? This could be in window boxes or on some of the countless hectares (acres) of land currently idle or devoted to the mass production of cash crops destined to feed factory-farmed meat and poultry.

Fifty years on from the Green Revolution, modern intensive farming has failed to alleviate poverty and environmental degradation. Revolutionary changes are needed in the mechanisms that control the food chain and which see agriculture as just another industry, subject to the same conditions as factories. As Dr John Zarb (writing for the *Ecologist*, vol. 30, no. 9, Dec 2000–Jan 2001) has pointed out, problems for farmers in rich and poor countries alike have arisen from a global agricultural trade that subsidizes and promotes large intensive systems that are good for agribusiness and supermarkets, but generally bad for the countryside, smaller farms and the communities they could support. It undermines the ecological, social and cultural factors essential to sustainable agriculture, forcing farmers into a state of technological dependence that further weakens their cultural links with the land. Whilst in developed countries there is an urgent need to help people understand the reality behind the production of much of our food, there is ample evidence that sustainable ways forward that care for the land, provide enough food and support farms and communities really work.

In various parts of the world crippled by economic, environmental and political conditions, such as Cuba, Chile and Bolivia, sustainable technologies, such as agro-ecology, are achieving real improvements for some poor farmers. Integrating indigenous and modern farming methods, agro-ecology is a productive, resource-conserving and socially equitable agricultural system, which puts the land and the people who depend on it

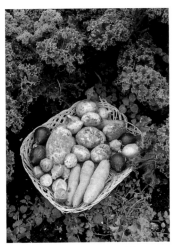

Seasonal organic vegetables, grown in Dorset, UK.

Sprouted seeds

Sprouted seeds of various kinds have recently been hailed as a 'perfect food' and an extremely **healthy** alternative to diets high in processed food known to produce health problems of many kinds. The process of sprouting – into tiny leafy greens – means that chemicals known as **enzyme inhibitors**, which can otherwise make nuts, seeds, pulses and grains difficult to digest, are broken down.

Simultaneously, complex, indigestible proteins, fats and starches are turned into easily absorbed and **highly nourishing** amino acids, essential fatty acids and simple sugars. Equipped only with a simple plastic tray, nutritious leafy greens can be obtained by simply **watering seeds** – such as alfalfa, red clover, radish and sunflower, mung beans and green lentils – for just a few days.

155

first. It emphasizes the importance of biodiversity, nutrient recycling between crops, animals and micro-organisms, and, using local skills and resources, creates conditions for stable local economies. It has been proposed that such systems could be even more successful in Europe or in the USA, which currently has the fewest farmers since the middle of the 19th century.

Organic agriculture

With a market that has increased ten-fold in the last decade, and with demand outstripping supply, Britain's organic food movement can be regarded as a modern success story. The UK has the fastest-growing organic market in Europe, with a retail value of £900 million in 2002. Based on the principle that no species is irrelevant but that all life on earth is inter-dependent, the idea of working with nature rather than against it to produce high quality nutritious food has been embraced by many people. Just as organic production promotes local distinctiveness in crops and animal species, and aims to maintain rather than weaken soil structure, farms are not just providers of food. Through valuing human relationships with the environment, they aim to support the economic fabric of real communities. But despite the popularity of organic farms and the benefits to human and environmental health, the UK government has set a target for only 4.7% of farmland to be organic by 2007.

The growing movement towards self-sufficiency is not just limited to the countryside. Greater London, for example, boasts some 30,000 active allotment holders and has some 500 ha (1,230 acres) under fruit and vegetable cultivation, producing about 8,400 tons a year. Food growing and gardening is happening in unexpected places, too, including city rooftops. Part of the hugely popular initiative nationwide in which producers sell

Permaculture

Based on the idea that we should value diversity rather than reduce it, permaculture is a method of plant management that tries to work in harmony with nature, using **natural ecosystems** as models. It allows a wide variety of different species of food and other plants to be grown side by side, occupying different ecological niches and maturing at different rates and times.

Rather than the adversarial approach adopted by much modern gardening, and most Western farming, in which anything that is not a friend is eliminated, permaculture respects the **interconnectedness** of species and doesn't depend on large inputs of fertilizer, pesticides and herbicides. Fewer plants are treated as weeds, and **companion crops** of many kinds enable species to be grown that will, for example, enrich the soil, have an insecticidal effect or attract useful birds or insects. Permaculture has the benefit of needing far less hard work than traditional methods, inspired by the **balance** found in nature.

According to Ken Fern of the charity Plants for a Future, over 5,000 edible species of plant can be grown outdoors in Britain. By using those **best suited** to different habitats, whether woodland, meadow, bog garden or lawn, our gardens can be productive in unconventional edible plants from the daisy to the fruit of the wild service tree.

direct to consumers, London's farmers' markets contributed £2 million to the economy in 2001.

Making demands

But biotechnology continues apace, changing flavours, textures and the nutritional content of familiar crops (such as vitamin A rice and caffeine-free coffee plants), modifying oils, fats and starches and developing new processes and products, ranging from low-calorie beers that depend on newly developed yeasts to sugar-free sweeteners. Fruit and vegetables (including bananas, avocados and tomatoes) have been engineered so that they suppress the production of ethylene, and therefore take longer to ripen. Scientists recently identified the gene that turns on and off hormones which instruct a plant to wilt – that is, to start dying. Hailed as a discovery that could lead to vegetables being kept 'fresh' for months after being cut or harvested, and therefore help perishable goods to be moved long distances to those in need, these new discoveries may be exciting, but what sort of society do they really promote?

Demanding foods that are produced by environmentally and socially responsible methods and insisting that industry use them should be priorities for us all. The modern, engineered varieties of our major food plants may suit the balance sheets of agrochemical companies, but the world's rich store of fruits, vegetables, tubers, grains and pulses are of ancient lineage. In their endless variety, they represent the miracle of evolution and the combined knowledge and skills of millions of people, mostly 'the poor', whom we never see. All the world's people should surely be free to reap the benefit of the plants of their ancestors.

Sustainable production of food and goods can take many different forms. In the extensive *dehesa* system in Spain, mixed farming centres around cork and holm oak woodland, interspersed with areas of scrub, grassland, orchards and cultivated fields.

Plants that **house** us

For as long as houses have been built, plants, chiefly trees, have played a major part in their construction. Whatever the age or style of your house, wood is almost certainly supporting, lining or adorning some part of it.

As the most versatile of all building materials – combining flexibility with durability and strength – and often the most readily available, wood continues to be indispensable for housing us and furnishing our homes. Over 70% of the population of the developed world (which includes North America, Scandinavia, Australia and Japan) lives in timber frame housing, and in the USA and Canada the proportion of low-rise buildings based on timber frames is over 90%. Though a much lower percentage of British buildings (13% in 2001) made use of wood in this way, the popularity of timber framing is rising rapidly, and it is now the fastest-growing method of construction in the UK.

Around the world some of the simplest but nonetheless effective dwellings have been based on a wooden frame. The North American Indian wigwam relied on a conical framework of poles to support a weather-proof covering of hides, and the felt-covered yurt used today by nomadic peoples of Mongolia and Afghanistan is constructed from a wooden lattice often cut from willow trees. Both homes have the advantage of being very strong, yet are easy to dismantle and reassemble as and when required.

In Europe, timber-framed buildings such as the Medieval 'cruck' and 'post and truss' styles appear to have developed from simple rectangular ridge-tent constructions in which poles were bent together in the form of an inverted 'V'. At that time, immense mixed forests blanketed the continent – conifers predominating in the northern and Alpine regions, whilst sturdy oaks prevailed in central Europe and at its western edge, in Britain. Above all other native trees, oak with its tremendous strength and resistance to decay was soon recognized as the most suitable wood for making the framework and supporting structures of substantial buildings.

In Europe's 'cruck' framed houses, massive timbers, often the matching halves of a single split tree trunk joined together by a ridge beam running the length of the house, took the roof load straight to the ground. As an understanding of the principles of timber framing and basic forms of bracing evolved, buildings became more complex. The 'post and truss' framed house developed as a secondary box-like structure built around the cruck but with roof supports constructed separately from the wall frame. Where ground space was restricted, the great strength of the oak timbers allowed cantilevered upper stories or protruding jetties to be built above a narrow first floor. The gaps between the supporting wall posts were filled with wattle and daub (flexible twigs plaited together and covered with a mixture of mud and straw) or lathes (thin strips of wood) and a coating of plaster.

In the 18th century, the use of the timber frame declined in Europe. But the tradition was continued in North America and Australia by European colonists who took advantage of vast native forests. Distinctive architectural styles arose – often characterized by weatherboard cladding – relying generally in North America on softwood species and in Australia on a wide range of eucalypts.

Previous page: Bamboo stems.
Opposite: Rainforest in Udzungwa National Park, Tanzania.

160

The hemp house

An interesting experiment in the use of environmentally friendly building materials got underway in Haverhill, Suffolk in 2001, as a young family moved into the first house in Britain to be made substantially of **hemp**. It is one of two houses made this way (the other unoccupied); it is supported by a timber frame and hemp – in combination with **hydraulic lime** – has been used to form the walls (having been poured into wooden shuttering, which was subsequently removed), and also the ground-floor screed. The two houses are being closely **monitored** and compared with two identical buildings built of traditional brick and block construction to measure qualities such as insulation, energy efficiency, sound proofing, structural stability, resistance to water and condensation factors.

They have already been found to need foundations only half the size of those of traditional buildings, and the whole effect of the 'hemp homes' – enhanced by other **natural materials** such as wooden window frames and clay floor tiles – is reported to make them much warmer than that of many of today's modern houses.

Looking to the future

Influenced by historical building styles and with the development of standardized, prefabricated frames and panels, timber-framed housing is now being encouraged at government level in Britain: a shortage of skilled 'brick and block' workers, the speed of timber frame construction and the warmth of the resulting homes all being contributory factors to the encouragement of timber framing by the government.

But wood has never been superseded as a building material in the conventional brick-built house and remains indispensable for the construction of rafters, joists and roof trusses, as well as skirting boards, staircases, traditional door and window frames, and flooring. Almost 6 cubic m (8 cubic yd) of timber is likely to be used in this way in a standard masonry-framed house.

The demand for wood and wooden finishes within our homes continues to grow. Though fashion has played a part in its appeal, very practical considerations such as the excellent capacity of wood to insulate both heat and sound and, where timber framing is concerned, its 'dry' construction are important attractions. Over 6,800 litres (1,500 gallons) of water may be used in the walls of an average brick house (in the mortar holding the bricks in place and in the plaster that covers them), and this can sometimes cause cracking if the house settles and dries too fast. But in a typical modern timber-framed house, prefabricated units that are easily assembled and then covered and reinforced with other wood-based sheets or panels replace supporting walls of masonry and plaster.

Other, less familiar plants and plant products are now proving their value for innovative modern house construction. One of these is hemp (*Cannabis sativa*) (see also page 76), whose strong stem fibres, mixed with hydraulic lime, are being used to form walls and a component of the ground-floor screed (see left).

Houses in Haverhill, UK. Hemp has been used in the walls and flooring.

Woods for timber frames and joinery

Roughly 180,000 new houses were built in Britain in the year 2000–2001, and with all of them using wood to a greater or lesser degree plus continual additions and embellishments to existing homes, the Forestry Commission has calculated that the UK's consumption of timber (for wood and wood products) reached 48.5 million cubic m (64 million cubic yd) in 2000. A recent WWF report has also estimated that despite accounting for under 1% of the world's population, the UK requires at least 6.4 million ha (16 million acres) of forest each year to supply its demand for timber products.

Oak, in demand for centuries for the construction of ships and all manner of buildings as well as interior fittings and furniture, is now a luxury timber, too scarce and expensive to be used on a large commercial scale for housing. Instead, most timber used today for joinery, framing, flooring, and general construction work in Europe comes from softwood tree species chosen for their straight grain, low production cost and regular availability.

Four main 'types' of softwood are currently commercially available in the UK: 'pine' or Scots pine (see right), spruce (Norway and Sitka) (see right), larch and Douglas fir (see overleaf). The silver fir has also been widely used as a joinery timber.

Large, managed plantations in Scandinavia and other northern European countries, as well as Russia and the Baltic region, and the immense natural forests that cover the northwestern seaboard states of North America, provide most of these softwood construction timbers: Scots pine, spruce, birch and larch often come from the northern European region, while Douglas fir, western larch and Sitka spruce are often from the USA. Almost 16 million ha (40 million acres) of commercial timberland are currently forested with Douglas fir and western larch in the western USA, with millions more hectares (acres) now protected from harvesting in various reserves or wilderness areas. Sharing almost identical structural characteristics and working properties, these species are interchangeable for many end uses. A number of other softwood tree species used locally in North America for construction work include western hemlock (see overleaf) and various pines, including lodgepole pine, Virginia pine and 'Southern' pine.

The pines

■ There are about 109 true pine species belonging to the genus *Pinus*. The pines are distinguished in general by their long slender needles and hanging cones. Their timber is divided in the trade into 'soft' and 'hard'. Hard pine is usually darker in colour, hardier, heavier, stronger and more resinous than soft pine. Whilst all except one are native to the northern hemisphere, the genus is most diverse in Mexico and the USA: *P. palustris*, for example, ranging from Virginia to eastern Texas, whilst *P. contorta* grows in the Yukon Territory. The Scots pine (*P. sylvestris*) is the only pine species native to the British Isles.

The spruces

■ The spruces are amongst the lightest of timbers used for construction and joinery work. There are 34 northern temperate species belonging to the genus *Picea*. They are characterized by needles that grow in spirals and are most diverse in western China and the eastern Himalayas. They produce timber of a creamy white to yellowish-brown colour with a straight grain. Some species have a lustrous texture – notably western white spruce (*P. glauca*), one of the most widely distributed softwoods in Canada. In Europe, Norway spruce (*P. abies*), also classified as white or silver fir, is an important species, also commonly used as a Christmas tree. The North American native, Sitka spruce (*P. sitchensis*), has been widely planted in Britain.

163

Douglas fir

■ Much used for general construction work in its native North America and traditionally imported into Europe from Canada, Douglas fir (*Pseudotsuga menziesii*) – native to the region of the North American Pacific coast – is now widely planted in the British Isles and other European countries. The tree, which produces one of the strongest and stiffest softwood timbers, can reach heights of 90–100 m (295–330 ft), with a diameter of 4.5 m (15 ft).

Western hemlock

■ Western hemlock or Alaskan pine (*Tsuga heterophylla*), which is native to the coastal regions of western North America, is one of 14 *Tsuga* species. A large proportion of all the western hemlock comes from large, managed plantations in Canada's British Columbia. It is the largest of the hemlock species, growing to around 65 m (213 ft) in height. Its light, uniform, knot-free timber, referred to as 'clear grade', is highly prized and is achieved by the very close planting of trees so that the lower branches, deprived of light, die off, leaving no knots. The strong, pale brown wood, which does not splinter, is one of the most valuable of North American timbers.

Which wood? Timber terminology

As do-it-yourself enthusiasts will know, the labelling of joinery timber, in Britain at least, has not tended to be specific as to the species used. Instead some very general terms have been employed by the timber trade, making it difficult for all but those growing the trees to identify them. For example:

● 'SPF'(Spruce, Pine, Fir) may be used to label a softwood timber from any of the commonly grown spruce, pine or fir species.

● 'Hem-Fir' generally refers to western hemlock (*Tsuga heterophylla*) or any of five true 'fir' species: *Abies magnifica*, *A. grandis*, *A. procera*, *A. amabilis* and *A. concolor*.

● 'White deal', 'whitewood' or 'spruce' indicates a *Picea* species, usually Norway spruce (*P. abies*) or Sitka spruce (*P. sitchensis*).

● European redwood, 'red deal' or 'pine', meanwhile, are terms used to describe Scots pine, which is also known as southern or Norway fir (*Pinus sylvestris*).

To further complicate matters, the terms 'pine' and 'fir' are often applied very generally and botanically incorrectly to a range of coniferous trees, and different names are used in different areas of the world for the same species. A case in point is Douglas fir (*Pseudotsuga menziesii*), which has also been called Oregon pine and British Columbian pine. The tree is, in fact, neither a true fir nor a pine species, but belongs to a separate genus altogether (*Pseudotsuga*). For much construction work, then, the precise naming of the tree species has not been regarded as necessary and common trade names do not generally distinguish between timbers from unrelated genera or places of origin.

Different rules and standards govern the felling, labelling and sale of commercial timber, and much that is destined for building work is processed in the country of origin. Entering the world market as relatively anonymous sawnwood – with labels that may simply state 'Scandinavian' or 'Baltic', for example, has made identification by the layman almost impossible.

The anonymity and lack of accountability that have characterized the timber industry in general have helped facilitate the destruction not just of tropical forests, but of some of the unique temperate forests of the world, which are still providing softwoods used in the construction industry. An example of this is the indiscriminate and highly destructive old-growth logging currently being inflicted on the Russia's boreal forests, eliminating not just trees and the habitat of many rare species, but the traditional lands and resources of local people. A proportion of this timber is currently being 'laundered' through countries such as Sweden and Finland, whose own image of good practice obscures its true origin.

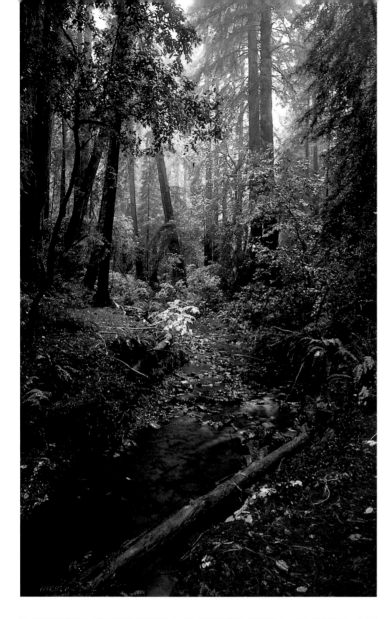

Californian or coast redwoods are graceful giants, growing to over 91 m (300 ft). Much of the great forest they once formed has now been logged.

Californian redwood

■ The most remarkable and perhaps best known of native North American trees are the trees known as the sequoias, including the Californian redwood (*Sequoia sempervirens*) and the giant sequoia (*Sequoiadendron giganteum*), also called Wellingtonia. The giant sequoias are famous for being the largest and amongst the oldest trees in existence – some are over 3,000 years old, with a height of over 80 m (262 ft) and girths of over 16 m (52 ft) at breast height. But their wood is soft and brittle, however, and of little commercial value.

■ The Californian redwood, on the other hand, one specimen of which is thought to be the world's tallest living tree, has been extensively exploited recently for shingles and exterior cladding for buildings. It is also being used for the construction of wooden pipes, tanks, vats, coffins and posts, for which the great durability of its rich reddish-brown timber makes it naturally well suited. The timber has also been used for plywood, and its thick soft bark made into fibreboard. The trees, once abundant, are now restricted to a narrow tract of land extending from southern Oregon to Monterey Bay on the Californian coast, with a few stands on the western slopes of the Sierra Nevada mountains. They are, however, still being logged.

Softwood or hardwood

The terms 'softwood' and 'hardwood' can be somewhat confusing, since some **softwood** timbers – produced by gymnosperms (conifers or cone-bearing trees that are mostly ever-green) – are physically harder than some of those produced by **hard-woods** (broad-leaved, dicotyledonous trees, characterized by seeds that are contained in an enclosed case or ovary). **Balsa** (*Ochroma pyramidale*), for example, is one of the softest and lightest woods in the world but it is technically a hardwood. Species of *Pinus*, meanwhile, belong to the softwood group, but some, such as *P. ponderosa,* are referred to as 'hard' pines. The two kinds of wood are distinguished anatomically by their water-conducting cells and tissues.

In search of good wood...

In the face of this obviously unsatisfactory position, with nearly half the planet's original forest cover now gone and with the world's 'old-growth' forests still disappearing at an alarming rate, the need for control has never been greater. Tropical rainforest is calculated to be disappearing at a rate equivalent to one football pitch every two seconds.

Many concerned individuals and some campaigning groups believe that the solution lies in a complete ban on the commercial felling of such forests (no matter what 'management' scheme is used). Others argue that 'certification' and 'sustainability' initiatives, which aim to assure the public that participating companies and landowners are committed to 'good forest stewardship', are the best way of bringing the rapacious destruction of our natural forests under control.

Various regional certification schemes and many certifying bodies are currently in place, but only one organization has rules that apply internationally. The Forest Stewardship Council (FSC) has drawn up principles and criteria intended to ensure that forests are managed in environmentally appropriate and socially beneficial ways, whilst being economically viable. Timber producers must adhere to these criteria to have their timber 'certified' by an accredited certification body. The individual tree species and country of origin are not listed on the label, but products originating from forests certified by FSC-accredited certified bodies are eligible to carry the FSC logo if the 'chain of custody' (tracking timber from forest to shop) has been checked.

While the uptake of FSC criteria by providers and users of timber for construction and joinery, for example, is encouraging, the suppliers of tropical hardwoods have been much less willing to adopt these standards. Nearly 2,000 companies (from sawmills to wholesalers), in some 56 countries worldwide, were FSC certified in June 2002, with over 400 forest certificates having been issued. But these figures represent less than 1% of tropical timber production. It has been estimated recently by Friends of the Earth that between a half and two-thirds of all tropical hardwood entering the UK has been harvested illegally. To make matters worse, 99% of the remaining imports are almost certainly from unsustainable sources.

The Good Wood Guide, produced by Friends of the Earth and Fauna & Flora International, provides information and advice designed to help consumers make the best environmental choice when buying or using wood in the UK. It lists companies selling certified products and reclaimed timber, and gives information about commonly used timbers and countries of origin. Friends of the Earth recommend that consumers should only buy timber products if they are recycled or have an FSC logo, and is calling on the British government to make it illegal to import and sell illegally sourced timber in the UK.

Turning the taiga into planks and paper

Making up one-third of the world's total forest area, much of it still mostly undisturbed, the **taiga** (or boreal forest) is a belt of forest that encircles the northern hemisphere. It represents the single largest terrestrial ecosystem on earth. The largely **conifer-dominated** forests extend across the far northern regions of North America, Russia and Europe. Russia alone contains nearly three-quarters of all boreal forests and about a fifth of the world's forest – the European region holding the continent's last remaining areas of old-growth forest of any size.

As well as containing a vast pool of biodiversity, including many unique or rare animal and plant species, this **vast region** is the home of hundreds of indigenous peoples with distinct cultures and identities, who have lived amongst and depended upon these forests for thousands of years.

The northern **boreal forests** already provide some 60% of the world's supply of industrial round-wood (including sawn softwood and plywood). But continued large-scale clear cutting and plantation forestry are now a **significant threat** to these forests, already almost transformed in Scandinavia to intensively managed secondary forests.
■ Much of the remoter forest cover in

Siberia (home to 26 distinct indigenous peoples), parts of Alaska and Canada is still intact, but untouched areas are increasingly being cleared.
■ In **Finland**, the tiny area of natural old-growth forest that remains (less than 5%) is still being logged.
■ In 2000, two sites in **Lapland** were cut down, in violation of the traditional rights of Saami reindeer herders who use the area as a winter grazing ground.
■ In 2001, over **50 more areas** of old-growth forest were also found to be destined for felling.

In Russia, rapid expansion of logging operations – most of it export orientated – and widespread corruption has led to an estimated 20% of timber being logged illegally.

Old-growth forests cover around 34 million ha (84 million acres) in European Russia, most of which is not protected. As much as 50% of the wood supplying sawmills (as well as pulp and paper mills) in this **enormous area** is currently coming from these forests. The Taiga Rescue Network – an international network of NGOs, indigenous peoples and nations working for the protection and sustainable use of the world's boreal forests – is campaigning to safeguard the last areas of old-growth forests in European Russia. While logging for timbers used in the construction industry is a **threat** to the taiga, the ever-growing demand for pulp for paper is a major cause of the current destruction of the boreal forests.

Some good news

In 2002, after years of campaigning by local and international organizations, six areas of wild forestland covering 1.7 million ha (4.2 million acres) in the Amur river watershed in eastern Russia were designated as protected. Now out of bounds to all major logging activity, some land is designated for the traditional use of the region's Evenk peoples.

Popular products: dangerous glues

While markets for wood-based panel products have been increasing for some uses, **concern** has arisen over the formaldehyde resins commonly used (in chipboard, plywood and MDF,

for example) as a bonding agent. Emitting formaldehyde gas, which is **toxic**, adverse health effects include irritation of the mucous membranes, such as eyes, lungs and nose, even at

very low levels, and cold-like symptoms. European manufacturers have, however, been steadily **decreasing** the formaldehyde content of their products.

Panel products from trees

New ways of using our most versatile building material are constantly being found. An example is 'glulam', glue-laminated timber, which comprises boards made up of small cross-sections of wood glued together in layers under pressure so that the grain of adjacent boards is parallel. First developed in Germany at the beginning of the 20th century, large pieces of any size or shape can be made. Important features of glulam include its resistance to chemical attack and its impressive strength to weight ratio as it is able to take greater stresses than solid timber boards. Popular in Scandinavia for beams, columns and rafters, it was recently used to construct the girders, deck and piers of a motorway bridge in northern Denmark, as well as the new terminal building at Oslo airport. With a combined length of 136 m (446 ft), sections of glulam joined together on the ground and then hoisted into place form what may be the longest piece of wood ever positioned in one piece for building work. Glulam is just one of an enormous range of products available for use in house construction and interior work that are made by sticking or bonding strips, veneers, chips, flakes or fibres of wood together in different combinations.

Plywood

Perhaps the best known panel product is plywood, which is made by sticking thin sheets or veneers of wood together in layers so that the grains are at right angles to one another, producing a very strong, inflexible board. The Egyptians made crude forms of plywood by bonding wood veneers together with glues made from animal bones and blood albumin. Today, standard plywood veneers are produced with the help of a peeling machine, which peels logs rather like a pencil sharpener. The finished product – available in various strengths and thicknesses and with different specifications – is useful for large,

More plants for panels

It is not just trees that are providing panelling for our homes and other buildings. A number of agricultural by-products, as well as waste paper (recycled newsprint), are being put to **good use** in some parts of the world, for wall panels, ceilings and door interiors. In the USA, **straw** is being used by several companies for boards. *The Good Wood Guide* reports that it has been calculated that if just a quarter of the USA's wheat straw were to be used in this way, it could satisfy the **entire** American demand for particle board. Hemp, flax fibres and shive (waste woody matter separated from flax fibres during processing), kenaf, bagasse (fibrous residue left after sugar has been pressed from sugar cane stems) and sugar cane residues are also being turned into panel products. Flaxboard, for example, is **cheap** to produce, lightweight and fire-retardant.

uninterrupted surfaces. Resistant to splitting, its thermal properties are similar to those of timber, and it can be bent without cracking to form smooth curves.

Numerous softwoods such as Douglas fir and pines from the USA and Canada, as well as tropical or temperate hardwoods, are used in different combinations to make plywood. A significant proportion of the plywood used in Britain, the fifth largest importer of plywood in the world in 1999, is made from tropical hardwoods, the largest supplier to the UK being Indonesia, which is then followed by Brazil.

Chipboard and particle board

As its name suggests, chipboard is made from chipped wood residues including forest trimmings and shavings from board mills, and post-consumer waste. These wood particles are dried and blended with synthetic resins before being pressed into mats, in a highly automated process. Chipboard is increasing in popularity for panelling, flooring, cladding and furniture – sometimes coated with a melamine foil veneer.

Cement-bonded particle board also uses woodchips, but in a mixture with either Portland or magnesite cement. Oriental strandboard (OSB) is similar to chipboard, but stronger, and is increasingly being used instead of plywood in flooring, roofing and beams.

Amazon rainforest timber. The Amazon contains the largest area of rainforest in the world, but destruction continues at an alarming pace. Around 15,000 sq km (5,800 sq miles) is currently being lost each year, much of it through logging.

169

Fibreboards

The main types of fibreboard include hardboard, medium board, soft board and medium density fibreboard (MDF). They are differentiated by the size and type of wood fibres used, their method of drying, the bonding agent used and the way in which they are pressed into shape. MDF and hardboard are the two most commonly used types.

To make hardboard, softened wood fibres are suspended in water, to which a number of other ingredients may also be added (such as resins, drying oils, preservatives and fire-resistant chemicals). The water is removed by gravity, suction and special rollers, which, by exerting high pressure at a very high temperature, produce a mat in which the fibres are interlocked or felted. Less dense than hardboard, MDF is made using a drying process at lower temperatures and with synthetic bonding agents rather than the natural resins in the wood. Around 67% of Britain's fibreboard is imported from European producers, the rest coming from a number of other countries, including the USA, Canada, Malaysia and the Baltic States.

Wood versus the elements

Best house for future

The building judged Best House for the Future in the 2001 National Homebuilder Design Awards was an oak timber-framed building comprising reclaimed, **recycled and natural** building materials, with cellulose from recycled newspapers and wool as insulation. Incorporating the latest environmental technology and uniquely energy efficient, it was described as an outstanding example to the house-building industry.

Straw houses

Compressed into a bale, **straw** makes an excellent building material. In the USA, where settlers in **Nebraska** first made use of it at the beginning of the 20th century, straw has been used to construct a great range of buildings, from homes and offices to workshops and community centres. Straw bales can be used in **various ways** to build a house which, when finished, gives greater energy savings than conventional building systems and needs less heating. Once covered in a lime-based render and with a roof overhang, it will **last indefinitely**. In the USA alone, 40 million tons of wheat straw are unused every year.

Timber cladding attached to the outside of timber-framed walls is becoming more popular in Britain for long-term weather proofing and for decorative appeal. But the USA, Canada and northern and Alpine regions of Europe have a long history of the use of timber weather boarding as an integral part of house structure and design. Many American and Antipodean homes are also complemented by timber decking, raised wooden platforms generally at ground-floor level, which provide an open leisure area adjoining the house.

Living in a log cabin or wooden alpine chalet makes sense where winter temperatures can fall dramatically. Timber cladding, particularly that with an internal air cavity, has far greater insulation properties than brick or concrete and will hold precious heat much more effectively. Where the climate is harsh, timber that grows slowly acquires qualities of strength and density that make it an ideal construction material. In Europe and North America, the trees most commonly used for cladding include Douglas fir, western hemlock, spruce, pine and larch species and – most popular in Britain – western red cedar. European oak and West African iroko may also be used. Pressure impregnation with chemicals means that timber no longer needs to be planed and treated with oil-based paints to protect it from the weather and insects.

Wooden shingles, small rectangular pieces of wood produced by sawing against the grain, have also been used for cladding and roofing and are an impressive feature of some traditional eastern European buildings. In America, shingles are often made of western red cedar, though oak and Californian redwood (see page 165) have also been used. Local availability will determine to a great extent which woods are utilized. In Australia, eucalyptus has been the traditional building material for much internal and external work and different species have been introduced to many other countries with similarly harsh or dry climates. Both these and radiata pine – originally from California – are now major building materials in South Africa and Chile as well as New Zealand and Australia. In southeastern Brazil and parts of Paraguay and Argentina, parana pine (*Araucaria angustifolia*) is still producing wood for all kinds of construction and joinery work, as well as for the plywood that is available in Britain. One of Brazil's most important timber species, the tree is now considered to be at high risk of extinction, as logging and forest clearance has reduced its natural range – originally covering some 200,000 ha (494,000 acres) by over 80 per cent.

Opposite: A roof thatched with reed stems will last for many decades and can be continually renewed. These reeds were grown in Turkey.

Plants on the roof – for homes of the future

Roofs thatched with wheat, straw or reeds are a familiar sight in various parts of Europe, such as France, Spain and the UK. Indeed, the thatching of houses with these materials is enjoying a **revival** in southwest England, particularly in Dorset, where council planners are recommending that one in every six new houses be thatched. Various other dried plants, including heather, bracken and gorse, are still in use on a very small scale in some areas, but for the ultimate in 'green' roofing, how about a house covered with living **sedum plants**?

Known also by their generic name of 'stone crop', the succulent *Sedum* genus comprises some 280 species found in habitats as diverse as the tropics, the Arctic and the Mediterranean coast. As succulents they are known for their **drought tolerance** and are often used in alpine gardens.

Following initiatives with green roofing systems already established in

Germany, living green roofs are being installed in the UK, using sedums – such as *S. oreganum* and *S. spathulifolium* – herbaceous perennials, grasses and bulbs. Though designs may vary, the basic principle involves either introducing seeds and cuttings (which will then establish themselves), or **plugging** ready-grown plants into a substrate suited to their needs – itself resting on a drainage and other sheeted layers. Alternatively, a ready-planted mat system is used, which can be rolled up and taken to wherever it is needed to **cover a building**. Needing little maintenance and adaptable for walls, gables and steeply pitched roofs, some of the benefits include:

■ Improved **sound-proofing** and thermal performance.
■ Reduced **rain run-off**. Green roofs can retain up to 60% of the rain water that would normally run off from the roof surface, a major problem in urban areas. In some German cities green

roofs on new buildings are already obligatory, chiefly because of rain run-off, and in the UK the Environment Agency requires that rain run-off must be minimal for certain buildings.
■ The removal by the plants of carbon dioxide and some pollutants from the atmosphere.
■ A reduction in the roof's capacity to act as a 'heat sink', trapping and intensifying the heat in urban areas.

As the reality of **global warming** becomes increasingly uncomfortable, and with predictions of great difficulty in keeping traditionally constructed urban buildings in particular acceptably cool, authorities in Tokyo (where temperatures are predicted to reach 43°C/109°F in the next 50 years) are asking companies to plant bushes, trees and lawns on their roofs to help **cool down** the city. All new buildings with a floor area of over 1,000 sq m (10,800 sq ft) will be required to put plants on a fifth of their roof area.

Hardwoods in our homes

Teak

■ Teak trees (*Tectona grandis*) produce small white flowers and fleshy fruits. The aromatic, golden-yellow heartwood becomes brown when seasoned and is sometimes patterned with darker markings, the grain being straight or wavy.

A **teak sapling** being replanted near Dar-es-Salaam, Tanzania.

In their countries of origin, tropical hardwoods have been very widely used for building work of all kinds; timbers with special properties particularly so. Teak in India, Burma and Thailand, iroko and afrormosia in West Africa and jarrah in Australia, for example, have all been favoured because of their resistance to attack by termites.

Efficient preservative treatments and modern building techniques, however, mean that a great range of tropical timbers have come to be used in temperate regions. In Europe, tropical hardwoods – especially those with high density and strength – are valued for some heavy construction work in industrial and institutional buildings, often replacing steel or concrete for components such as joists, beams, portals and lintels. We are more likely to find tropical hardwoods in our homes, however, as flooring, doors or kitchen fittings, for example, and, of course, as furniture.

Indonesia is currently the largest exporter of tropical timber in the world, but at immense cost to both environment and local people. In 2001, the International Tropical Timber Organization (ITTO) concluded that the Indonesian forestry industry was in disarray, that illegal logging was out of control, with profits concentrated in the hands of a few corporations, and that the remaining forests were likely to be completely destroyed within 15 years.

Around 200 species of tropical hardwoods are estimated to be available for use in the UK, but about 20 of these make up 80% of those used. The relentless felling and demand for many of these timbers has brought an alarming number of them to the brink of extinction, and eliminated vast areas of forest in which they grew. Teak, rosewood and mahogany are just three of those at risk.

Teak

One of the finest of all tropical hardwoods, teak (see opposite) enjoys worldwide fame because of its strength, durability, stability in fluctuating atmospheres, the ease with which it can be worked and its beautiful appearance. It continues to be used for many purposes in Western homes and offices, including all types of furniture, top-class joinery, panelling, flooring, doors, window frames, staircases and veneers. Outdoor furniture, such as patio tables and chairs, flooring and decking, is, however, the major end use. Other important applications include boat- and shipbuilding – especially expensive floorings and veneers in luxury sailing boats and yachts – dock and harbour work, piling, bridges and sea defences. Teak is much favoured for decking and for the trimming of boats because of its unique resistance to the corrosion of metals in damp environments. Its oils also make it resistant to insect attack.

Holding most of the world's primary teak forests, and the principal exporter for many years, Burma currently provides about 80% of the world's supply. In the financial year 2000/2001, production of teak, the country's top export product, was said to have increased by some 39% over the previous year. Trade, however, is even greater than official statistics indicate, since a large black market in the timber is known to exist between Burma and its neighbours. These include Thailand (along whose border about 80% of the remaining teak forest is situated, but where logging is officially banned), China (with recorded imports of some 500,000 cubic m/17.6 million cubic ft of teak from Burma in 1995) and India. This trade is thought to be providing financial support for Burma's military regime, which is responsible for serious human rights violations. Cambodia is also aggressively logging its ancient teak forests, having signed contracts with many foreign companies allowing them to operate.

173

Most of this timber is still being harvested – unsustainably – from natural forests, to supply major markets within the region alongside western Europe, the USA, Singapore and Hong Kong. Teak is distributed very sparsely in the wild – with only one to five trees occurring per acre. The relentless felling – including undersized trees in recent years – has caused massive devastation, and harm to indigenous tribal people and wildlife alike. The FAO believes that as much as 607,000 ha (1.5 million acres) may have already been destroyed; the annihilation of most of the world's remaining old-growth teak forest appears to be imminent.

Numerous teak plantations have been established within the tree's natural range, which includes Burma, Laos, Cambodia, Thailand and much of peninsula, India, as well as in Indonesia (mostly in Java), Central America (mostly Costa Rica), Sri Lanka and Brazil. Despite the advantage of being cheaper, since timber is harvested from much younger trees (only 20 years old or less), wild harvested teak is still preferred by many users. To make matters worse, some dealers and exporters are mixing old-growth teak with Javanese plantation teak for sale to Europe and America, making it impossible for the user to determine its true origin.

Rosewood

Rosewood takes its name from the rose-scented timber that is produced by a number of tropical evergreen trees of the genus *Dalbergia*. Although many are traded under a variety of different names, some of the best-known species include Brazilian rosewood (see below left) and kingwood or tulip wood from South America; Honduras rosewood, granadillo and cocobolo from Central America; East Indian rosewood or Bombay blackwood from southern India; and African blackwood from tropical Africa.

Most of these members of the rosewood family have been so widely sought after and felled that they – and their native forests – are now in serious trouble in the wild. Brazilian rosewood, for example, which has been used as a decorative wood for some 200 years, has been so relentlessly harvested that the tree is considered to be at high risk of extinction. Fewer than 5,000 individual trees now remain within a restricted area in Brazil.

Although the name 'rosewood' is shared with trees of the *Dalbergia* genus, fragrant rosewood oil – which is highly prized by the perfumery trade (see page 48) – is extracted from trees belonging to the genus *Aniba* (of the unrelated *Lauraceae* family). The presence of linalool in *Aniba rosaeodora*, native to Amazonia, has led to the extensive exploitation of this species and the trees are now also seriously endangered in the wild.

A plantation-grown teak tree.
Since teak timber will sink in water unless dried, in India bark and living tissue is removed near the base so that the tree dies and stands for two years, to dry out, before felling.

Rosewood

■ Brazilian rosewood (*Dalbergia nigra*) timber may vary from chocolate or violet-brown to a rich purple-black, marked with irregular black streaks. It has a rather oily appearance and when worked releases a distinct though mild and fragrant odour. Still being illegally traded, its uses include veneers for furniture and cabinet making, musical instruments and billiard cue butts.

Mahogany

The name mahogany has been used for a number of tropical hardwoods. Today two 'true' mahogany groups are recognized: those describing several *Swietenia* species (the 'original' mahoganies) from central and South America and the West Indies, and the related African *Khaya* species.

The best-known Cuban mahogany, *Swietenia mahogani*, was first shipped to the United Kingdom and Europe some 400 years ago. The most cherished cabinet wood in the world, this species has been so widely exploited that its significance is now largely historical. Despite its great value for furniture making, enormous numbers of smaller trees were cut from forests and used to fire the boilers of Caribbean sugar mills and steam trains, whilst many larger trees were felled for fence posts and railway sleepers. This indiscriminate use of the timber finally led the government of Cuba to ban the export of mahogany logs in 1946. Along with Honduran mahogany (*Swietenia humilis*), today the wild populations of Cuban mahogany are highly fragmented, representing a major loss of genetic diversity. However, small quantities of timber from plantations remain occasionally available on the international market.

Another species, *Swietenia macrophylla*, also known as big-leaved mahogany or baywood, which has come to fill the position of *S. mahogani* as the commercially important species, is now, tragically, facing a similar fate. Despite a wide natural distribution from southern Mexico, through Central America and large areas of the Amazon Basin, intense demand has caused this majestic tree, which can reach a height of 45 m (148 ft), to suffer a serious decline.

175

Killing for mahogany

In Brazil, Indian lands are – in theory – protected from **illegal logging**, but for years this practice has been commonplace and gone unpunished. Mahogany logging has now extended to all 15 Indian lands in the state of Para. Indians generally have no choice but to let the loggers (who may iden-tify mahogany by flying hundreds of kilometres over dense forests in search of **scattered trees**), destroy their forest. Illegal access roads are then bulldozed through previously untouched forest, paving the way for colonists, ranchers, plantation owners and illnesses to reach isolated tribes. In some cases, the Indians have been obliged to sell their trees themselves, but for a pittance, or, in many cases, have ended up in **debt bondage** to the sawmills.

Indians have been forced to take **direct action** to try to halt the illegal loggers; many violent conflicts have been reported on Indian lands and an unknown number of Indians have been murdered as a result. The Greenpeace report *Partners in Mahogany Crime* has revealed the extraordinary **profits** involved for those who deal in the timber and sell the final end products. An Indian may be paid only $30 for a tree that will be resold for about $1,400, and which may eventually be converted to 15 large dining tables, worth some $8,500 each when sold in London or New York. This makes one tree possibly worth $128,250 by the time it reaches its final destination.

FSC logo spraying. Most of the tropical timber that enters Europe is logged illegally and unsustainably. This timber, however, is FSC certified, which means that it complies with rules intended to ensure that forests are managed in environmentally and socially acceptable ways.

Often referred to as 'green gold' (1 cubic m can fetch over $1,600), the exploitation of this mahogany, fuelled by international demand, has been a major factor driving the destruction of the Brazilian Amazon, where the tree occurs across some 80 million ha (198 million acres).

While official figures are unreliable because of the illegal nature of most of the trade, the main exporters of mahogany are currently Peru, Brazil and Bolivia. *Partners in Mahogany Crime*, a recent report published by Greenpeace, which focuses on the Amazon, has revealed the alarming truth about the mahogany trade in Brazil. According to Brazilian government figures in 2001, at least 70% of the mahogany harvested had been stolen from public and Indian lands, much of it in the state of Para, where the largest remaining concentrations of mahogany are to be found. This has resulted in catastrophe for the indigenous people, animals and forest (see opposite).

In 2001, following initiatives in recent years by Brazilian agencies to restrain the illegal mahogany trade and pending investigations into illegalities and corruption exposed by the Greenpeace report, Brazil announced that it was suspending indefinitely the trading of mahogany and banning its export. But despite these measures, illegal logging and export are still continuing. The report provides evidence that much of the Brazilian trade is controlled by just two key figures – one of whom is calculated to make as much as $1 million a day from his activities during the logging season – and that these are connected to at least five export companies who control about 80% of the exports from the state of Para, the major supplier to importers in Europe and beyond.

In countries bordering Brazil, the situation appears to be equally serious. According to a recent TRAFFIC report (a central clearing house that collects and distributes information about the global trade in wildlife; jointly run by WWF and IUCN), Peru exported over 51,000 cubic m (1.8 million cubic ft) of mahogany in the year 2000, with significant illegal harvesting occurring in regions such as Madre de Dios. In Mesoamerica, over-exploitation has aided a similar decline both in mahogany and the forest that supported it. Having now lost over 80% of their broad-leaved forest, the tree is now considered commercially extinct in El Salvador and Costa Rica, while illegal trade continues in Belize, Guatemala, Honduras, Mexico and Nicaragua.

The USA was by far the largest importer of mahogany during the period 1997–1999, having imported over 253,000 cubic m (8.9 million cubic ft), 90% coming from South American countries, followed by the EU, which imported over 28,000 cubic m (988,000 cubic ft). Within Europe, most mahogany was consumed by the UK, following the Dominican Republic to become the third largest importer in the world. Much of this has been used for high-class furniture, panelling, doors, window frames, toilet seats and guitars.

THE REAL PRICE of tropical hardwood

Buying a piece of tropical hardwood for a new door or window frame may be hard on the pocket, but it is much harder on the rainforest and people from which it has been taken. Despite the warnings and predictions made by experts over the last 10–15 years about alarming rates of forest loss, and the pledges made by various governments and some trading bodies to take effective action, the destruction of tropical forests has been steadily increasing. It is now worse than ever before. The latest assessment (made by the World Resources Institute and Emily Mathews in 2000 and supported by WWF) is that natural forests in the tropics are currently being cleared for timber, pulp and plantations at the rate of nearly 16 million ha (39.5 million acres) per year.

Illegal logging
Around 200 different tropical timbers are currently available for use in British homes but, according to a recent study by Friends of the Earth, over 99% of all tropical timber being imported into the UK is from uncertified sources. Furthermore, about 60% is logged illegally, in violation of national laws. Illegal logging includes cutting down more trees than permitted (including those that may be under- or oversized) as well as obtaining concessions and processing timber corruptly. This horrifying finding means that hardwoods are still routinely being taken from areas which belong to local people, are designated as reserves or which have some other category of protection. The result is immense environmental damage and human misery.

The UK currently obtains 90% of its tropical timber from just three countries: Brazil, Indonesia and Malaysia. Friends of the Earth research has brought together findings that show the catastrophic level of logging in these countries in recent years. They conclude it has made a significant contribution to:
■ The loss of 53 million ha (131 million acres) of forest between 1972 and 1998 in Brazil (an area the size of France), most of it in the Amazon, where illegal logging was assessed to have reached 90% in 1998. Around 1.7 million ha (4.2 million acres) are now being felled each year.
■ Rainforest loss in Indonesia: nearly 75% of its original forest is now destroyed.
■ The destruction of Malaysian forests: 50% has been cleared.
■ Massive forest loss in Cameroon (the fourth largest supplier to the UK): 81% of unprotected forest has been allocated to logging companies.

Illegal logs reaching the UK are 'laundered' to obscure their origin, making it almost impossible to identify them on arrival. A recent Environmental Investigation Agency report revealed that, in 2001, three ships apprehended by the Indonesian authorities in the Java Sea were found to be carrying 25,000 cubic m (882,500 cubic ft) of illegal timber, much of it meranti from central Kalimantan, worth about $4 million.

Poor management
As if this weren't bad enough, the overwhelming majority of the rest of Britain's imports are, in any case, being harvested in ways that do not meet basic criteria for 'sustainable'

management. Logging in many tropical regions is still carried out indiscriminately, with huge areas – including small islands in Malaysia and Indonesia – being denuded of their trees. The World Bank estimates that there will be no natural forest of any quality left in Sumatra by 2005 and none in Kalimantan by 2010.

Hardwoods are often used for general construction work in their countries of origin and for specialist building work and interior fittings and furniture elsewhere. But the introduction of machines that can reduce almost any timber to wood chips in a matter of minutes has exacerbated the situation. Hardwoods can still end up as prefabricated panel products, or plywood.

Temperate hardwoods around the world are a cause for concern, too, as many are being illegally logged. The pool of large trees, which have often taken hundreds of years to reach maturity, is being rapidly depleted, and many are suffering from the effects of atmospheric pollution and acid rain.

In Chile, large areas of the remaining monkey puzzle (*Araucaria araucana*) forests have recently been set on fire. It seems that this has been done so that timber companies can fell these trees and use their hard and durable timber. Monkey puzzles being protected by law has not stopped those determined to bring about the trees' demise.

Since it is currently impossible for the consumer to distinguish illegally logged timber, including tree species threatened in the wild, the best we can do is to buy only recycled wood or timber which displays the FSC logo.

177

Furniture from plants

Birch

■ Whilst the silver birch (*Betula pendula*) is distinguished by having rough silvery bark with drooping branches and smooth, short, pointed leaves, the downy birch (*B. pubescens*) has a smoother, often reddish bark at its base with branches that are closer together and more horizontal. Its twigs and leaves are covered with soft hairs.

■ In very cold areas, birch trees may never grow beyond the size of small shrubs, or will branch just above the ground, but in kinder climates they can reach heights of 18–20 m (59–66 ft). Birch timber is whitish to pale brown in colour with no distinct heartwood. It is fairly straight-grained and fine textured and when dry is similar to oak in terms of strength, and superior in rigidity. The biggest uses of birch timber are marquetry, veneers and strips for plywood, dowelling rods, interior fittings and furniture.

178

Furniture and household accessories – both modern and antique – made wholly or partially from tropical hardwoods are to be found in countless homes and places of work in Britain. Mahogany, teak and rosewoods of different kinds, for example, though seriously depleted in the wild, still occur very frequently in solid and veneer forms in tables and chairs, wardrobes, shelving, cabinets, work tops, lamp stands, chopping boards, saucepan handles and toilet seats.

Temperate forests have also been used for centuries to furnish our homes. Whilst walnut, one of the most important woods for traditional British furniture, is too expensive now for wide-scale use, oak, beech, ash, birch and sweet chestnut, in various combinations and often with other natural and man-made materials, are widely available today.

Other temperate woods often used by individual craftsmen, with a long history of use for decorative carving, veneer and inlay work, include:

● Pear, a striking terracotta brown or reddish-gold in colour.
● Sycamore, a milky-white wood that has a natural lustre and that polishes to give an ivory-like surface.
● Box, a fine even-textured wood, ranging in colour from pale yellow to bright orange (see page 284).
● Crab apple, generally a soft pink.
● Lime, pale creamy yellow in colour and relatively soft but dense.

Softwoods commonly used for furniture in Britain include a range of pine species, not generally specified individually by manufacturers but popular in the form of 'antique' pine – timber chemically treated to look old – and yew, Britain's densest softwood, which has a rich red-brown timber. The pines are likely to be grown in Scandinavia and other northern European countries such as Finland. The more northerly the origin and the slower the growth of the tree, the better the timber is said to be – of a finer, denser quality.

We tend to take our furniture for granted, but design and availability have progressed only gradually over the centuries. Until Elizabethan times, only people of high social rank in Britain would have possessed chairs, which were solid constructions made of a heavy frame of squared timber. Those of lesser rank had benches and stools; stools were in fact the forerunners of what was to become the first chair available to the cottager, the Windsor chair, distinguished by its turned, lightweight legs and back supports (not added until the 16th century) set into the seat. Many different woods have been used: beech, birch, apple or pear for the legs; elm or oak for the seats; and hooped backs often of steam-bent yew.

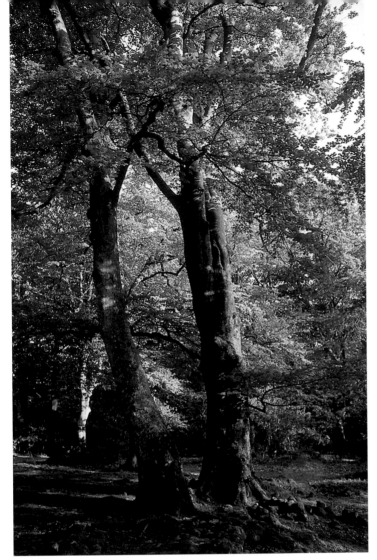

Mature beech trees in the New Forest, UK.

Beech

■ In dense forests, the beech tree (*Fagus sylvatica*) can grow to a height of about 45 m (148 ft), with a clear bole of around 15 m (49 ft), yielding a substantial quantity of straight-grained timber. The texture of the wood is fine and even but density and hardness vary according to the region of origin. When freshly cut, the timber is a very pale brown, turning to pink or reddish-brown on exposure to the air and a deeper colour after the steam treatment commonly given.

■ Whilst 'green' (before drying-out in the open air or by kiln), its strength is similar to that of oak but, after drying, beech becomes 20% stronger in bending strength and stiffness, and considerably stronger in its resistance to the impact of heavy loads.

179

European beech

With its fine veil of soft translucent green in springtime and leaves of copper-gold in the autumn, the beech tree (see right), with its elegant grey trunk, is one of Europe's most impressive native trees.

There are innumerable applications of beech wood in the home and it is now the most commonly used native hardwood in Britain. Interior construction and joinery, mass-produced furniture, flooring and all sorts of domestic woodware use large quantities each year.

Since beech wood bends so well and can be turned so easily, it is an ideal wood for making furniture, especially chairs. In the Chilterns, where beech chairs have been made for centuries by 'bodgers' (traditional chair makers), some beech woodlands are still managed to supply this ancient industry. Sadly, much of Britain's native beech woodland, especially in southern England, has now been felled and replaced with quick-growing conifers. Large quantities of beech wood are now imported into the UK, in particular from France and Germany.

Oak ecology

The oak, for so long part of the European landscape, fulfils an important **ecological role**. It is estimated that some 500 different species of insect and other invertebrates depend to a greater or lesser extent on the foliage, the deeply furrowed bark or the tree's timber for their existence, thereby providing an important source of food for many birds. The tree has also been directly beneficial to people and their domestic animals as a **source of food**. In Saxon times, poor people ate acorns during periods of famine and towards the end of the 7th century special laws were made relating to the feeding of pigs in woods, on acorns and beech nuts. These 'rites of pannage' as they are known today are still observed by the commoners of the New Forest.

Oaks are also beneficial to the **soils they grow in** (often heavy clays), in that they play a major part in draining them. A mature tree will draw up through its roots around 91 litres (20 gallons) of water a day – a natural water pump for soggy ground.

In recent years, the health of oaks has **declined** all over Europe, with the most serious deterioration in Germany, the Czech and Slovak Republics, and Poland.

Birch

Found throughout Europe from Lapland southwards to central Spain and from Ireland to Russia, birch (see page 178) is one of our most versatile hardwoods. It grows further north than any other broad-leaved tree and yet will tolerate great heat. Two birch species out of 35 are those primarily used for timber in the UK: *Betula pendula*, the silver or warty birch, and *B. pubescens*, the downy or hairy birch. Both are major timber species in Scandinavia.

Oak

Of all the trees native to the forests of Europe the mighty oak (see right) has held a special fascination since the earliest times. The tree was sacred to the ancient Greeks and to the pre-Roman Celts, who emerged from the Rhine lands of central Europe. This king of British forest trees, believed to live for over 1,000 years, became very early on the foremost construction material of Europe, and remained the most important furniture timber from the Middle Ages to the 17th century, when it began to be displaced by walnut.

Not only very strong but durable, the natural formation of its branches allowed timber to be cut from the tree in curved shapes suitable for the cruck frames of Medieval houses (see page 160) and the frame supports for ships. Oak timber had become so important to British shipbuilding by the 16th century that King Philip of Spain ordered the Armada to burn and destroy every oak that grew in the Forest of Dean – the source of much of the British Navy's timber at the time. Although the Armada was defeated, so many oaks had already been felled that extensive laws were passed by Queen Elizabeth I to protect them for future use, and extensive planting was subsequently carried out in the royal forests.

Despite centuries of use for iron smelting, tanning and construction purposes, much of southern England still had a large amount of woodland cover up to 1945. Between 1945 and 1985, over half of the remaining hardwood forest was clear-felled or severely degraded – an amount equal to that destroyed during the preceding 400 years. Scarcity of good oak has now made it, by and large, the most expensive of British timbers and it is therefore often supplemented by faster-growing, imported American oaks (often *Quercus rubra* and *Q. alba*), although these may be less attractive to carpenters.

Oak timber has a long history of ecclesiastical use – for the interior fitting of churches and cathedrals – and for marine construction, the building of boats, docks and sea defences, etc. Its water-tight timber has also made it eminently suitable for barrel making and, with the help of rush to seal the joins, it remains essential for storing precious liquors such as whisky while they mature.

Left: An ancient common or pedunculate oak.
Overleaf: Oak woodland with hazel trees.

Oak

■ The oaks of Norse and Celtic legend classified today by the timber trade as European oak refer to just two of the 400 or so separate species in the genus *Quercus*: *Q. robur*, known also as the pedunculate oak, and *Q. petraea*, the sessile or durmast oak, which tends to inhabit warmer sites and lighter soils. Both of these may also be called English, Polish or Slavonian oak by traders, according to the country of origin, and they are both classified on the basis of their wood structure as 'white'.

■ Since there is little difference between the appearance of the woods of *Q. robur* and *Q. petraea*, the species involved is not usually distinguished by the timber trade. To complicate matters, many hybrids between the two species exist. Growing conditions may affect the character of the timber, however, and trade names chiefly mark these differences.

181

182

Plants for padding and polishes

Lacquerware

The application of a natural resin made from the grey, syrupy sap of lacquer trees to furniture and other household items had become an **art form** in China before the 1st century AD, but was perfected later in Japan. Still practised today, up to 300 layers of filtered resin may be applied, each one rubbed with charcoal powder then polished after drying to give a final **glass-like finish**.

The natural strength, resistance to heat and imperviousness to water of the lacquer once dry have made it perfect for protecting the **surfaces** of household utensils and furniture. Fine designs in gold leaf or rice paper may be laid onto the surfaces, sometimes also painted, inlaid or carved, and coated with additional layers of lacquer.

Japanese lacquerware – mostly derived from the sap of *Rhus verniciflua* (used also in China and Korea) – is considered to be of the **highest quality**, but various other lacquer-producing trees are also tapped in Vietnam, Korea, Burma, Taiwan, Thailand and Indonesia.

Other plants have helped to make our seating arrangements more comfortable. Until the 1980s many upholstered armchairs and sofas were stuffed with vegetable 'horsehair' consisting of fibres obtained from the leaves of the dwarf palm, *Chamaerops humilis*. Found in North Africa and Mediterranean regions including Italy, Spain and southern France, this is one of only two palm species native to the Mediterranean. Although the practice is in decline, vegetable horsehair is still being produced on a limited scale in Morocco and used as a packing material, as well as a stuffing for outdoor furniture and mattresses.

Spanish moss

Spanish moss or old man's beard (*Tillandsia usneoides*), the air plant found hanging in festoons from trees and overhead wires – with a natural distribution from southern Virginia to Argentina – has also been much used for padding upholstery (including the car seats of the model T Ford), mattresses, cushions and packaging. Also used as a binder for mud or clay in house plaster and in the construction of chimneys, Spanish moss was harvested on a large commercial scale in the states of Louisiana, Georgia and Florida until the mid 1970s. It is still used for packaging flower bulbs and in floral arrangements.

A shine from plants

Brazilian wax palms (*Copernicia prunifera*) from South America and candelilla bushes (*Euphorbia antisyphilitica* and *Pedilanthus* spp.) (see page 40) from Mexico help us to keep our wooden furniture in good condition in the form of furniture polish. Carnauba wax, which forms on the palm leaflets as they develop, constitutes 5–20% of some household furniture polishes, which generally have a base of paraffin wax and beeswax. An important attraction is its hardness, since only small quantities are needed to give good protection and an attractive shine to the surface being polished. Candelilla wax, also brittle and hard but with a higher resin content, has strong adhesive properties, which make it a very useful ingredient in polishes for floors, furniture and leather upholstery.

Plants have also been involved in the production of a number of wood 'finishes', used partly to seal the grain against the atmosphere and so limit the wood's tendency to warp and shrink, and partly to protect its surface from becoming marked or dirty, and also to enhance the beauty of the grain. Linseed oil, for example, is still found in some oil-based varnishes. Valued for its 'drying' properties, it has been widely used to give preliminary protective coatings to oak, mahogany and walnut surfaces. While tung oil, as well as pine and natural gum resins, may still be found in some varnishes, most now use synthetic resins and cheaper oils. In the past, this oil base was sometimes darkened by the immersion of alkanet roots (*Alkanna lehmannii*), which were a traditional source of deep red-brown dyes.

The original base of the various types of French polish is shellac, the secretion of a parasitic insect, *Laccifer lacca*, that feeds on dhak trees (*Butea monosperma*), naturally distributed throughout Southeast Asia from India to Burma. Shellac is processed by heating and dissolving in industrial alcohol. On application of the French polish, the alcohol dissolves, leaving the shellac deposited on the wood.

Though true French polish is, in general, being superseded by synthetic finishes that are more resistant to scratching, heat and staining from dampness, Oriental lacquerware (see opposite) still depends on the sap produced by plants of the *Anacardiaceae* family. The most important species is the Chinese or Japanese lacquer tree (*Rhus verniciflua*), native to China.

Inside a Maori meeting house at Waitangi, New Zealand. The floor is polished with gum from the kauri tree (*Agathis australis*).

Trees for the home: too good to waste

Spoilt for choice when it comes to timber, few of us think of the environmental **impact** of furnishing our homes, or of what will happen to our furniture and fittings when we move on to something new.

Currently over 8,000 tree species – that is about 10% of the world's total – are threatened with **extinction**; over 1,000 of these due to logging. Many of these trees, including species of birch, cedar, pine, oak, maple, meranti and mahogany, will have been used as construction or furniture timbers around the world.

To make matters worse, huge amounts of wood are routinely wasted on their way to becoming the items we **take for granted** in our homes. Friends of the Earth and the Global Trees Campaign (an initiative of

Fauna & Flora International and the United Nations Environment Programme–World Conservation Monitoring Centre/UNEP–WCMC) are concerned not only that so many of the world's trees are in serious trouble, but that as little as **half** of a typical piece of sawn timber actually ends up as a manufactured item, such as a chair or door – representing **just 14%** of an entire tree.

In addition, they estimate that 750,000 tons of reclaimable timber is thrown away in the UK each year from the demolition of buildings alone. Much of this, along with the wood that is discarded by furniture makers, builders, pallet users, other industries and householders (often as old furniture, doors, windows and DIY off-cuts) is ending up in **rubbish tips**; timber

is now estimated to comprise 7–10% of household waste sites. Finding new uses for this waste timber is an important goal:

■ Some **off-cuts** can be used by schools or recycled into chipboard, for use in place of new wood.
■ Other waste may be turned into **shavings** for animal bedding, **shredded** for compost or mulch, or used to make fuel briquettes.

Effective action by government and industry is crucial, both to halt un-sustainable logging and to make standard the recycling of all that we already consume.

Repairing our furniture and using **reclaimed timber** to make much of what we need – whether floorboards, doors or garden chairs – will help save the world's threatened trees.

185

Plant stems for furniture

The stems of plants belonging to two very different families – one native to the tropics and the other to temperate areas – have given us distinctive cane and wicker furniture.

Willow

■ During the Second World War, food and ammunition hampers made from willow were used by the British since they could be air-dropped without breaking (commercial use of the basket willows for any other purpose at that time was banned), and willow is still regarded as the best material for the baskets of hot air balloons.

Rattan

■ The stems of the moderate-sized rattan *Calamus caesius* grow high into the forest canopy, reaching about 100 m (328 ft) in length. *C. caesius* is widespread in the lowlands of Sabah, where it is known as *rotan sega*, and elsewhere throughout Borneo, Sumatra, the Malay Peninsula and Southern Thailand, where it may have been introduced. In Kalimantan it has been widely cultivated by small-holders for over a century, often in abandoned rubber plantations, and also incorporated in traditional forest gardens.

■ Acknowledged to be the best quality cane of its size and class, *C. caesius* is considered ideal for all kinds of furniture work. Rattan is a very versatile material. While the larger-diameter canes make stable frames for furniture, others may be split into fine strands for chair backs and seating, for basketry and also for decorative screens.

Willow

Wickerwork is a very ancient craft. The Roman scholar Pliny recorded the Etruscans reclining upon wicker couches and the Romans them-selves favoured wicker furniture. Then, as now, the material involved was willow. The supple young shoots of these water-loving trees have been made into a range of household items, chiefly baskets, chair seating and backing.

Though the shoots of most willow species are flexible enough for weaving, one in particular, *Salix viminalis* – often referred to simply as 'osier' or 'common osier' – has been most commonly used in Britain. *S. purpurea*, the purple willow or purple osier, and *S. triandra*, the almond-leaved willow, are also widely used for wickerwork. These trees are found wild in marshy areas across central and southern Europe and into Asia, but a large number of hybrids and cultivated varieties also exist today, producing coloured shoots ranging from pale yellow to a rich purple.

Instead of being pollarded like some other willows to produce straight poles, osiers are coppiced, that is cut down to ground level once a year, so that a mass of long, pliant stems or 'withies' is produced. Much in demand in former times for chair seating and for a number of other purposes including fish traps and lobster pots, the cultivation of withies for commercial purposes is now practised only on a small scale in Britain, having suffered greatly through the intro-duction of mass-produced furniture. Many individual craft workers or small-scale enterprises, however, are using willows today for baskets, furniture, woven panels and plant climbing frames, while willow stems or 'spilings' are also used to form fences to restore river banks and prevent erosion.

Rattan canes

Used for chairs and many other light but sturdy items of furniture, and usually described as 'cane', are rattans, the rope-like stems of various thorny climbing palms from the tropical forests of southern and Southeast Asia. Rattans belong to a subfamily of the palms, the *Calamoideae*. Two genera supply the bulk of all rattans used in the furniture trade –

Weaving strong, supple strips of **rattan cane** in Indonesia.

Harvesting rattan

Mature rattans grow upward with the aid of **backwardly curved** spines attached to 'whips' borne at the end of the leaves or on leaf sheaths. The stems grow singly or send out a number of suckers from the base of the plant. Once a stem has reached the maximum diameter for the species, it grows longer, **towards** the source of light.

To extract rattans from the forest, accessible stems are cut close to the ground and dragged from the supporting branches. After removing the **leaves and spines**, the stems are cut into more manageable lengths of 3–5 m (10–16 ft). To make them more durable (by removing gums, resins and water), larger stems are usually cured with a hot oil mixture. They are then **rinsed or scrubbed** before being bundled and left to dry in the sun. Smaller canes are often dried without curing. Some species must be rubbed with sand or pulled through bars or pulleys to remove the silica from their epidermis.

Stems are also **graded** according to dimension, hardness, defects and colour. The canes are then split, often into three or four separate pieces. **Secondary processing** involves steaming, bending, splitting (of the larger canes), dyeing, sanding and finishing.

Calamus (of which there are some 370 species) (see page 186) and *Daemonorops* (115 species). While many of these are useful at the local level, not all rattans have commercial value: about 27 species provide the most sought-after, fine-quality canes

With an export value of hundreds of millions of dollars each year, the rattan trade is extremely important in Southeast Asia; rattans are, in fact, the main commercial product of its tropical forests after timber. Highly labour-intensive, it is estimated that over a million people are employed in the manufacturing of rattan furniture. About 500,000 people are involved in the manufacturing sector, and around 700,000 in the collection, primary processing and transportation of the stems, mostly on a seasonal basis, providing a vital source of income in many rural areas. Furniture is the biggest end use of rattan canes, and many European, Japanese, American and Australian companies import the finished products or the canes to make them. However, a vast array of other items are made from them – important either at local, national or international level – which include cordage for general tying and binding, rope, baskets, matting, panelling, hats, walking sticks and sports equipment.

Estimates of the enormous socio-economic value of rattan are hard to make because of its huge cultural importance and the fact that the rattan industry is largely based on small-scale enterprises. But it is believed that over 700 million people trade in or use rattan in some way, and the global rattan industry – including both local usage and external trade – is estimated to be worth around $6.5 billion per year. Indonesia, Malaysia, Thailand, the Philippines and China have all been important exporters of furniture and furniture canes, the industry having expanded rapidly between the 1970s and the early 1990s. In Indonesia alone, for example, the annual allowable cut has been estimated at about 700,000 tons.

To stimulate the development of their rattan-based industries, in 1988 the main exporters banned the export of rattan canes unless in the form of ready-made articles. Indonesia, however, which dominates the world trade (accounting for around 80% of raw rattan), lifted its ban in 1998 and is said to have been flooding the market with relatively cheap supplies of cane.

Bulrush

Another traditional material used for making the seats of chairs, this time in temperate areas, is *Schoenoplectus lacustris*, the true bulrush or clubrush. A freshwater rush, native to the swampy areas and waterways of Europe (with close relatives found also in North and Central America, Asia, Africa and Australia), it has been used as the traditional rush for chair seating in Britain since the 17th century.

THE REAL COST of rattan furniture

The colossal volume of the rattan trade has come at a very high price indeed in environmental and human terms. Uncontrolled harvesting – mostly for export – as well as massive deforestation means that many of the most desirable species have become exhausted. This is a situation, of course, that poses a serious threat not just to the rattan industry, but to the countless poor families (involved in a highly complex, unregulated trade) whose livelihoods have come to depend upon it and for whom it plays a critical role as a form of income. Many Asian countries have had declining exports in recent years, those most affected being China and the Philippines, though canes from Indonesia and parts of Indo-China are finding their way to manufacturers there. Of the 600 or so rattan species, 117 were recorded in 1997 as being threatened to some degree. Since the taxonomy of and threats facing some are still poorly known, their true conservation status is difficult to assess. Species thought to be new to science may become extinct before they are studied. Wastage, estimated to be about 30%, has contributed to the problem of over-harvesting.

Across Southeast Asia, numerous plantations have been set up, some as part of agroforestry systems in areas controlled by local communities. But many are owned and managed by private companies. Around 31,000 ha (76,600 acres) have been planted in Malaysia with *Calamus manan*, for example, while *C. caesius* and *C. trachycoleus* have been planted in another 10,000 ha (24,700 acres), and other species have also been tried. Though returning a profit (not, however, generally shared by local people), such plantations have not met demand and have supplied only a very small part of the market.

Most recently, other land uses, in particular African oil palm plantations, have been considered more lucrative and many rattan plantations have been or are being converted. In the face of the widespread loss of traditional management practices, fuelled by the goal of meeting an insatiable international demand, the need for a controlled and sustainable harvest of rattans, which respects local rights and knowledge, could not be greater.

Dredging, pollution and the use of herbicides in land adjacent to the rivers in which they once flourished has, sadly, led to a considerable decline of rushes in many areas of Britain. Only one major collector currently exists, harvesting bulrushes chiefly from the Great Ouse in Cambridgeshire and Bedfordshire. The craft of rush work has not been lost, however, and a number of individual rush weavers are still active around the country. As well as providing the seating for chairs, rushes are woven into floor matting, baskets, screens, bags and shoes. A less well-known use is as a sealant in the manufacture of some Scottish whisky casks, strips of rush being inserted between the curved oak panels to make them water- (or rather whisky-) tight.

Growing to a height of around 3 m (10 ft), the soft, rounded bulrush stems are harvested in Britain in late June or July, being cut on or just below the surface of the water, and carefully dried for some weeks. Before use they must be moistened once again to make them pliable enough to weave.

A number of British weavers make use of rushes grown in Holland, where they also occur naturally and are managed on a large scale in filter beds for water purification.

189

Dried harvested stems of the true bulrush or clubrush.

Plants on the floor

Jute

■ Jute (*Corchorus* spp.) is an annual plant that flourishes in monsoon climates and which grows to 2.5–4.5 m (8–15 ft). Bangladesh is the world's leading exporter, followed by India. To extract their fibres, tall jute canes are first cut by hand close to the ground and left lying in the fields for a day or two to allow the leaves to fall from the stems. Bundles of stems are then steeped in water for several days so that their fibres become loosened. Whilst standing in water, farmers strip the fibres from the plants, beating them with wooden flails and washing them to remove the remains of the outer bark. The fibre is finally dried in the sun, baled and sorted, ready for export.

■ Extremely strong and hard-wearing, jute has been used extensively not just for carpet backing but for sandbags, sacks, ropes, cables and twine and for the manufacture of roofing felt, damp courses, plaster board, caulking for the decks of ships, lining materials for vehicles and the soles of 'rope' sandals. Grown several thousand years ago in Asia, jute leaves were used first as a vegetable and source of medicines, and only later were the fibres utilized for rope and cloth.

■ Today, there is no jute spinning industry in Europe, but major end uses of the woven yarn in the UK include carpet backing, roofing felt, upholstery and furniture webbing, sacking for tree roots in preparation for planting, and sacks for storing seed potatoes.

190

In many parts of Europe, long before the days of carpeting or other mass-produced flooring materials, rushes were amongst the principal plants used to make simple coverings. Simply strewn on the bare earth or stone floors, they could be replaced when dirty. In country churches and chapels, this practice (which was widespread in Britain before the boarding of floors) had its focus in annual 'rush-bearing' ceremonies in late summer – a tradition which still survives in some areas today. Sweet-smelling flowers and herbs such as meadowsweet, sweet rush, lavender and fennel, which released their fragrance as they were trodden on, were also added to the loose rushes, which later came to be plaited or woven into mats.

A number of other water-loving plants, often simply described as 'rushes', are also used for seating and mat-making in various parts of the world. *Juncus acutus*, the traditional mat and basket rush of Italy, has long been used to make the soft, padded covering for Italian wine bottles and receptacles for olive pulp before pressing, and is still a raw material for chairs and matting. *J. maritimus*, known simply as the sea rush or sparto, is similarly woven into floor coverings, especially in Spain and Morocco, whilst the material known colloquially as Chinese seagrass (*Cyperus malaccensis*) has been widely made into the 'rush matting' sold as a floor covering in Britain. It is exported in large quantities from China, where it has long been used. In India, *C. pangorei* or Calcutta mat rush is an important raw material for mats.

The carpets and rugs that most of us possess depend on plants but in quite different ways. While many carpets are made of synthetic, oil-based materials, a large number are made from viscose or similar fibres, derived from wood pulp. These carpets also depend on plants in the form of gums and starches for sizing solutions (see page 70); potato and maize starch and sodium alginate from seaweeds being commonly used. A much more surprising source of carpet material, however, is now being provided by maize. A new plant-derived polymer, polylactic acid (PLA), derived from maize starch, is being made into recyclable and biodegradable carpets available in Europe and the USA.

Jute for floors and furniture

There may be plant materials beneath the carpet too. The backing or scrim, a loosely woven cloth stuck to the underside of many tufted and woven rugs and carpets, has traditionally been made of cotton or jute (see left). Jute in particular has been of very great use to flooring manufacturers for its exceptional strength and durability, helping to keep products stable and give them a better 'handle'.

Imports of jute into western Europe (now only in processed form) have greatly diminished in recent years as the fibre has been replaced by longer-lasting synthetic polypropylene. But many of the more traditional uses of jute cloth in the home – usually as hessian – continue in the form of linoleum manufacture, backing for carpets, upholstery fabrics, furniture webbing and bedspring coverings.

Jute is the fibre separated from the stems of *Corchorus* species. In the mid-1980s, Bangladesh derived half of all its foreign exchange earnings from jute – at that time its most important export crop. By 1991, however, the jute trade represented just under a quarter of export earnings and, by 2000, only 7 per cent. A number of factors contributed to this decline – chiefly a sharp drop in world demand due to competition from synthetics, made worse by procurement and processing problems and 'natural' disasters, such as the recent drought in major jute growing areas. Meanwhile, the domestic consumption of jute in producer countries has, conversely, been rising, to try to boost exports once more, India and Bangladesh have been making great efforts in recent years to diversify the range of products made from this fibre. These include packaging for food-grade products such as cocoa, coffee and shelled nuts, bags of many kinds, paper, a variety of blended yarns for clothing, blankets and floor coverings, footwear, jute-reinforced composite boards for construction use, and moulded jute 'plastic'.

Jute is cultivated by millions of very poor farmers, who grow the crop on small parcels of land (1–2 ha/2½–5 acres annually), in rotation with rice, to provide extra income. After cotton, more jute is consumed today than any other vegetable fibre.

191

Linoleum from plants

Whilst jute has been the traditional material for backing linoleum, the bulk of this tough yet pliant floor covering comprises several other substances from plants. Linoleum takes its name from the Latin *linum* for 'flax', and *oleum* for 'oil' – referring to linseed oil – since this was, and still is, the product's traditional base. The processing method and materials used have scarcely changed since it was first made in 1863. After thickening with oxygen, the linseed oil – which forms about a third of the mix – is combined either with ground cork or 'wood flour' (finer than sawdust, and chiefly obtained from softwood plantation trees), accounting for another third, and a filler: chalk or limestone. Pine rosin is added as a binder and sometimes 'tall oil', a by-product of pulping pine wood, as well as pigments, before pressing onto a jute backing.

After a decline in production, roughly from the 1960s to the 1980s, during which time plastic floor coverings – mostly of PVC – displaced it in many households, the movement away from PVC (on health and environmental grounds) has helped revive the fortunes of lino. Today's biggest users in the UK are hospitals. Its natural properties make lino not only very durable but very hygienic, as it is so easy to clean. Furthermore, a reaction between the oxygen in the air and the oil in the lino means that the surface molecules continue to harden after it has been laid and this prevents the growth of certain staphylococcus bacteria.

Ban on plastic bags boosts jute

In February 2002, Bangladesh banned the use of **all polythene bags** in the capital city of Dhaka, a measure that will, it is hoped, help the environment, the jute industry and millions of Bangladeshi people. Although only introduced in the early 1980s, around 10 million polythene bags have been used **each day** in Dhaka alone. The bags were found to have been a major cause of the blocking of the drainage systems during the 1988 and 1998 floods, and have also been clogging up farm land and reducing fertility. The production and recycling of the bags, meanwhile, has also caused **health problems**. A return to the use of jute rather than plastic for bags should bring many benefits.

192

Coconuts from doormats

A different tropical fibre – coconut – may well be the first thing that welcomes you as you cross the threshold of your home. Better known as coir, it has been the traditional fibre used to make doormats for many years. India is the chief producing country – the industry employing around 500,000 people, mostly women. Because over 50% is home-consumed in India, however, a larger amount is exported from Sri Lanka.

Much in demand for its natural resilience, durability and resistance to wetness, coir is the coarse fibre stripped from the outer husk and shell of the coconut fruit, which surrounds the nutritious white flesh of the inner seed. Two kinds of coir are distinguished:
- Pliable white fibres, extracted from immature, green coconut husks, which may be spun into yarn.
- Brown fibre (used for doormats) obtained from coconuts that are fully mature and whose white flesh is ready to be processed into copra and desiccated coconut. Classified as hard industrial fibres, coir is available in its raw state, as spun yarn or ready-made floor coverings.

Tree bark on the floor

The bark of an evergreen oak tree, *Quercus suber*, native to southern Europe, has provided us with a flooring material with very special qualities: cork. Composed of a honeycomb of tiny, water-tight cells held together by a strong resinous substance and filled with an air-like gas, the great value of cork as a flooring and panelling material lies in its absorption of sound and vibration. It is also light, has a unique elasticity and is resistant to the passage of liquids.

Cork has lined floors, walls and ceilings in southern Europe since Medieval times, affording protection from the intense summer heat, yet providing a surface always warm to the touch in winter. With a cushioning effect on feet, legs and back when walked upon, it is very hard-wearing and resistant to fire – a natural feature of the bark, designed to

Harvesting cork

It is not until the tree has reached 25 years old that the first harvest of cork is taken from the cork oak. In a **highly skilled** operation, this is done during the months of June, July and August, by making careful vertical and horizontal cuts into the tree's bark with a special axe or curved saw and then prizing off

large pieces. As long as neither the **inner layers** of bast material, which carry the sap up through the tree, nor the **cambium cells**, which produce new wood and regenerate new protective layers of cork bark, are harmed, this operation can be performed every nine years without permanently

injuring the tree. The bark renews itself **naturally**, with fresh layers growing each year.

Cork from the first two harvests, which is less smooth and even, is used to make products such as floor tiles and insulation. By the third harvest, it is of the right quality for wine stoppers.

protect the tree against forest fires as well as drought. Today, cork is being used in modern flooring in the form of tiles and as a constituent of linoleum. In tiles it may form the surface and other layers (interspersed with other panel materials), or provide a 'core' to wooden-panelled flooring. The world's biggest cork processing company – which takes some 68% of world production – currently manufactures in the region of 7 million sq m (75.3 million sq ft) of cork flooring a year.

Cork tree, Sardinia, Italy, showing area harvested of bark.

To make flooring and other cork products, slabs of bark cut from the tree are first seasoned to allow the cork to stabilize and develop a uniform moisture content. After this they are boiled in purified water to remove solids lodged in the pores of the bark and, in the most modern processes, any polluting chemicals. Boiling also dissolves the tannins and softens the cork, thus making it more pliable. The planks are now stacked and allowed to dry, before grading.

Cork for flooring is commonly made from the material left over after corks – which require the best-quality bark – have been punched out of the slabs, or from slabs that are considered too poor for corks in the first place. Where cork trimmings are used, these are compressed so that the natural resins bond the individual granules, leaving a firm but cushioned surface that is very durable. In combination with linseed oil, granulated cork is still used to manufacture some linoleum.

Today, over 50% of the cork harvested is used for the production of stoppers of various kinds, but much is used for thermal and acoustic insulation, pressed into boards and pipe coverings. In this form it stops the pipes from freezing up in countless cold-storage and meat processing plants, and in oil refineries. Cork is used also in buoys and fishing-net floats, gaskets, sports equipment (such as hockey and golf balls) and sound proofing in shotgun cartridges. When burnt, cork shavings produce the black pigment used by many artists.

The world's supply of cork comes to us courtesy of an ancient system of management that benefits the millions of us who use cork directly, and the thousands of people involved in its production. Its

193

production also benefits a great variety of wildlife – some of it rare or endangered – and the regional climate of a significant portion of the western Mediterranean.

Known as *dehesas* (in Spanish) or *montados* (in Portuguese), forested regions characterized by cork and holm oaks, interspersed with grassland, scrub vegetation, orchards and cultivated fields, extend across 3–4 million ha (7.5–9.8 million acres) in western Spain and eastern Portugal and other western Mediterranean countries. Created over many centuries by selectively felling the oak forest of the region, and now supporting one of world's few truly sustainable agricultural systems, including cork production and the grazing of bulls and wild pigs, the *dehesas* provide a habitat for a rich diversity of plants, mammals, insects and birds. Notable examples include the very rare Imperial eagle, the black stork, Egyptian mongoose, wild cats and the Iberian lynx – now rarer than most tiger species.

More cork in our kitchens

The first recorded use of cork as a **stopper** was by the Egyptians, thousands of years ago. The ancient Greeks, too, used cork to plug vessels of olive oil and wine. It was not until the 1600s, however, when the French Benedictine monk Dom Pérignon began using corks to stopper sparkling wine, that cork came to be considered essential for the bottling of wine in general.

The only known solid that can be compressed in one dimension without expanding in another, the **unique cell-like structure** of cork means that it can be condensed to about half its width without losing any flexibility. With its 'elastic memory', cork constantly tries to return to its original size and is still the only material that can **naturally accommodate** any faults in the necks of bottles, forming an air-tight seal. A multi-million-pound industry now revolves around the production of cork for bottles.

In recent years, concern about the problems of 'cork-taint' (a misnomer,

since it is not the cork itself which affects a small percentage of bottled wine, but the contaminant TCA, now affecting many food and drink products) led to a major move to substitute cork with **composite** (plastic) stoppers. This was further justified by predictions that despite a huge increase in production over the last 50 years, the ever-increasing demand for cork might outstrip supplies in the future.

Many feared that a move to plastic corks would lead harvesters to abandon the ancient **cork forests**, which might then be cut down in favour of much less appropriate land uses. One recent estimate was that the cork industry could crash within 15 years because of the rapid take-up of plastic stoppers.

Despite this trend, many wine producers are now **moving back** to cork, as problems associated with its usage are overcome. Recent major studies have confirmed that natural corks are

still **hard to beat** in terms of ease of extraction and re-stoppering, preservation of the wine's character, and low incidence of leakage and taint. The worldwide revenue generated from the production of nearly 13 billion wine stoppers each year is estimated to be £625 million. The world's **biggest cork manufacturer** claims that there is enough harvestable cork in Portugal's cork forests alone to meet market demand for over 100 years.

In the year 2000, around 250,000 tons of cork bark were produced in Portugal and Spain, involving some **80,000 people** in some stage of its production, harvesting or processing.

It takes nearly 50 years for a cork oak to produce its first wine cork, but just a few seconds to throw the cork away. Though it cannot be used again for wine, recycled cork has many uses: floor tiles, dart boards, hockey balls, gaskets and place mats being just a few ... so, **save and recycle** all your corks.

Needing large amounts of sunshine – summer temperatures can reach 45°C (113°F) in the shade – and with their primary growth period coinciding with the wet season, the cork oak forests flourish like desert oases whilst the treeless land around them is baked dry.

Most of the world's cork is currently produced in Portugal (which accounts for over 50%) and Spain. In Portugal, rules protecting the cork trees go back to the 13th century and it is illegal to fell a healthy cork oak. Most other producing nations, which include Italy, Algeria, Morocco, France and Tunisia, also have rules determining how often cork can be harvested. The Portuguese government recently helped preserve its cork oak forests by requiring a new cork oak to be planted for every old one cut down; the area under cork forest is now said to be growing by about 4% per year.

In various regions, however, these rules have been ignored and cork oak forests have been felled, largely to increase cereal production or grow eucalyptus trees for paper production. In Portugal, a large area containing cork oaks was recently flooded due to the construction of the Alentejo Dam, containing Europe's largest reservoir. In some regions, too, young people have been abandoning their rural way of life and migrating to cities.

The cork oak forests have also suffered a reduction due to climatic factors, such as more frequent and severe droughts, which they have played an important part in ameliorating. Protecting the ancient oak forests through our continuing use of cork should help offset the worst excesses of climate change now threatening the Iberian peninsula, and maintain a truly sustainable way of life.

Freshly harvested cork bark ready for processing.

195

Plants on the wall

Quick-drying plant oils such as linseed and tung, together with mineral pigments, thickeners such as China clay or talc, and other ingredients, including resin from various pine trees, have been traditional components of many cans of household paint. Thinning agents, such as turpentine, extracted from pine oleoresin and processed pine wood, have also been used extensively in paint manufacture and as cleaning substances for paint brushes.

The household paints most commonly available today fall into two categories: solvent and water based. A major component of modern solvent-based paints is generally an alkyd resin, derived from commercial vegetable oils, such as soya. Although turpentine was widely used in the past (and is still used in artists' paints), the solvent found in these household paints is white spirit – a cheaper petroleum derivative. Common water-based paints, described as acrylic or vinyl, are now formulated with synthetic resins, but some comprise water-dispersible alkyds, which, like the solvent-based paints, are also derived from vegetable oils.

Many paints also contain plant starch or cellulose in soluble form to give them stability and to make application easier. Cellulose gives an even consistency to the mixture, helping the pigment to disperse rapidly. Alginates from seaweeds are similarly employed, improving the paint's flow, suspending pigments and helping good coverage to be achieved. In some vinyl emulsions, gum arabic (*Acacia senegal*) controls the consistency of pigments and any settling that occurs.

Wallpaper

Most wallpaper, like ordinary writing paper, is made from a base of wood pulp. Some companies are using increasing quantities of recycled paper as a base for their wallpapers, but trees will has been the starting point in most cases. A great range of woods has been used over the years, with different sources of supply for different countries and the companies within them. The price of pulp from different regions may determine which tree species are ultimately involved, and as this can vary greatly and different pulps are mixed, it is not easy to say what a particular roll of wallpaper is made from. In Britain, softwoods such as spruce and pine species have a long history of use, as well as hard-woods such as birch, beech and aspen, often grown in northern European countries. North American forests producing a mixture of soft- and hardwood trees have also been a traditional source of pulp, though today's supplies could be made from eucalyptus trees grown in South Africa or Brazil or other fast-growing plantation species. (See page 82 for a description of pulp production.)

The wallpaper paste or glue that sticks each sheet to the wall comprises many plant-derived ingredients too. Most of these are thickeners and binding materials and include potato and maize starch and powdered plant cellulose.

Household accessories from plants

The average house is full of articles made from plant materials; examples include everything from wooden bread boards, cork place mats and rattan dog baskets to household brooms. A number of brooms and brushes on sale in hardware shops and stores in Britain are made directly from plants. In Britain, in the past, before brush making became a commercial industry, locally available materials such as twigs and small branches, grasses and reeds or shrubs like heather and broom would have been used in different regions for sweeping floors, but the main source of broom fibre was hog bristle. Today, despite the impact of synthetic substitutes, which are far more widely used than their natural counterparts, plant fibres – especially those from palms – remain an important raw material for brush and broom making. For certain applications – in particular those involving heat or friction – various natural fibres still cannot be matched.

Good broom fibres should have plenty of natural spring in them and, especially for heavy use, should be able to recover their normal position immediately after sweeping. Palm fibres are particularly well suited to the job. They are extracted from the portion of the palm leaf stalk that remains on the tree after the leaves have been removed or have fallen and are extremely resilient and strong. Many will withstand constant use and immersion in hot or corrosive liquids. They make excellent scrubbing and nail brushes as well as yard or outside brooms and are generally long-lasting. For more information on broom fibres, see the box to the right.

Traditional houses

Traditional architecture almost everywhere in the world relies on plant materials to some degree or other. Some of the simplest huts and shelters, most extensive family houses and elaborate ceremonial buildings have used timbers, branches, stems, leaves or fibres so skilfully and successfully to fulfil design requirements and to suit the needs of people using them that their form has stayed virtually unchanged for centuries.

However, as many people's capacity to support themselves has been undermined by the global economic system, there has been a massive shift in the world's population. By 1999, 45% of the population were living in urban areas, the majority in shanty towns or slums, making the best of anything to hand (much of it post-consumer waste). Moreover, the UN project that by 2030, 60% of the world's population will be living in urban areas.

Where traditional houses do remain, however, whether in the foothills of the Himalayas or the Australian 'bush', they incorporate local materials tried and tested over long periods. In areas of tropical forest especially, the range of building materials used by different peoples reflects the abundance and diversity of plant life around them – but two sets of forest plants in particular stand out: bamboo (tropical grasses) (see overleaf) and palm tree species. Between them, these plants have provided an extraordinary variety of timbers and fibres, poles and thatch to suit almost every need.

Broom fibres

The most common natural broom and brush fibres in the UK include:

Coir from Sri Lanka, for soft domestic brooms.

Palmyra fibre (*Borassus flabellifer*) from southern India, used in stiffer brooms and scrubbing brushes (normally in combination with a much smaller quantity of fibres made from PVC or polypropylene).

Tampico or Mexican fibre (*Agave lecheguilla*), used in high-quality bath and nail brushes.

Kitool fibre (*Caryota urens*), has remarkable elastic properties and was widely used until the mid-1970s. It was later eclipsed by synthetics, but various other brush and broom fibres are still available. These include:

Bahia bass (*Attalea funifera*), which is stiff and resilient and was once widely used for chimney sweeps' and outdoor brooms.

Arenga or gumati fibre (from *Arenga pinnata*), which has a heat resistance of up to 150°C (300°F), thus still unrivalled for the manufacture of rotary floor scrubbing and buffing brushes. Its other uses include roof construction and concrete reinforcement.

The dried flowering stems of cultivars of the cereal plant *Sorghum bicolor* (see page 121), which provides broom corn or Italian whisk.

197

Rainforest homes

A great variety of buildings and structures – all appropriate to the culture and needs of their builders – have graced the world's rainforests, displaying the skill, ingenuity and knowledge of local materials held by indigenous peoples. Whether small, temporary dwellings, single family homes or the huge communal houses of various tribal groups in the Amazon rainforest and in Indonesia, tragically, almost all now face some degree of threat, as the frenzied destruction of forests, largely fuelled by commercial imperatives determined thousands of miles away, continues unchecked.

In the northwest Amazon region, a number of Indian peoples, including the Tikuna, Barasana, Witoto, Yukuna and Yanomami, have all traditionally lived in large communal houses, generally referred to

198

Bamboo

A huge and complex group, whose members have not yet all been classified, around 1,000 species of bamboo, belonging to 80 genera, are thought to exist. Members of the **grass family**, they are native to tropical, subtropical and some warm temperate regions of the world, but are especially diverse in South and Southeast Asia.

Some bamboos grow taller than many **trees** and have stems that are stronger than a number of timbers. *Dendrocalamus asper*, for example, can grow to 30 m (98 ft) and has extremely strong and durable stems. Bamboos grow exceptionally fast (the shoots of some, such as *Bambusa tulda*, increasing by around 3 cm/1¼ in an hour) and many species reach their full height in only **6–8 weeks**. In the wild the larger species form dense, often impenetrable, forests, while the smaller species, equally dense, remain as undergrowth.

Bamboo is probably used by more than **half** the world's population every day. The plants can serve almost the entire needs of a commu-

nity and some 1,500 different traditional uses have been documented. The **stronger stems** (or culms) are used for house construction, scaffolding and bridge-building as well as for water pipes and storage vessels, the **canes** for furniture, the **shoots** of many species are eaten as a vegetable and the leaves used for thatch. Other uses include musical instruments, containers of various kinds, cooking pots, rafts, fish traps, mats, hats and bags. **Fibres** peeled from the split culms – themselves very important basket materials – are used to make rope and twine (useful for fastening panels and poles into place), and these fibres are three times stronger than hemp.

As a **building material** bamboo plays an immensely important role in almost every country in which it flourishes. In Bangladesh, for example, 73% of the population live in houses made almost entirely from bamboo, the sturdy stems providing walls, pillars, window frames, rafters, ceilings and roofs; whilst in Java woven bamboo mats and screens commonly

fill timber house frames. Elsewhere in Asia rural and forest dwelling peoples make use of the plant in one form or another – often for the **supporting structures** of their homes, for flooring made from the split and flattened culms, or for making separating walls and screens. The square culms produced by placing a wooden mould over the shoots as they develop are often used as decorative **alcove posts** for the traditional raised recess, the tokonoma, built into Japanese houses.

The utility of bamboo culms extends further, to the manufacture of high quality **veneers**. Laminated furniture and flooring made from bamboo that has been peeled and processed, then stuck together to make plywood sheets or onto other boards, for example, is now very popular in China and Japan.

As an alternative to our fast depleting wood supplies, and to man-made materials, bamboo offers endless possibilities: in Australia, even **surfboards** are being made of bamboo.

as *malocas*, immense buildings that serve a number of functions that go beyond our own conception of a family house. Within a given region, different materials and variations in building techniques, such as the way in which roofing leaves are woven into thatch, are a reflection of cultural identity, but the many differences in style are variations on a common theme. The *malocas* of the Yukuna people, for example, are round; those of the Witoto oval, and the Tikuna's rectangular, but all are built around four pillars, which support a steep thatched roof.

Different kinds of *yano*, the communal house of the Yanomami, one of the most numerous of South America's forest peoples, whose lands straddle the Brazilian/Venezuelan border, are distinguished by their roughly circular shape, surrounding a large open central area. Sometimes up to 40 m (131 ft) across, a *yano* will house a whole village of between 25 and 400 people. While individual families have their own hearth and area where hammocks are slung beneath the slanting roof, the centre of the *yano* is a large space open to the sky, where dancing and ceremonies take place. The building materials, gathered and erected as a joint activity by the men, consist of strong palm and other timbers, vines for tying, palm leaf thatch for roofing and often flattened sheets of bark for making walls.

A rubber tapper's (individual) family house in Acre, Brazil, built using local rainforest materials: wall screens are made from split palm trunks, and the roof thatch from palm leaves. Stilts raise the house away from animals and wet ground during periods of rain.

199

Several thousand kilometres to the southeast, the peoples of the Upper Xingu river also live in vast houses, but of a rectangular shape, with roof and walls that curve continuously like an old-fashioned aircraft hangar. Majestic and very beautiful in construction, a typical Mehinaku house will be some 30–40 m (98–131 ft) long, 16 m (52 ft) wide and 9 m (30 ft) high. Its framework, traditionally made of very flexible pindaíba wood (*Duguetia lanceolata*), forms a graceful oval dome, which is thatched with bunches of dried satintail grass (*Imperata brasiliensis*). Each house is an architectural masterpiece, with a perfect ventilation system created by the overlapping grasses of the roof; the interior is cool in summer yet warm during the cold winter nights. About 30 people may live in each of these Mehinaku buildings, which are generally separated from each other by distances of several kilometres.

For each person the communal house is the centre of his or her world. It serves as a family home and kitchen (each family will have its own private space and fire), dining room, meeting place, workshop and storehouse, dance hall and spiritual focus. The house and each feature of its design represent both the structure and divisions of Mehinaku society and a model of the cosmos and spirit world, which links the people intimately to their forest home.

Inside the Mehinaku house, hammocks used for sleeping are made from fibres extracted from the leaves of the buriti palm (*Mauritia flexuosa*). The silky fibres make a tough cord used not only for weaving hammocks but for many other household purposes. Only the most tender buriti leaves are chosen by the women for hammock making. They are gathered by canoe from the swamp lands and river banks on which they flourish and shredded to make the cord. A strong but soft

Kalimantan long houses

In Kalimantan, the Indonesian part of Borneo, **Dayak peoples** (a name given to many different groups of people who have lived on the island since the earliest times) have traditionally lived in long houses.

Described as the 'ultimate in communal living', these houses – raised on stilts and with pitched roofs – have accommodated up to **100 families**. With a wide veranda running the length of the building at the front,

for social and communal activities, each family has its own front door and private area inside, comprising sleeping accommodation, a storeroom and a kitchen. Long houses were often built almost entirely of **bamboo**, using the stems for structural support, walls and flooring.

Though still used in Sarawak and elsewhere, unsympathetic administrations in Kalimantan have worked to break up communal houses,

regarding them as a barrier to 'progress'. The **logging** of the Dayak's forests and the ill-conceived transmigration scheme imposed by the Indonesian government, which, since 1969, has been transferring families from the overcrowded islands of Java, Bali and Madura to the traditional lands of indigenous peoples, has also forced many to **abandon** their traditional way of life and live in poverty.

and comfortable hammock is a prized possession amongst the Mehinaku. On the island of Sulawesi in Indonesia, the Toraja people have also developed a spectacular and beautiful style of architecture. Tall, narrow, family houses, known as *tongkonan*, with extraordinary curved roofs, stand on sturdy wooden piles cut from tree trunks. Carved and painted wall panels and doors are made of planking cut traditionally from wood that is both rot-resistant and finely grained, whilst roofs are composed of split bamboo tiles, which are threaded together using strips of palm fibre and tied in place. A great number of layers (around 60 in some houses), the upper overhanging the lower in succession, form this very dramatic and distinctive structure. The upper curve of the roof is shaped to resemble the outline of a boat, in reference to the traditional importance of this means of transport. The Toraja believe that their ancestors came from southern China or Cambodia by boat and their houses are also sited facing north to mark the event.

A combination of factors have contributed to the erosion of the traditional Torajan way of life and the modern *tongkonan* now has a more symbolic than functional significance. New technology has also allowed the houses to become taller and less practical, so that they tend now to be reserved for use on special occasions.

Long house, on the island of Siberut, Indonesia. An entire Mentawi family group live in this house, constructed from local plant materials. The relative isolation of the Mentawi Islands, off the coast of Sumatra, has helped ensure the survival of their traditional houses.

Reeds for ancient architecture

Reeds, it seems, may have been the material used to make the earliest **pointed arches** in the world. In Iraq and elsewhere in the Middle East where wood is scarce, reeds have been used as supports for various mud constructions, and notably for arches. Long after the reeds have **rotted away**, the arch, made of layers of baked mud, still stands.

In Britain, reeds of a different species are more familiar to us as the thatching material that has been used on many traditional buildings, particularly in eastern and southern parts of England, now enjoying a revival in popularity.

In **drier regions**, a well-made thatched roof will last for some 80 years, and can, of course, be continually renewed.

Disappearing homes

The unprecedented destruction of the world's tropical rainforest in the last decade, particularly, has caused suffering and hardship to millions of people. Amongst the casualties have been the Baka pygmies of Cameroon.

Until relatively recently, many of Cameroon's Baka pygmies were able to live their traditional, semi-nomadic lives in the rainforest. Although typically also involved in complex and unequal relationships with Bantu cultivators for whom they have generally been obliged to work, the Baka, with their sophisticated understanding of its rhythms and resources, have traditionally spent several months each year, during the rains, travelling deep in the forest, hunting and gathering seasonal fruits.

In order to do this they use simple but effective temporary homes that can be easily constructed on the move. Dome-shaped huts (*mongulu*) and short-term, three-sided shelters (*lobembe*) are made from an interwoven frame of springy saplings. Various species are used including *Diospyros canaliculata*, which are then covered with the giant leaves of *Megaphrynium macrostachyum*, the stalks of which are hooked onto the framework by bending them double. The leaves lap over one another like shingles, working from the bottom upwards, thus making the hut waterproof. Piles of branches are then usually placed on top to hold the leaves down in winds or heavy downpours. The doorway of the hut usually faces those of friends or close relatives but can be repositioned as relationships between the members of the group change.

After a month or two, when local game is depleted and other forest foods such as nuts, fruit and honey have become scarce, the pygmy camp moves on. Important social events such as births, deaths, marriages or the start of puberty also prompt a change of location. Like other rainforest peoples, the Baka are expert plant-users. They generally make their beds, for example, from the bark of two trees, *Triplochiton scleroxylon* and *Annickia chlorantha*. After being prized from the tree, the bark is pounded flat and left to dry to a hard, rigid board. This is then raised off the ground on stakes and covered with leafy bedding. To illuminate their evenings, the Baka make resin torches, moulding the sap of several tree species (such as *Canarium schwein-furthii*) into shape before it sets hard.

Today, the Baka's way of life has been devastated. In the mid-1980s, a fall in the price of cocoa and coffee induced a severe economic downturn for the country and logging remained the easiest option for earning foreign currency. Since then it is estimated that Cameroon has lost 90% of its frontier forest and most of the rest is now threatened by European and Asian logging companies. For most of Cameroon's Baka people, their ancient way of life is almost at an end. Although trips into the forest are still made, these are ever more restricted, and many have been forced to settle at the edge of the roads along which the logs from their forest are transported.

A desert home

Another tragedy – shocking because of the malicious intent behind it – has befallen a quite separate people, who made their homes amongst marsh and shallow swampland in southern Iraq. For centuries, the Marsh Arabs built their houses – entire villages, in fact – from the huge reeds that dominated the landscape they occupied. Sturdy and often very beautiful, the framework, walls, roofs and flooring of these houses were made entirely from these perennial plants with their tall, flat, tapering leaf blades, chiefly *Phragmites australis* and *Arundo donax*. Each year large areas were burned to encourage new growth, and the young reeds that sprang up provided pasture for the Arabs' water buffaloes and other livestock. The tall stems, which may grow to 3 m (10 ft) in height, were bound together in bundles after cutting and bent into arched pillars to provide the framework of houses, storage buildings and animal pens, whilst walls and roofs were made from reed matting.

The area provided some of the main battlegrounds for the war between Iran and Iraq in the 1980s, and the Marsh Arabs (predominantly Shi'ite Muslims) were then driven from their homes in the early 1990s as Saddam Hussein unleashed a plan of genocide against them. Suspecting them of being opposed to his regime, he imposed a deliberate strategy of damming both the Tigris and Euphrates rivers and thereby draining the marshes, in order to destroy their self-sufficient way of life and their identity. The result has been acute suffering, malnutrition, a health crisis and the mass exodus of people from the area. Around 100,000 Marsh Arabs have fled to refugee camps in Iran, while possibly 400,000 are displaced elsewhere in Iraq. The land that the Marsh Arabs had occupied for so long – some 15,000–20,000 sq km (5,800–7,700 sq miles) – has almost completely dried up and regressed into desert; an entire ecosystem and way of life destroyed.

Plants that **cure** us

When we reach inside the medicine box we are often reaching indirectly for a plant. For numerous minor ailments as well as many of our more serious complaints that need hospital care, plants have given us effective modern treatments.

The sophisticated, brand-named medicines we use to help us combat everything from hayfever to heart disease, however, tend not to advertise any connection, where one exists, with plants. But one in four modern prescription drugs has been estimated to contain at least one compound currently or once derived from or patterned after active substances found in plants. In 1997, it was estimated that 57% of America's top 150 proprietary drugs contained at least one major active compound derived from or modelled after natural compounds – with a global retail value of some $120 billion.

The global value of plant-derived medicines is very difficult to establish, but a figure of $30 billion has been estimated for 2002. Just one drug, Taxol, an anti-cancer agent derived originally from the Pacific yew, was worth $941 million in 1997.

Plants are living chemical factories, and it is the complex substances they produce that have proved so useful in curing, or keeping our illnesses at bay. The vast range of botanical chemicals that have evolved over millions of years is used by plants in many different ways – to protect them from disease, for example, or stop them from being eaten (by everything from insects to elephants), to help regulate growth, to act as storage compounds in roots or seeds or, in some plants, to counter solar radiation.

The array of conventional medicines derived from plant chemicals is impressive, and includes heart drugs, analgesics, anaesthetics, antibiotics, anti-cancer and anti-parasitic compounds, anti-inflammatory drugs, oral contraceptives, hormones, ulcer treatments, laxatives and diuretics. But a surprisingly small number of plants have yielded the plant-based prescription drugs currently licensed for use worldwide.

Very few of the world's plant species (a figure of 5% is generally estimated) have had the full spectrum of their pharmaceutical potential tested in laboratories. But the technological advances of recent years mean that thousands of plant chemicals can now be screened for activity against particular targets, very rapidly. It is recognized, however, that such screens almost certainly miss many important modes of action. Those plants that have historical or contemporary usage in the treatment of illness have tended to be those most closely investigated – species such as the sea pink (*Armeria maritima*) and wood anemone (*Anemone nemorosa*), once used in treatments for tuberculosis and leprosy, have recently been investigated in the UK). But it is now recognized that any plant is of potential interest. It may contain compounds never dreamt of by pharmacologists, which could eventually initiate the formulation of a powerful new drug.

Previous page: Opium poppy.
Opposite: The areca nut and betel catkin have stimulant and other properties.

Using modern genetic fingerprinting and chemical isolation techniques, many familiar plants are now under scrutiny. A large number of previously unknown biologically active alkaloids and unusual glycosides have been discovered in the bluebell (*Hyacinthoides non-scripta*), for example, while the bulbs of daffodil cultivars (*Narcissus pseudonarcissus*) are now the source of the compound galantamine, which is being used to help treat Alzheimer's disease. While some believe that micro-organisms are the most promising organisms for drug discovery today, plants offer an almost limitless array of novel chemicals.

Plants for medicines

Plant compounds are used in Western medicine by incorporating them directly in products or by using them as blueprints or starting points for the manufacture of new or synthetic drugs. Many biopharmaceuticals are now made by recombinant DNA techniques, followed by production using fermentation. It has been estimated that plants are the source of over 30 drugs of proven value in worldwide use and that over half of today's 25 best-selling pharmaceutical drugs either directly incorporate the chemical compounds of plants (grown for this purpose as crops) or are the result of the synthesis or manipulation of a plant chemical. Globally, sales of all plant-derived medicines are worth an estimated $30 billion. But Western (pharmaceutical) medicine treats only a small proportion of all the people on earth. According to the World Health Organisation (WHO), up to 80% of the world's population relies entirely or in part on locally produced medicines, almost all of them produced from plants. The number of species that have been used medicinally by different peoples is unknown, but as Western drugs continue to be unaffordable or unavailable for use by one-third of the world's people (a figure rising to over half of all people in the poorest parts of Africa and Asia), safe and effective traditional therapies are critical to the provision of basic health care.

Traditional plant-based medicine is officially accepted in a number of countries today, including China, Thailand and Vietnam. In China alone – with its population of 1.3 billion – over 10,000 plant species are used in traditional and folk medicine, of which it is estimated that up to 500 species are commonly employed in contemporary practise.

Ayurvedic and unani medicine, practised on the Indian subcontinent, and which, like many other belief systems, treats disease as a state of disharmony in the body, also uses plants extensively to treat both physical and spiritual ill-health. According to the WHO, in the year 2000, 25 countries reported having a national traditional medicine policy for primary health care, and in countries as far apart as Ethiopia and Ecuador, the use of medicinal herbs is encouraged by local health care officials. Though such knowledge is increasingly threatened by

Poison or panacea

Grazing mammals have been eating plants for 180 million years. It is among the plant chemicals that are poisonous to mammals that we find many **useful medicinal compounds**. Part of the $570 million or so that it costs to turn a newly discovered plant compound into a Western drug is spent on **safety testing** – the difference between miracle cure and damaging poison often being simply a matter of dose.

Daffodils are helping sufferers of Alzheimer's disease.

the displacement and social change caused by economic or other pressures, ordinary people everywhere in rural areas are likely to inherit medicinal plant lore, which helps to keep them well.

In industrialized countries, a strong and increasing interest has recently emerged in complementary and 'alternative' medicine, in parallel with conventional (or 'allopathic') medicine, particularly for treating and managing chronic disease. In the UK, consumers are now spending £100 million a year on alternative medicines – including herbal remedies – to treat everything from depression to memory loss. In Germany, Europe's biggest consumer of plant-derived medicines, up to half of all medicines in use are based on plants and their constituents.

The global market for herbal products has grown rapidly in the last decade and in the USA (where one-third of all adults are reported to use them) annual retail sales of herbal medicines were estimated to be nearly

Bluebells may be a source of future drugs.

$4 billion in 1998, while in the UK the figure was £50 million. (In the UK, estimates of the value of herbal medicine use are difficult to gauge, as many products are considered to be food supplements.) In Japan, where nearly three-quarters of 'Western-style doctors also use traditional medicine, the herbal medicine market was worth $2.4 billion in 2000. The world trade in raw materials for botanical medicines, vitamins and minerals was estimated at some $8 billion in 1997.

Many conventional medicines use raw materials – plant or otherwise – as the source of single, active compounds to effect the treatment or cure. However, herbal medicines (as used by medical herbalists) generally take the form of extracts or material in which a much greater range, if not all of the plant's natural compounds are still present – dispensed as much for preventative purposes as to effect a cure.

For most of our history and all around the world, it seems that plants have been used in this way (often alongside varied supernatural or spiritual beliefs) to cure and prevent illness. Many of Europe's oldest botanic gardens, such as that in Pisa (opened in 1543) and at Oxford University (founded in 1621), were 'physic gardens', created as a teaching aid for the medical students of the day. Now the wheel has almost come full circle, as drug companies consult historic herbals to find leads for the development of new pharmaceuticals, and ever more people become interested in the healing power of plants.

Natural cures

On the shores of Lake Tanganyika both chimps and local people use **Aspilia species**, members of the daisy family, to get rid of intestinal parasites. South American Capuchin monkeys, meanwhile, rub their bodies with **pepper** (*Piper* sp.) leaves to help prevent skin infections and repel insects.

THE REAL PRICE of wild plant medicines

When plants used by traditional peoples are 'discovered' for wider use in herbal or conventional medicine, the consequences can be disastrous for natural plant populations due to unsustainable collection.

At least 2,000 medicinal and aromatic plants are used commercially in Europe (which accounts for about one-quarter of world imports), and about 90% of these are still collected from the wild. While many of these plants supply herbal medicines, a study conducted in Germany showed that the conventional drug trade derived 70% of its plants from the wild, and only 30% from cultivated sources.

As a result, many species are now endangered. Examples include arnica, orchid and liquorice species, cowslip, yellow gentian, oregano and thyme. In many cases, claims by those who market them that wild-harvested plants are more effective as remedies have amplified the problem. In addition, species whose roots or rhizomes are harvested, such as American ginseng (*Panax quinquifolius*) and Devil's claw (*Harpagophytum procumbens*), are particularly vulnerable.

Many trees used medicinally also face increased threats, as their active ingredients are often harvested from their bark – the removal of which can be fatal to the tree.

The endangered *Prunus africana*, for example, is a slow-growing member of the cherry family whose bark contains a remedy for prostate disorders. Soaring demand has created a market now worth £137 million a year. In Kenya, where one tree can be worth £125 (about a year's income), trees are often felled for their bark. At the current rate of harvesting, the tree is predicted to be extinct there in 5–10 years. A research programme by the International Centre for Research in Agroforestry is developing faster-growing varieties to plant in orchards, where regular rotation harvesting should enable growers to take some bark without destroying the whole tree, whilst providing valuable income for poor farmers. In some parts of Europe, such as Albania and Hungary, the collection of wild plants still provides important income for poor families. According to WHO, the International Union for the Conservation of Nature (IUCN) and WWF the best way to satisfy the medicinal plant market, address conservation and aid local economies is by cultivating medicinal plants in the regions where they grow naturally. It is recognized, however, that this is not always feasible or economic and there is an enormous lack of knowledge about cultivation techniques. The next best solution is to have realistic controls to ensure that harvesting is sustainable.

Cuts, bruises and skin conditions

Our skin is one of the body's first lines of defence against disease. Left untreated, scratches and cuts can lead to major infections, particularly in the tropics or where general levels of nutrition and health are poor. A plant that has been of major importance in the past for dressing wounds, now being investigated for modern use, is sphagnum moss (see right). A variety of other plants are more likely to come to our rescue, however, including cotton and several different species of pine tree, as well as seaweeds.

The fluffy white mass of cotton wool is comprised of the long, silky hairs that cover ripe cotton seeds. Naturally absorbent, when clumped together as cotton wool they provide a huge surface area within a small volume of material, ideal, once sterilized, for staunching minor wounds. The high price of cotton has led to the development of substitutes, but plants are still involved as much 'cotton wool' is now made from viscose, derived from wood fibres. Viscose that has been charred at very high temperatures to form a charcoal is also the base material for activated charcoal cloth, which is very effective at absorbing bacterial toxins and odours from dirty or infected wounds. Cotton fibres also help us in the form of bandages and 'lint' dressings (cloth that has a raised nap, originally made from linen).

Most fabric-backed sticking plasters are an amalgamation of materials traditionally made almost entirely from plants. The fabric is often cotton, while the base for the adhesive may be latex from the rubber tree (*Hevea brasiliensis*) (see page 240), which provides elasticity, with the addition of gum rosin, a 'tackifier', obtained from the oleo-resin (comprising rosin and turpentine) produced by coniferous trees.

Much of the gum rosin used today comes from China, Indonesia and Brazil and it is extracted from a wide range of conifers, though Scots pine (*Pinus sylvestris*) is said to produce the best yields and the highest quality resin. China alone produces about 400,000 tons a year, accounting for two-thirds of world production. As both gum rosin and latex can cause allergic reactions in people with sensitive skins, many manufacturers now produce plasters with synthetic acrylic adhesives.

Seaweeds, meanwhile, supply the calcium alginates used in some dressings applied to wounds that are very difficult to heal. Gauze-like in appearance, these highly absorbent dressings minimize bacterial contamination while promoting coagulation of the blood flowing from a wound. Any alginate fibres trapped in the wound are absorbed into the body as it heals, dissolving slowly. The seaweeds used for the production of alginates (also used by dentists as the principal components of most dental impression pastes) are brown algae (*Phaeophyta*), and

Sphagnum moss

■ Able to absorb up to 20 times its own weight in water (and more than twice as much moisture as cotton), sphagnum moss (*Sphagnum* spp.) helped staunch the bleeding of injured soldiers during both World Wars. Long famous for its antibiotic activity and wound-healing capacity, analyses of the moss have revealed the presence of several associated micro-organisms, including *Penicillium fungi*, which seem to be responsible.

■ Bronze Age people in Britain made use of the extraordinary properties of sphagnum moss for healing wounds, and the plants have a variety of traditional uses – including nappy padding – among the Lapps, some native North American and Inuit peoples.

211

Opposite: Herb bunch from medicinal plant garden, Cuenca, Ecuador.

Ephedra

■ Ma Huang, as ephedra is known in China, has been used in medicine there for the last 5,000 years. It is still used to treat coughs, bronchitis and asthma, as well as hayfever and other allergic conditions.

■ The introduction of ephedrine to Western medicine is relatively recent, but it is now a very important multi-purpose drug. Concerns over side-effects, including heart conditions and strokes, means that *Ephedra* and ephedrine should only be used under medical supervision.

■ Over-collection of the ephedra plants from wild populations has begun to threaten the most popular species but the drug is now mainly produced synthetically.

large quantities are gathered off the coasts of North America, Britain, China, Japan and Norway.

Brown seaweeds have also been a major source of the iodine widely used in the past as an antiseptic. Iodine tincture for general medical use has not been derived from seaweed ash for some 50 years, but there is still a market for seaweed-derived iodine products (from various species of kelp), marketed as nutraceuticals (see opposite) or herbal remedies.

A more familiar household antiseptic is wych hazel (*Hamamelis virginiana*). When distilled with alcohol the leaves and bark yield an astringent extract, which helps prevent inflammation and controls bleeding. Wych hazel tincture or decoction is also widely found in eyewashes as well as gargles for mouth ulcers, gum problems and sore throats.

Tea tree oil (from *Melaleuca alternifolia*) has become famous for its powerful bactericidal properties, whilst arnica (from *Arnica* spp.) is well known for its effectiveness in treating bruises.

The distinctive smell of some familiar household antiseptics derives mainly from the inclusion of terpineol or pine oil. Though now considered to have very little antiseptic effect, these oils contribute antibacterial properties, whilst acting as solvents and coupling agents for other ingredients.

Eczema

The properties of many plants – including chickweed (*Stellaria media*) and herbs used in tradtional Chinese medicine – have been experimented with over the years in the attempt to cure or at least control the symptoms of eczema. Steroidal creams developed from compounds first discovered in yams have been a standard treatment in conventional medicine. A plant product that has attracted much attention for its reported benefits in various other conditions – including rheumatoid arthritis, various cancers, heart, renal and liver disease, multiple sclerosis, pre-menstrual syndrome and hyperactivity in children – and now licensed for the treatment of atopic eczema is evening primrose oil. Extracted from various *Oenothera* species, especially *O. biennis*, the action of evening primrose oil is attributed to its high linoleic and gamma linolenic acid content.

Treating allergies

As all hayfever sufferers know, plants can make us ill as well as healthy. Tiny pollen grains that irritate the membranes of nasal passage-way, provoke a reaction in many people, which can cause weeks of misery. Plants, however, can also bring relief. While eyebright (*Euphrasia officinalis*) is sometimes used in homoeopathic and herbal remedies, an extract from the roots of butterbur (*Petasites hybridus*), taken only when needed, rather than in anticipation, is also said to clear nasal congestion effectively, it is also reported to be useful for migraine sufferers, although extracts must be carefully prepared to be free of harmful alkaloids.

Many allergic reactions are treated through the sympathetic (involuntary) nervous system and one of the most widespread anti-allergic medicines is a plant compound that is chemically similar to the human hormone adrenalin, which plays a key role in the transmission of nerve impulses. Sufferers of various allergies including hayfever and urticaria (nettle rash), as well as asthmatics, whose condition often occurs as an allergic reaction, may well have been treated with ephedrine, extracted from the green twigs of various species of *Ephedra* shrubs, native to China and Japan (see opposite). As a sympathetic nerve stimulant, ephedrine has valuable anti-spasmodic properties. It acts promptly to relieve swellings of the mucous membranes and clear both bronchial and nasal passages and has thus become a common ingredient of medicines used to treat allergic conditions.

Evening primrose oil is a rich source of GLA for eczema and PMS.

Healthy eating

Nutritionists are helping people in Western societies to **rediscover** what many other cultures around the world have always known: the line between plants that form part of a **healthy diet** and that stop us becoming ill and medicinal plants used to treat specific disorders, is a very fine one.

One recent study involved young offenders in UK jails. Inmates tend to **avoid fruit and salads** on the prison menu so the study looked at the effect of supplementing their diets with minerals, vitamins and fatty acids. Those receiving the supplements behaved better and committed fewer offences.

Nutrition experts are increasingly talking about **nutraceuticals**. This term refers to foods or elements of foods, the main benefit of which is prevention (and sometimes the cure)

of illness, rather than nutrition. Perhaps the best known is **vitamin C**, which is essential in forming collagen fibres in tendons, skin and blood vessels. It is present in most fresh fruit and vegetables but the lack of it leads to the disease **scurvy**, a link that once led naval surgeons to order sailors to drink an ounce of lemon or lime juice every six weeks at sea.

Folic acid is a natural growth regulator especially important to pregnant women as it prevents spina bifida – spinach is a rich source. Fruit and vegetables are also good sources of **essential fatty acids**, which are used to create many compounds in our bodies. Gamma linolenic acid (GLA), for example, is usually created from precursors in the fruit and vegetables in our diet by a particular enzyme. In some people

this enzyme doesn't work properly, however (drinking, stress, eczema and pre-menstrual syndrome can all affect its function), leading to various disorders. **Evening primrose oil** is a rich source of GLA and is now licensed as a pharmaceutical for the treatment of eczema and PMS.

Modern technology is being used to help us get the most out of some plant nutraceuticals. Sulphoraphane, in broccoli, for example, stimulates enzymes that **detoxify chemical carcinogens**, but modern broccoli has been bred to contain little of it because it tastes bitter. Plant breeders are looking to create new strains of broccoli that contain analogues of sulphoraphane that are less bitter tasting.

In the USA, sales of nutraceutical products are projected to reach $8.6 billion by 2006.

213

The sap of parsnips, rue and giant hogweed can cause irritating rashes or even blisters if in contact with the skin on a sunny day. But the same chemicals that cause this reaction, known as psoralens (first discovered in *Cullen corylifolium*, which make skin more sensitive to ultraviolet light), are used in curing psoriasis and vitiligo, a condition in which patches of skin lose their melanin pigmentation and turn white. Today, bullwort (*Ammi majus*), which belongs to the carrot family, is the source of methoxsalen, the photosensitizer used to treat these conditions, together with ultraviolet light.

Plants that help kill pain

Those who suffer from migraine may well be familiar with feverfew (see left). This strongly aromatic member of the daisy family, which grows wild in many parts of Europe, has a long history of use for the treatment of fevers and general aches and pains.

But aspirin, the most frequently consumed drug in history, owes its existence to willow trees. Though aspirins are made today from synthetic compounds, it was substances first extracted from the leaves and bark of the white willow (*Salix alba*) and later from the perennial herb meadowsweet (*Filipendula ulmaria*), which gave chemists the blueprint for this famous headache cure.

The willow tree has a very long history as a source of pain-killing compounds. By the time of Dioscorides (1st century) it was also being used to treat gout, and over the following centuries many uses were ascribed to the tree, including the alleviation of rheumatic and labour pains, toothache, earache and headaches.

During the 18th century, willow was widely used as a substitute for cinchona bark (for curing fevers, see page 218), but it was not until 1828 that salicin, the active pain-killing ingredient, was isolated and named. Shortly after this both salicin as well as salycilic acid (also found in many other plants) were extracted from meadowsweet. While salicylic acid was found to be helpful in cases of rheumatic fever and for arthritic conditions, it was in the search for an alternative to sodium salicylate (which had come to be considered a superior treatment, but which was unpleasant to take and caused gastric irritation) that acetyl salicylic acid was produced in Germany in 1897. Named aspirin, by taking the 'A' from 'acetyl' and 'spirin' from the Latin name for meadowsweet in use at the time (*Spiraea ulmaria*), it proved to be much more effective, free from unpleasant side-effects, and able to give quick relief for all kinds of pain.

Much more recently, aspirin has come to be regarded as a life-saving drug, thanks to its blood-thinning, anti-platelet effects, helping

Feverfew

■ Feverfew (*Tanacetum parthenium*) was recorded by John Parkinson in his 17th-century herbal as 'very effectual for all paines in the head'. Modern research, including clinical studies into the effects of the leaves when taken by migraine sufferers, has suggested that feverfew may be very effective for the prevention and treatment of migraine. Further research is considered necessary, however, to establish the full activity of the plant and it is advised that it should be taken under medical supervision.

■ The name feverfew is said to derive from 'febrifuge' – in reference to its tonic and fever-dispelling properties. It has also been used for arthritis, stomach ache and toothache, amongst other things.

prevent strokes and control angina. It has been estimated that if those at risk from heart attack took an aspirin at the onset of the pain, the risk of death in the first 48 hours would be greatly reduced.

Some long-haul airline passengers now take aspirin as a precaution against deep-vein thrombosis: blood clots in the legs caused by long hours of inactivity.

Killing pain with poppies

As an analgesic, aspirin brings relief to millions of people every day, but the greatest natural painkiller of all is produced by the opium poppy (*Papaver somniferum*).

Morphine is just one of some 25 alkaloids present in opium, the milky latex exuded by the seed heads of the poppy when cut. It is also extracted from the stems and leaves of the plant.

Able to relieve pain and resulting sleeplessness, and calm anxiety, morphine takes its name from Morpheus, the Greek God of Dreams. The Greeks regarded sleep as the greatest healer and the poppy was celebrated in their art, literature and religion. Neolithic peoples in the western Mediterranean, however, were the first to cultivate the opium poppy, from 6,000 years ago. Able to induce a state of euphoria, it became widely regarded as a general panacea – and its medicinal use spread via numerous trade routes around the Mediterranean and beyond.

Today, medical practitioners prescribe the powerful analgesic properties of morphine to relieve severe pain. Cancer sufferers, people with heart or lung failure and those suffering traumatic shock are all likely to be given the drug, and it has been used routinely for pre-operative analgesia and relaxation.

Papaverine, another of the opium poppy's many powerful alkaloids, is used to treat intestinal colic, while the drug verapamil, structurally related to papaverine, is used to treat high blood pressure, angina and irregular heartbeat.

Codeine, meanwhile, also an alkaloid (although only one-fifth as strong as morphine), is familiar as a 'household' painkiller, especially as a cough suppressant, in cold remedies and for the relief of all kinds of bodily aches and pains.

Some 400 tons of opiate raw materials used for medical and scientific purposes were produced in 1999 and since morphine still cannot be synthesized chemically, opium poppies are the only source. Traditionally, opium latex has been collected by scraping it from the surface of the unripe seed pods in the form of a brown gum. About

Cannabis

■ *Cannabis sativa* has a long history of medicinal use, and had some very well-known beneficiaries, including Britain's Queen Victoria. In more recent times, its uses have included the reduction of pressure in the eyeball in glaucoma patients, but its anti-vomiting properties have proven to be especially useful and have been found to be helpful for some patients receiving chemotherapy for cancer as well as for treating AIDS. Cannabis has recently been trialled in the UK (in the form of an under-the-tongue spray that delivers a controlled amount of its active ingredient, tetrahydrocannabinol – THC) for the relief of cancer pain.

■ It is also under trial for the relief of severe pain caused by nerve damage and spinal cord injury, as well as multiple sclerosis. While a synthetic copy of THC was licensed for use in 1982, modern trials may lead to the wide adoption of cannabis-based prescription drugs.

18,000 poppy heads are needed to produce 9 kg (20 lb) of raw opium in this way. For modern medicinal use, however, opium poppies are cut by combine harvester and morphine is extracted chemically from the poppy stems and seed capsules. Codeine is extracted from morphine and isolated directly from harvested plant material.

For medicinal use the most important opium-producing countries today include Australia, Turkey and France, but India is by far the largest producer. According to the UN, 70% of the heroin consumed as an illegal drug in western Europe originated in Afghanistan in the year 2000. In some rural areas, smoking opium is considered an acceptable pastime, and the country does not have a major drug abuse problem. Afghanistan has, however, announced a concerted strategy to eradicate the opium poppies that produce the raw material for heroin. Some 20% of illegal opium was produced in Myanmar.

Native to Asia Minor, this attractive perennial with red, pink or white flowers is a prime example of the dangerous duality of some of nature's most useful medicinal plants. Opium is highly addictive, and, as the source of heroin, can be a deadly and destructive drug when misused.

British physicians who prescribed morphine during the 19th century in medicines such as laudanum (now called tincture of opium) for such famous patients as Samuel Taylor Coleridge and Elizabeth Browning caused or contributed unwittingly to their addiction to the drug. Apart from its soothing action, a significant attraction lay in its effect as a mental stimulant, creating the impression of alertness without any inhibition.

British policy also played a distressing role in the addiction of thousands of Chinese people to opium in the early 19th century. The British East India Company used opium, produced from poppies grown in Burma and India and smuggled into China, to obtain silver bullion with which to pay the Chinese for their coveted tea and silk; a situation that finally led to the Opium Wars between 1839 and 1842. Heroin, discovered in 1898 as a result of chemical changes made to the morphine molecule, was found to be more effective than morphine as a painkiller and more effective as a cough suppressant than codeine. It was openly sold in cough suppressants in North America until 1917. It was not until the Second World War, however, that morphine and heroin addiction became serious social problems in the USA and Europe.

For modern medicinal use opium poppies are cut by combine harvester and morphine is extracted chemically from the poppy stems and seed capsules.

Coughs, colds and fevers

In 2001, the British alone spent more than £600 million on 'over-the-counter' remedies to help alleviate coughs and colds. Though it's unlikely that science will ever find a cure (so diverse and fast-evolving are the viruses that produce them), a large number of plant extracts are available in the form of syrups, pills, powders and sweets, as decongestants for the head and soothing, counter-irritants for sore throats and coughs. Lemon (which has anti-microbial and antiseptic activity), eucalyptus and cinnamon leaf oils, thymol from thyme (strongly antiseptic) and menthol from mint species are all widely used.

One of the most common ingredients in cough sweets and syrups is liquorice (see right). Its saponin-like glycosides, such as glycyrrhizin, have both expectorant and anti-inflammatory properties, and its distinctive taste also helps disguise more bitter ingredients in these and other medicines. A treatment for mouth ulcers is also based on a semi-synthetic derivative. Liquorice (see right) is obtained from the underground stems of several *Glycyrrhiza* species, but in particular *G. glabra*, still gathered from the wild in Turkey, Afghanistan, Azerbaijan, Iran, Pakistan, Russia and Syria. European countries imported about 6 million kg (13.2 million lb) of roots from this species in 1996. In the 1950s, Turkey alone was exporting around 20 million kg (44 million lb) of wild collected liquorice a year. This had fallen to 1.56 million kg (3.4 million lb) by 1995, a huge decline, suggesting that *Glycyrrhiza* has become scarce because of over-exploitation.

Camphor (*Cinnamomum camphora*), an evergreen native to China, Japan and Taiwan, has been used for centuries to help alleviate colds. Steaming the leaves, roots and wood and distilling the product yields both camphor and white camphor oil, both of which are slightly antiseptic and have expectorant and analgesic properties. Synthetic camphor, derived from turpentine, is now a common substitute for the plant-derived compound (now banned in the UK because of its toxicity), which has also been used to treat rheumatic pain and inflammation and neuralgia.

Though long appreciated by Australian aborigines, eucalyptus oils – with their powerful antiseptic and expectorant properties – are a more recent addition to our armoury of remedies against coughs and colds. The fresh leaves of various *Eucalyptus* species (see right), largely grown in plantations in China (the main exporter of oil), Australia, South Africa, Brazil, India, Portugal and Spain, are steam distilled to produce oils used for medicinal, perfumery and industrial purposes worldwide. Of the medicinal oils, *E. globulus* is the most important,

Liquorice

■ China is now the largest liquorice producer but is almost wholly reliant on unsustainable wild collecting. In the 1960s the plain of Nenjiang river was famous for its *Glycyrrhiza uralensis* population. By 1980 there were no bushes with any commercial value left in this area.

■ About 6,000 tons of *G. uralensis* are exported annually and the domestic consumption amounts to several thousand tons. The reliance on wild-collected material to supply the world's cough remedies cannot continue and cultivated crops have become established in southern Europe, Australia, Brazil and parts of central Asia.

Eucalyptus

■ Eucalyptus oil, which has powerful antiseptic properties, is distilled from the fresh leaves of a number of eucalyptus species, particularly *Eucalyptus globulus*. It is against the oil from this species that the quality and productivity of others are judged.

■ Several hundred *Eucalyptus* species are known to produce volatile oils, but only six currently account for most of that produced. Whilst some have found industrial applications, the aromatic eucalyptus oils are important in the perfumery trade. Oil from the leaves of *Eucalyptus citriodora* emits a lemon scent, and is used both in its whole form and as a source of citronellal, a major source of aromachemicals.

Garlic

■ Intended by nature to provide food for the growing garlic plant, this pungent bulb (*Allium sativum*) has a long history of use both as a human food and as a medicine. By at least 3000 BC, Mediterranean and Far Eastern peoples had discovered its virtues and the plants (which originated in central Asia) were soon distributed around the world.

■ The active ingredients in garlic – responsible also for its tenacious smell – are sulphur-containing compounds, of which allicin is particularly important. The exact way in which garlic works in producing its many beneficial effects is still unclear, however.

■ Nearly all the members of the *Allium* genus, to which garlic and its relatives belong, and in particular the onion, leek and chive, have been reported to have medicinal properties.

Quinine

■ Quinine is only one of 35 different alkaloids – including quinidine, cinchonidine and cinchonine – present in cinchona bark which, together with other natural compounds, have a broad-spectrum, anti-malarial action. Several of these have applications in other areas of medicine, too, such as the treatment of heart arrhythmia. While around 60% of cinchona alkaloids is used for pharmaceuticals, much of the rest is used in the food and drinks industry.

both in terms of the amount produced and its high cineole (or 'eucalyptol') content, the principal constituent. It is much used as a soothing anti-irritant in throat pastilles and cough syrups as well as an antiseptic in gargles and as an ingredient of inhalants used to relieve bronchitis and asthma.

Many remedies are available containing ingredients from plants to help us cope with colds or flu, but one plant in particular stands out for its ability to help protect against these and other illnesses. Garlic (see left), a member of the lily family, has been used in many parts of the world both to treat and prevent all manner of infections, including those affecting the nose, ears, chest and throat. With its strong antimicrobial activity (particularly effective in the gut and lungs), chest congestion, sinus problems, tonsil infections, whooping cough and bronchitis as well as intestinal disorders, high blood pressure, indigestion, acne and asthma are all reported to have been relieved with the help of garlic. In herbal medicine, the modern uses of garlic have focused on its reputed cancer-preventative, anti-microbial, anti-hypertensive, anti-thrombotic, antioxidant and cholesterol-lowering effects.

Malaria

To those of us who are used to them, colds and flu are rarely life threatening, but fevers such as that caused by malaria are quite a different matter. In any given year, nearly 10% of the global population will suffer a case of malaria, a disease which contributes to the death of about 3 million of us each year. In sub-Saharan Africa, it is estimated that malaria kills 3,000–4,500 children every day. To make matters worse, as an effect of global climate change, scientists predict that the mosquitoes that carry malarial parasites could return to regions from which they have been eradicated for many years, including Britain. It is more important than ever that we have ready access to simple and cost-effective treatments.

That's why one of the most precious of all medicines continues to be quinine (see left). The effectiveness of the bark of the small evergreen cinchona trees (*Cinchona* spp.) for curing fevers was known to native South Americans long before Europeans reached the continent. When Jesuit missionaries working in Peru at the beginning of the 17th century began to fall ill with malaria, native doctors, with their vast knowledge of local medicinal plants, used cinchona bark to treat them. In 1633, Father Calancha, a Jesuit priest, wrote of the 'miraculous cures' effected by a powdered bark 'the colour of cinnamon', but it was not until 1639 that the 'Peruvian bark', as it was described, reached Rome and Spain. However, the treatment was scorned and rejected by the medical profession, who had much to gain financially from treating malaria sufferers inadequately themselves. It was not until the end of the 17th century when a London apothecary, Robert Talbor, successfully cured Charles II of the 'ague', as malaria was then known, that Peruvian bark was finally accepted, and the quest began to discover the source of this 'miraculous' new cure.

The subsequent story of quinine's introduction into Western medicine is a long and complex one, involving the work and research of many people. It was not until the early part of the 19th century that the first serious study of *Cinchona* species was eventually published, by Alexander von Humboldt and the French botanist Aimé Bonpland. Then, in 1820, the alkaloid quinine was isolated from the powdered bark and named by two French doctors (from an Amerindian word for the cinchona tree, *quinaquina*, which literally means 'bark of barks') and demand for the drug soared. By the mid-19th century, reckless over-exploitation of the wild trees was threatening the survival of some species.

With malaria rife in their African, Indian and East Indian colonies, the Dutch and British tried to produce their own supplies by employing explorers to collect seeds and seedlings with which to establish colonial plantations. The Dutch raised trees in Java and the British in Sri Lanka and India, but the quinine content was extremely low. (Concentration of the alkaloid varies widely between different species and even between populations of the same species.) A British collector, Charles Ledger, managed to obtain what was probably a batch of *Cinchona officinalis* seeds, which produced trees containing more than 13% quinine, compared with the less than 7% typical in plantation stock. Refused by the British government, the Dutch government bought a pound of seeds for £6 7s in 1865. Described by some as the best investment in history, the Dutch successfully grew 12,000 trees in Java and for almost 100 years controlled nine-tenths of the world's supply of this vital medicine.

In 1944 the first synthetic quinine was made by two American scientists and synthetic drugs proved very effective in the control of certain strains of malaria. But malaria is not a simple disease and strains of malarial parasite, particularly *Plasmodium falciparum*, developed a resistance to some of them.

Cinchona contain several alkaloids, such as mohave importan.

Scientists predict that mosquitoes that carry malarial parasites could return to regions from which they have been eradicated for many years, including Britain.

Despite the considerable medicinal attributes of natural quinine, its use to treat malaria has largely been overtaken by semi-synthetic drugs. But the effectiveness of quinine against strains of parasite that are resistant to chloroquine has maintained a considerable market for

Other plants for malarial cures

Attention continues to be drawn to many other plants already used by native people to treat **fevers** and which could be used in the development of malaria drugs. A recent study conducted in Roraima, Brazil, found that seven indigenous peoples used more than 90 species from 41 different plant families as sources of **malaria treatments**. Only 24 of them had previously been known to have anti-malarial properties. Latin America's malaria problem is most acute in Amazonia and has seen a huge increase in reported cases in recent years. In tropical Africa and southern and Southeast Asia, meanwhile, species of *Ancistrocladus* and *Triphyophyllum peltatum* have been found to contain alkaloids that appear to be **highly effective** against *Plasmodium falciparum* and *P. berghei*, yet with no significant side-effects on the patient. These plant-derived anti-malarials are not yet on the market, but the Public Health Service in the USA is seeking licensees to develop them.

Another promising species is *Eurycoma longifolia*, native to the rainforest-clad slopes of the Malaysian peninsula. Its roots have been used traditionally as a cure for fevers and malaria and three compounds with potential anti-malarial activity were recently discovered.

The race to counteract the **deadly work** of the mosquito, by developing cheap and effective drugs, is largely dependent on the willingness of the world's major pharmaceutical companies to invest, and also on that of developing countries to implement good health-care systems.

the natural drug. Around 600 tons of cinchona alkaloids are currently produced each year, from the bark of trees grown mostly in Congo Kinshasa and Indonesia. A traditional Chinese malaria treatment, however, has provided a powerful drug, artemisinin, for use against chloroquine-resistant cerebral malaria. Its discovery has been described as 'monumental' because of its nearly 100% effectiveness when used alongside established drugs (such as mefloquine). The Chinese have been using this treatment for many years (based on their own discovery in the 1970s of the active properties of the plant *quinghao*, closely related to sweet wormwood, *Artemisia annua*), but Western governments and companies have acted slowly in promoting its adoption. A cheap, all-in-one tablet is urgently needed.

Intestinal/digestive conditions

Laxatives and purgatives have probably been the most easily discovered of all plant medicines and they have also been among the most ubiquitous. One of the most useful and widely used ingredients in proprietary laxatives is cascara, obtained from the bark of *Frangula purshiana*, a North American species of buckthorn (see opposite). Perhaps even better known than cascara is senna, obtained from the leaves and pods of *Senna alexandrina* (also known as *Cassia senna*), a member of the pea family. These contain glycosides, which, like cascara, stimulate peristalsis. India is the largest source, exporting around 6,000 tons a year, but about 700 tons are still exported each

Bio-pharming

'Just one mistake by a biotech company and we'll be eating other people's prescription drugs in our corn flakes.' Larry Bohlen, Friends of the Earth, USA (2002).

Having so far concentrated on making crops easier to grow, a major new focus for biotechnology is bio-pharming. Exploiting the ability of plants to make medicinally important proteins by splicing them with human genes (at far less expense than fermentation factories), researchers are experimenting with maize, canola, potatoes and tomatoes, amongst others, to engineer vaccines, enzymes, antibodies and hormones. More than 300 field trials of GM bio-pharmaceutical crops had already been conducted in the USA by mid-2002.

Maize, for example, has already been genetically modified to produce the gastric lipase enzyme, used to treat digestive problems caused by cystic fibrosis, as well as trypsin, a protein used to make insulin, and one company now has a permit to grow maize spliced with a herpes-fighting human gene. Bananas, meanwhile, have been developed to produce an edible vaccine against hepatitis B and potatoes to produce a vaccine against cholera.

Risks and benefits

Such technology, used successfully with bacteria (in laboratories) since the 1970s, could have profound benefits. For example, it is theoretically possible to modify a common fast-growing species to produce large quantities of compounds otherwise available at almost prohibitive expense. Most antibody drugs available are currently grown in Chinese hamster ovary cells and turned into drugs in bioreactor plants costing millions of dollars each. By contrast, edible vaccines in bananas offer the means to vaccinate large numbers of vulnerable people at very low cost.

However, the concerns expressed by those alarmed by GM food crops are magnified when it comes to bio-pharming. For example, it is feared that pollen from plants designed for pharmaceutical purposes may impregnate crops intended for food uses. What would happen to people or animals if they unwittingly consumed powerful drugs that should only be given under strict medical supervision? Many scientists believe that current gene containment strategies, which might work in the laboratory, are unreliable in the field.

Bio-pharming also increases the potential for bio-piracy (see page 153), as it could be difficult to detect if genes have been appropriated from traditional medicinal plants. And bio-pharming in industrialized countries could have effects on the economies of those developing countries that cultivate and sell traditional medicinal plants or their products. However, much pharmaceutical synthesis is based on petroleum, and bio-pharming could be viewed as a significant option when supplies run out. One company has estimated that 10% of the US maize crop will be devoted to pharmaceutical production by 2010.

year from Sudan, much of which is gathered from the wild. Other plants used in laxatives include:

- The ripe seed and husks of two common plantain species produced commercially mainly in India – *Plantago ovata* and *P. afra*. Also containing a useful mucilage, they complement the action of cascara and senna by acting as bulk-forming laxatives. The seeds also absorb bacterial toxins, enabling them to be expelled from the body.
- Psyllium husks. The world's largest importer is the USA, where in 1998 the FDA approved the claim that their consumption reduces the risk of coronary heart disease.
- Rhubarb was a traditional constipation remedy in much of Europe and Asia, and traditional Chinese medicine still uses the roots of a native species (*Rheum palmatum*) as the source of a laxative.
- Castor oil and extracts from various aloe species are also used.

Buckthorn

■ For centuries, native North Americans have used various parts of *Frangula purshiana* for medicinal and other purposes; they chewed pieces of bark, or made infusions from it to use as a laxative and to treat intestinal parasites. In Europe there is evidence that Saxon peoples used the berries of buckthorns for similar purposes. The active compounds in cascara, which are often used to treat chronic constipation, are converted by gut bacteria into substances that stimulate the peristaltic action of the large intestine.

Heart and blood conditions

Pineapple

■ Pineapples (*Ananas comosus*) were first cultivated by the Guarani Indians of Brazil and Paraguay. Local people drank the juice to aid digestion – especially after eating meat – and as a cure for stomach ache. Pineapple juice was also used to promote the healing of wounds. Now bromelain, the anti-inflammatory enzyme currently extracted from the stems of pineapple plants (though it is also present in the fruit and leaves), may help thrombosis sufferers.

Perhaps the most important medicinal plant that grows wild in the UK is the foxglove (see opposite). Heart drugs based on digitalis – a mixture of glycosides found in its leaves – have saved the lives of countless people suffering from congestive heart failure and helped millions of heart disease sufferers to lead more normal lives.

Foxgloves have been used since ancient times in folk medicine, mainly to make preparations for treating cuts and bruises. It was not until the late 18th century, after the publication of Dr William Withering's investigations into the plant, that its value as a cardiotonic was realized. Withering had spent 10 years looking into the properties of the foxglove after finding that an old lady he had been asked to visit had cured herself of dropsy (an unnatural accumulation of fluid in the tissues) by taking a herbal cure containing it. 'This medication,' Dr Withering wrote, 'was composed of 20 or more different herbs, but it was not very difficult for one conversant in these subjects to perceive that the active herb could be no other than Foxglove.' The improved circulation that had resulted from the woman's ingestion of digitalis had boosted the performance of her kidneys, clearing the accumulated body fluids, which are a symptom of the complaint.

Digitalis glycosides, including digoxin, digitoxin and lanatoside c, have a powerful effect on the heart and circulatory system. They can strengthen and increase the muscular activity of the heart, stimulating more forceful contractions in one that is inefficient whilst regulating a dangerously fast heartbeat. Like many valuable medicinal plants, the foxglove is highly poisonous.

A rainforest vine, *Strophanthus kombe*, from West Africa, whose extracts have been used as a local arrow poison, has also provided a useful medicine for the heart. Strophanthin acts as a heart stimulant and has relatively few gastro-intestinal side effects.

While aspirin is now recommended by some doctors in order to reduce the risk of heart attacks (see page 214), the widely used anti-coagulant drug warfarin was originally developed from compounds in the fragrant herb *Melilotus officinalis*. More recently, a quite unrelated plant – the pineapple (see left) – has been of interest as the source of the anti-inflammatory enzyme bromelain, which has also been shown to reduce the clumping of platelets in the blood and the formation of plaques in the arteries. Bromelain digests the by-products of tissue repair and reduces inflammation caused by fatty substances inside the blood vessels, so it is thought to help maintain healthy cardiac tissue and reduce the risk of stroke. Some sufferers of arthritis, carpal tunnel syndrome and sinus congestion may also benefit from its use.

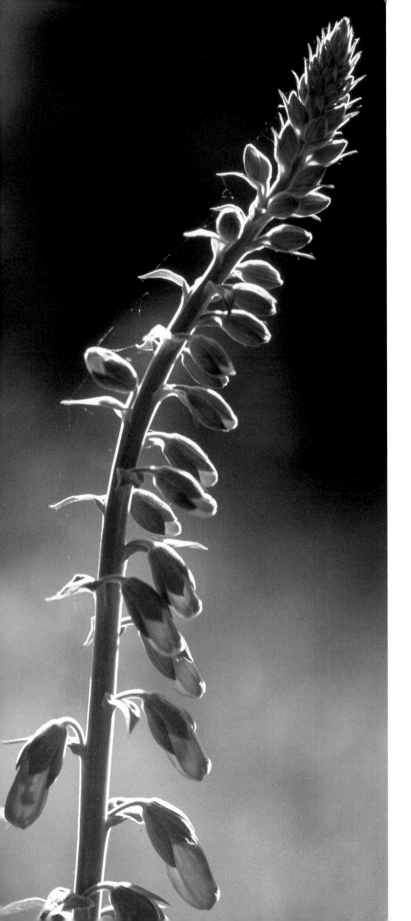

Foxgloves

■ Though the common British fox-glove, *Digitalis purpurea*, was the first source of digitalis glycosides, it was later discovered that *Digitalis lanata*, the woolly or Austrian foxglove from southeast Europe, contained an even greater concentration of these active principles.

■ Today, digitalis is obtained from the dried leaves of both species, grown commercially in several countries, including the Netherlands, Hungary and Argentina. Just one digoxin product, marketed by a major pharmaceutical company in the USA and Europe, is currently worth some $50 million a year. Many other companies sell cardiac glycosides. Neither digoxin nor digitoxin have yet been synthesized.

The common foxglove. Foxgloves contain digitalis, a mixture of powerful glycosides. An overdose of digitalis can lead to life-threatening symptoms, so getting the dose right is critical.

Screening and herbalism

Western medicine has had many spectacular successes in the last two centuries, including the discovery of **antibiotics** and the control of major **infectious diseases**. But many of the disorders industrialized countries are now trying to overcome have much more complex causes than infectious diseases. Cancers, heart attacks and rheumatism are caused by combinations of complex environmental and genetic factors that do not necessarily lend themselves to the 'magic bullet' approach of a single chemical molecule as a cure. The medical effects of **healthy diets** are acknowledged to play an important role in the prevention of such disorders (see Healthy eating, page 213).

Up to now, Western scientists looking for new plant-based drugs have 'screened' plants to look for individual chemical compounds that show some kind of **biological activity**. But new approaches to the study of plant medicines may lead the way to treatments that contain mixtures of the ingredients found in the plants themselves. Among the tools will be complex **computer programmes** that can help analyse the interactions between many different ingredients and symptoms. The resulting medicines could be much closer to the treatments prescribed by herbalists, whose extracts contain a mixture of compounds which they contend work synergistically, their **sum effect** being greater than that of any individual ingredient.

There is much debate between scientists about the efficacy of herbal treatments compared with drugs based on individual plant chemicals.

Although the very existence of herbal remedies that have been handed down over generations can in itself be regarded a screening process, **herbal medicines** should not be used without the supervision and direction of a qualified herbalist. As with pharmaceutical drugs, it is possible for a patient to develop an allergic reaction to a herbal remedy or it may react with or interfere with other treatments being used or conditions a patient may have. Herbal remedies are certainly not automatically safer than pharmaceutical drugs just because they occur naturally; many are poisonous if taken at the **wrong dose** or in the **wrong way**. Even popular herbal remedies need to be taken with care. St John's wort (*Hypericum perforatum*), which has become hugely popular in Europe for its antidepressant effect (sales were worth $6 billion in 2000), has been found to interfere with the activity of certain prescription drugs.

Ginkgo seeds on sale in South Korea. Ginkgo is used to treat many ailments.

High blood pressure/circulatory problems

Those suffering from high blood pressure or hypertension have also been helped by plant-based drugs. One of the most effective, which has helped many millions of people to lead reasonably normal lives, comes from a monsoon forest plant that has been used for more than 4,000 years by Ayurvedic healers in India to treat both snake bites and mental illness. The plant is *Rauvolfia serpentina*, commonly called snake root or rauvolfia (see opposite).

From some 24 different alkaloids contained in the plant, and found in the highest concentrations in its roots, three have traditionally been most important (rescinnamine, deserpidine and reserpine) in medicines to reduce high blood pressure and as tranquillizing drugs.

Using one drug to treat disorders of both the circulatory and nervous systems is not as odd as it sounds. Reserpine is considered a sympathomimetic agent, one that targets the sympathetic nervous system. It has been found to lower blood pressure in remarkably low oral doses and does so by acting on the parts of the nervous system that affect the heart and blood vessels, decreasing the heart rate and dilating the vessels. Reserpine has also been used in small amounts for the treatment of menstrual and menopausal problems, and in larger doses as one of the most effective plant-based tranquillizers. Another compound, ajmaline, contained in the roots in fairly large quantities, has been used as an antiarrhythmic.

Khella (*Ammi visnaga*), with a long history of medicinal use, produces the active substance khellin, which has been used to treat renal colic. In 1945 a medical technician noticed that khellin also eased his angina. Research subsequently led to the discovery of two major new drugs (both of which are now synthesized): nifedipine, used for angina and high blood pressure, and amiodarone, which restores abnormal heart rhythms.

Compounds from the leaves of the ginkgo tree (*Ginkgo biloba*), which has featured in Chinese medicine for thousands of years, are now being used in the West for a range of ailments that obtain relief through increased blood flow. These include chilblains, Raynaud's disease (a condition in which the blood vessels in the body's extremities react so much to cold – and sometimes to anxiety or emotion – that the blood supply to the fingers, hands, toes and feet fails), senile dementia and Ménière's disease (a disease of the inner ear).

The activity of ginkgo on the circulation is reported to be partly due to its vasodilatory effects, that is, helping to keep the blood vessels in the extremities dilated. Increased blood flow to the brain, enabling its cells to receive more oxygen and a better energy supply, is reported by some to be able to bring relief in certain cases of tinnitus (although this effect is now in doubt) and to slow the deterioration of memory in old age.

Snake root

■ Although reserpine has been successfully synthesized, it has been cheaper and easier to obtain the compound from plants. As a result, *Rauvolfia serpentina* is collected on a large scale in the wild from India, Thailand, Bangladesh and Sri Lanka, both for export and the traditional medicine industry. The international monitoring of trade in endangered species has revealed that this species could become threatened if trade is not regulated. Cultivated plants are already being grown in plantations in Nepal, India, Sri Lanka and Java, and experimentally in the United States. Scientists have also developed a way of propagating *R. serpentina* using cell culture.

■ A sister species, *R. vomitoria*, is much richer in rauvolfia alkaloids, but most roots are harvested from the wild in West Africa.

225

Anaesthetics and nerve conditions

A range of psychoactive chemicals and nervous system toxins developed from plants have found important uses in modern medicine to treat nervous disorders and as a source of anaesthetics. Surgeons everywhere and, of course, their patients, have benefited hugely from the leaves of the coca bush (see overleaf), which have made available some our most valuable and widely used anaesthetic compounds.

In its native South America, coca has been grown for thousands of years by indigenous peoples of the Andes, from Colombia to Bolivia. The Incas regarded the bush as sacred and the property of only the ruling caste and the leaves were used for a number of important

Coca

■ Today, the leaves of the coca bush (*Erythroxylum coca*) are chewed by thousands of poor and under-nourished Andean peoples to give them stamina. A wad of leaves kept at the side of the mouth and constantly renewed as the leaves lose their potency helps make the carrying of heavy loads and strenuous work of all sorts more bearable. The compound responsible is cocaine.

■ The illegal manufacture of cocaine as a 'recreational' drug for Western markets has led to the wholesale destruction of many large coca plantations in South American countries, adding to the suffering of those who grow or use the leaves not for profit or to reach a 'high' but to supplement their meagre diets.

ceremonial purposes, as part of religious observance, in the form of offerings and for the divination of omens. The dead began their journey to the afterworld in the company of coca leaves and the plant was revered everywhere. The reason for this special reverence lies in the chemical composition of the leaves. When chewed with a small amount of lime or plant ash, their active compounds are released, stimulating the nervous system in general, enhancing muscle potential, increasing stamina, depressing hunger and generally relieving pain. As native South Americans then, as now, did not ingest the cocaine alkaloid alone, but together with other compounds in the leaves, they did not suffer the addictive or mind-altering side-effects associated with the use of cocaine addiction in Western societies today.

Soon after their arrival in Peru and quick to oppress their newly conquered subjects, the Spanish made sure that coca leaves were available to their workforce. Its use allowed them to exploit Indian labour mercilessly, for little food and no remuneration, and helped the Indians to endure appalling cruelty and deprivation.

The cocaine alkaloid was first isolated from the leaves during the 1840s. Such was the interest and so glowing the reports of its effects that coca leaf extracts soon became a common addition to tonic drinks and powders in Europe and America. The discovery in 1884 of the power of cocaine to deaden nerve endings and its suitability therefore as a local anaesthetic brought about a revolution in surgery. It was found to be the most effective anaesthetic for eye operations, for example.

Coca leaves grown by a government-licensed company in Peru are currently the chief source of the anaesthetic alkaloids in Britain. A crude leaf extract is exported, from which cocaine is extracted for use in the formulation of a variety of drugs, mainly used in local anaesthetics for nose and eye surgery. Coca cultivation for illegal drug production is a recent development in South America, where Colombia is the largest single source. Before the 1990s, only a few indigenous communities grew small amounts of coca for ritual use.

Awareness of the dangerously addictive properties of cocaine led to research into the synthesis of non-addictive substitutes, but without the coca leaf many valuable drugs could not have been produced. Lidocaine (formerly lignocaine), used as a local anaesthetic in dentistry, is a synthetic derivative – by a circuitous route – of an alkaloid contained in barley.

Plants that relax us for the surgeon

Before certain operations can be performed, and to facilitate the surgeon's job, it is important that the muscles to be cut are in a relaxed state. For abdominal surgery and obstetrics, as well as a range of disorders affecting the voluntary muscles, plant chemicals from the

Amazon, made famous as ingredients of the arrow poison curare, have provided a vital adjunct to anaesthesia to secure muscle relaxation.

A group of Spanish conquistadores exploring the Amazon in the 16th century were the first Europeans to record the deadly effects of curare, when one of them was hit in the hand by an arrow, dying soon afterwards. Curare was widely used as an arrow poison at that time by many Amazonian Indian groups and some still use it today. There are many different sorts of curare, each one based on the compounds contained within a different plant or plants. In some cases, fruits are used – as with *Chlorocardium venenosum*, used by the Kofan Indians of Colombia and Ecuador – but some of the best-known ingredients come from species of *Chondrodendron* and *Strychnos* vines.

The bark of these vines is scraped off in sections and then pounded, before filtering in water to extract the ingredients. The processes involved are complex and have generally been kept a closely guarded secret. Up to 30 different ingredients may be used for each recipe and recipes vary considerably between different groups of people. For two centuries the exact contents of curare poison remained a mystery to Western observers and it was not until 1800 that Alexander von Humboldt, who had witnessed the preparation of one kind of curare by native people of the Orinoco river, gave the first accurate account.

A little later, in 1814, the eccentric Victorian explorer Charles Waterton performed some experiments to find out how curare worked. One of these involved injecting a donkey with the drug.

Coca harvesting in Peru. Cocaine is used to produce a variety of drugs, mostly used as anaesthetics.

Coca eradication destroys land and lives

Huge quantities of toxic glyphosate and granular herbicides have been continuously sprayed on community land belonging to some 58 peoples of Colombia's forested **Putumayo region**, including the Kofan, Quichua, Siona and Embera. Pleas to desist, and to honour previous agreements to give economic and cultural assistance to communities who have become obliged to grow coca to survive (and to help them substitute these illicit crops), have gone **largely unheard**.

The policy of coca eradication in rural Colombia, instigated and financed by the USA, has been widely condemned as an attack of terrorism on the thousands of people living in the region.

Despite evidence that spraying only worsens the problem, the massive new eradication campaign announced in July 2002 promises to drive thousands more people from their homes by indiscriminate poisoning of the crops, medicinal plants, animals and water on which they **depend**. By 1998, it had been reported that some 150,000 ha (37,000 acres) of rainforest in the Orinoco and Amazon basins had been destroyed. For every hectare (just over 2 acres) of coca sprayed, 4 ha (10 acres) of forest are **devastated**, and it is estimated that, by 2015, 70% of the region will have been converted to wasteland.

The Organization of Indigenous Peoples of the Putumayo Zone declared in July 2002: 'As we know from bitter experience, fumigation [spraying] will devastate everything: our farms, our vegetable gardens, our animals, our people. It will destroy our means of subsistence and threaten our very survival … The rich biodiversity of the Amazon forests will be poisoned by glyphosate.'

Within 10 minutes she appeared to be dead, but Waterton inserted a pair of bellows into her windpipe through a small hole he had cut for the purpose and started pumping to inflate her lungs. He must have been delighted to see that she soon 'held her head up and looked around' – though he had to keep up his means of artificial respiration for some two hours until the effects of the curare had worn off. The experiment demonstrated that curare killed by immobilizing the voluntary muscles, so that breathing became impossible. It did not, however, affect the (involuntary) heart muscle, which meant that as long as breathing was artificially maintained, curare could be used as a muscle relaxant.

Later experiments showed that curare blocked the transmission of nerve impulses to the muscles – taking maximum effect two to five minutes after injection. It was not until 1939, however, that the active principle of curare was isolated and not until 1943 that it was introduced successfully into routine surgery. Although hyascine from the mandrake root (see below) had been widely used in the past, at this time the three main components of anaesthesia – hypnosis, analgesia and muscle relaxation – were all brought about by the use of one anaesthetic agent, usually ether or chloroform. It was only after the patient had entered a deep stage of narcosis that sufficient muscle relaxation was obtained to enable the surgeon to operate easily. The introduction of curare meant that the patient could be kept lightly anaesthetized while obtaining complete muscle relaxation and without the risk of surgical shock.

Potato relatives for surgery and travel sickness

Some 2,000 years ago, the alkaloid **hyoscine** (scopolamine), obtained from **mandrake root** (*Mandragora officinarum*), was used to relieve the unimaginable pain of amputation and crucifixion. Now produced synthetically, it has been commonly used as a standard **pre-medication** before surgery. Relaxing muscle spasms, lessening pain and inducing sleep, hyoscine has also been an important component of some travel sickness remedies, and was used in

the Second World War by soldiers in the D-Day landings to prevent sea-sickness. The synthetic version is still a constituent of some over-the-counter travel sickness remedies.

Some other, **very poisonous** relatives of the **potato**, henbane (*Hyoscyamus niger*), thornapple (*Datura stramonium*) and corkwood (*Duboisia myoporoides*), also contain hyoscine, as well as hyoscyamine, while deadly nightshade (*Atropa belladonna*) and corkwood are sources of

atropine. Used by Greek ladies to enlarge the pupils of their eyes to make them more attractive, derivates of atropine are still used by opthalmologists to dilate the pupil for purposes of refraction or to treat iritis. Due to its sedative effects on the stomach, atropine has also been used for **travel sickness**, in a different capacity, to restore or control the heartbeat and as an antidote to the suffocation that can be caused by the inhalation of nerve gas.

Muscular blocking agents

The alkaloid tubocurarine is contained in the stem of *Chondrodendron tomentosum*, a liana found in the Brazilian and Peruvian rainforest. It has been widely used as a muscle-relaxant during surgery and in the treatment of multiple sclerosis, tetanus and stroke patients. Tubocurarine is the model for a series of synthetic neuro-muscular blocking agents, such as vencuronium and atracurium, used routinely today as anaesthetics.

Other plants traditionally used by Amazonian peoples as a source of arrow poison, such as *Strychnos guianensis* (strychnine), have been investigated as sources of other medicinally useful compounds. One of the most recent discoveries is guiaflavine, an alkaloid found in the bark of *S. guianensis*.

Psychiatry

The volatile oils of valerian (*Valeriana officinalis*) have a sedative activity – old herbals also mention the plant to treat certain forms of epilepsy – and it has been used in the past to treat shell shock and other neuroses.

In 2001, the drug Reminyl (galantamine), now synthesized but first found in several plants, including daffodil bulbs, was given legal approval in 22 countries. It is expected to increase the quality of life for sufferers of mild to moderate Alzheimer's disease, since clinical trials demonstrated a significant improvement in the cognitive performance of those who took part.

Cancers

In the last decade or so, the search for cancer treatments has stimulated increasing interest in screening plant chemicals for their potential as drugs. By 1997, almost two-thirds of all approved anti-cancer drugs were of natural origin or modelled on natural products. But our continuing struggle to combat the causes and effects of different kinds of cancer – greatly helped by the strategies adopted by plants to help them survive – is not new. Probably the most famous of all plant-based cancer drugs, the rosy periwinkle (see right), was discovered in the 1950s.

Already known to traditional healers in both Madagascar and the West Indies, where it was used to treat diabetes, research revealed the presence of two alkaloids – vincristine and vinblastine – which,

229

Rosy periwinkle

■ Grown by many people as a small pot-plant with striking pink or white flowers, the rosy periwinkle (*Catharanthus roseus*) is a member of the poisonous dogbane family and a relative of the hardy, evergreen creeping periwinkles that decorate many British gardens. Present in minute quantities, the rosy periwinkle alkaloids, used as anti-tumour agents, account for just 0.00025% of the dry weight of the leaves.

though not useful for diabetes, were active against certain kinds of cancer cell. Today, these alkaloids are still key drugs used in the treatment of leukaemia, sarcoma, Hodgkin's disease and other lymphomas, and cancers of the breast and testicles. The use of vincristine in combination with other drugs for cases of childhood acute lymphoblastic leukaemia has resulted in a first remission rate of 90% and a cure for over 75% of patients. Too complex to synthesize, these alkaloids are extracted today from plants (in which they are present in tiny amounts) cultivated chiefly in Texas, in the USA. Since the 1950s, over 100 other potentially useful alkaloids have been discovered in this single species. Some semi-synthetic derivatives of the vinca-alkaloids are in various phases of clinical trial.

Long before Western medicine discovered the rosy periwinkle, native healers had been using various kinds of plants to treat tumours and cancers. Native North Americans, for example, recognized the anti-cancer properties of plants such as mayapple (*Podophyllum peltatum*) and bloodroot (*Sanguinaria canadensis*). The Penobscot Indians used mayapple to treat warts and skin tumours while others used the red sap of bloodroot against various kinds of malignant growth.

The roots of *P. peltatum* and a Himalayan species, *P. hexandrum*, which has become endangered through over-collection, are the source of the semi-synthetic etoposide and teniposide drugs used against lung and testicular cancers, leukaemia and lymphoma.

Cures from yew and hazel

One of the most important plant-based anti-cancer drugs to emerge from screening programmes in recent years is paclitaxel, a compound first isolated from the bark of the Pacific yew (*Taxus brevifolia*) in the 1960s, and now the source of Taxol and Taxotere treatments for ovarian and breast cancer.

The development of paclitaxel as a drug was hampered because it is only present in the slow-growing yew in minute amounts and wide-scale felling in the 1980s and early 1990s (when several million pounds of bark were collected) became a conservation issue. In the UK alone, more than 15,000 women die from breast cancer each year, but the bark of one 60-year-old tree provides only just enough to treat a single patient.

The problem was partially overcome by the discovery of the precursors to paclitaxel (which could be used in a semi-synthetic method to create Taxol) in the leaves of the common yew, *T. baccata*, and in the UK and elsewhere programmes to collect the clippings from yew hedges were initiated. However, in the late 1990s, paclitaxel and similar substances were discovered in other, quite unrelated species, including the nuts, branches and shells of the hazel tree (*Corylus avellana*).

Hazel trees. A recently discovered source of the compound paclitaxel, first found in the Pacific yew, the common hazel may be able to help in the fight against ovarian and breast cancers.

Although paclitaxel can be synthesized, the method is said to be too complex and expensive to implement commercially, so an increased number of sources from which to make the drug should be good news. Certain species of fungi have also been found to produce the compound, suggesting the possibility that it may be 'brewed' on a commercial scale. Chemists have also been able to produce similar compounds synthetically to produce the drug Taxotere.

Other plant-based treatments

Plant cells can now be grown in industrial cultures to produce medicinal compounds. In trials, such cultures have been developed from leaf tissue of *Ochrosia elliptica*, for example, one of some 30 species in this genus found on islands in the Indian and Pacific Oceans, and which contains alkaloids with anti-tumour activity, opening further avenues for drug production.

Technology can also be used to modify compounds that show promise as drugs but which may have dangerous side-effects. Camptothecin is a quinoline alkaloid first reported in the Chinese tree *Camptotheca acuminata*. Clinical trials were conducted during the 1970s but development of the drug was stopped because the compound was too toxic. But scientists persevered with research on its unique mode of action and in 1985 discovered that it interfered with an enzyme involved in gene regulation, thus upsetting the normal pattern of cell growth and division. This paved the way for chemists to create camptothecin analogues with the same medical action but with greater solubility and lower toxicity. These are now found in drugs such as CPT-11 (active against leukaemia and lymphoma) and topo-tecan, both approved in the USA for treating ovarian cancer.

Plants and cancer prevention

While a great range of plants from mistletoe to the African bush willow are known to contain powerful chemicals that are effective at fighting certain types of cancer, the role of plants in helping to prevent many kinds of cancer is also becoming increasingly clear. As with heart disease, good diet is a key element of preventative medicine. Plants grow by harvesting light energy but the chemical reactions that harness the sun's light also produce as by-products 'free radicals', highly reactive molecules that can damage delicate cell structures such as membranes. Plant cells contain flavonoid compounds, which have evolved to mop up these free radicals, and many fruits and vegetables contain large amounts of them. Chemical reactions in our own cells also produce free radicals and they are thought to play a role in some cancers. The *American Journal of Clinical Nutrition* recently reported that flavonoids found in fruit, vegetables, chocolate, tea, wine and

Leaves of the common yew.
Yew leaves are highly toxic, but they contain compounds that are helping to make powerful anti-cancer treatments. Using leaves rather than bark has helped conserve Pacific yew trees, in which paclitaxel was first discovered.

231

grape juice 'reduce cellular oxidative stress and are associated with a reduced risk of heart disease and cancer'.

Soya, now the source of various health concerns (see page 149), is a source of flavonoids but it is also rich in phyto-oestrogens, chemicals that mimic mammalian sex hormones. Their presence in soya is believed to be linked to the low incidence of breast, colon and prostate cancers in some Asian peoples; it is known, for example, that phyto-oestrogens block the growth of cancer cells and slow the growth of new blood vessels into tumours. Why would plants contain such substances? Perhaps because they can reduce the birth rate of animals that graze on them.

Plants for small families

Our best-known form of modern birth control, the contraceptive pill, taken by an estimated 81.7 million women in the year 2000, comes to us courtesy of a family of climbing forest vines, the yams (*Dioscorea* spp.). Although the pill is now mostly synthetically made in the West, it was diosgenin, a steroidal sapogenin that occurs naturally in very high levels in some yam species, that enabled what was, in the 1960s, a revolutionary means of birth control to be produced.

The discoveries of an American chemist, R. E. Marker, who showed in 1942 how male and female steroid hormones could be made from diosgenin (first isolated from the Asian yam, *Dioscorea tokoro*), were at first ignored by the large pharmaceutical companies. But after his discovery of several yam species native to Mexico whose tubers were big enough to make commercial production of progesterone viable, these vegetables became the basis of the world's oral contraceptive industry. Of the world's 850 species, about 20 contain enough diosgenin to be useful for the production of progesterone, oestrogen and other hormones. These species include *D. nipponica* and *D. zinziberensis*, both widely cultivated in China (the biggest single supplier of diosgenin). Another species, *D. deltoidea*, was once common in the Himalayan foothills but has been so over-harvested from the wild that it is now a threatened species and in 1994 its export from India was banned. It has, however, been cultivated in other parts of the country and in China.

Steroidal sapogenins are the starting materials of one of our most important groups of drugs because, alongside oral contraceptives, their products include cortisone and hydrocortisone, sex hormones and anabolic agents. Although they can now be totally synthesized, various other plant species, including *Agave sisalana*, are still an important

source of some steroidal drugs. Included in these is the drug Crinone –
a progesterone drug used for maintaining pregnancy and for HRT.

Other plants for contraception

Many different plants have been used historically as aids to family
planning – to prevent or defer conception – and to abort or aid the
progress of a pregnancy. Ancient Egyptian medical papyri, for example,
mention pine and acacia species for their anti-fertility properties. Other
plants used for such purposes include:

● The pomegranate (*Punica granatum*), whose rind was recommended
as a contraceptive in the 2nd century.

● The date palm (*Phoenix dactylifera*), willow species and juniper
(berries), which have a very long history of use and Western research
has confirmed the activity of the compounds in these plants. The WHO
is reported to have recently investigated one of the willow's relatives, a
poplar species, for compounds that could be used to make cheap birth
control pills.

● Papaya seeds (*Carica papaya*), which are also under investigation.

Women in many different parts of the world are still using plants in
their natural state to control fertility. At least 4,000 species, half of
them from tropical forests, have been recorded in modern times –
mostly from information supplied by local people – to contain anti-
fertility compounds. Although only a fraction have been screened to
find their active chemicals, more than 250 plants from South America
alone have been confirmed as having birth control potential. The nut of
the majestic greenheart tree (*Chlorocardium rodiei*), for example,
better known for its yield of one of the world's most durable timbers,
is a source of contraceptive compounds (long used by the women of
several Amerindian groups in its native Guyana, see page 236).

Oral contraceptives for men

Oil from the seeds of the cotton plant could also become a source of
commercially available oral contraceptives – this time for men. In the
Chinese province of Jiangxi, cotton seed oil was used for cooking
during the 1930s, when the processing of cotton in the area made it
cheap and readily available. An unexpected result was that no children
were born there while the oil was being used. The oil contains gossypol, a bitter yellowish pigment, which was later found to interfere
with the cells responsible for sperm formation.

In China and Brazil, continued efforts have been made to develop
gossypol commercially as a male contraceptive pill. Concerns about
the incidence of permanent infertility and other side-effects caused the
WHO to abandon further research in 1980. But while the Chinese are
still pursuing the use of the compound (in some cases in combination

Plants and PMS

Hormone-like compounds derived
from plants may also be used to
help women who suffer from the
hormonal imbalances that
contribute to pre-menstrual
syndrome. A traditional herbal
remedy obtained from the chaste
tree (*Vitex agnus-castus*) has been
found to contain compounds that
mimic the effects of certain sex
hormones and neurotransmitters.
In trials, these have reduced PMS
symptoms such as depression,
mood swings, headaches and sore
breasts.

Cotton seed oil may provide us with
a contraceptive for men.

233

with other plant extracts such as *Tripterygium wilfordii*), an alliance of many countries is now investigating it as an alternative to vasectomy. Other plant species investigated for their potential as male contraceptives include the *Eugenia* and *Portulaca* species.

Infertility remedies

Many remedies for infertility also derive from plants. In Cameroon, Baka pygmies have traditionally used the young leaf shoots of *Palisota schweinfurthii*, a common herb of the forest understorey, which are chopped and boiled to make a special drink. Asama bark, scraped from the tree *Turraeanthus africanus*, is also traditionally made into an infusion for drinking. The soothing, sympathetic manner of the experienced older women dispensing these preparations and their reassuring words create the positive frame of mind that is an important facet of their medical prowess. Tragically, Baka society, like that of so many other forest peoples worldwide, is critically threatened by deforestation and other pressures.

Plant species that yield compounds that can be used to treat some kinds of organically caused male impotence are also under study. The alkaloid yohimbine, derived from the bark of the African yohimbe tree, *Pausinystalia johimbe*, may prove effective. One study found that 10 out of 23 men derived benefit from a drug containing the compound.

Plants fighting **HIV/AIDS**

Perhaps the most familiar and important of all plant-derived products used in family planning is not a drug at all, but the rubber condom, which plays a role not just in contraception but in preventing infection from sexually transmitted diseases – of which HIV/AIDS is the most serious.

All condoms are made from a base of natural rubber, the great utility of this compound being its ability to stretch whilst keeping most of its original shape. In its crude form, natural latex has many properties similar to cows' milk, and it must be treated carefully to stop it curdling and decomposing before processing and the addition of chemicals, organic salts and waxes. The slightly oily feel of some condoms is due in part to the natural migration of these waxes towards the outside of the rubber composition. To counteract this effect and help put on the condom, maize and potato starch often coat the outside surface.

The WHO estimates that 42 million people are living with HIV/AIDS around the world, 95% of them in developing countries. Africa, with around 30 million sufferers, is the most badly affected of all.

Seeds of the **Moreton Bay chestnut** are helping in the battle against AIDS.

The rubber condom plays a role not just in contraception but in preventing infection from sexually transmitted diseases.

The appearance of AIDS in the early 1980s provoked urgent research into potential treatments, and the screening of plants for possible anti-AIDS drugs formed a major element of the work. One of the first plant chemicals to show some promise was a novel kind of molecule discovered in the seeds of an evergreen Australian tree, the Moreton Bay chestnut (*Castanospermum australe*). Castanospermine was first isolated at King's College London in 1981, by a member of a research team working on plant toxins. The unusual structure of this alkaloid – one of a number of previously undetected 'sugar mimics', so-called because of their resemblance in size and shape to simple sugar molecules – had caused it to be 'invisible' in the classic screening procedures used previously.

Plant chemists at the Royal Botanic Gardens, Kew, found that many of these sugar mimics had powerful deterrent or toxic effects on insects and on the digestive systems of insects and mammals. The Kew scientists were also collaborating with a team at St Mary's

Natural rubber condoms protect against pregnancy and infection.

235

Bio-prospecting and patenting

Knowledge of how medicinal plants are used by peoples in countries from which they originate has helped make **large profits** for the international pharmaceutical industry. It has been estimated that nearly 75% of 119 drugs derived from higher plants were discovered as a result of ethnobotanical research. The appropriation and use of such knowledge – the intellectual property of different peoples – is causing **increasing conflict** between developing countries, which possess most of this kind of 'intellectual capital', and industrialized countries, which have the money and infrastructure to make drugs based from it.

Compounds found in two medicinal plants traditionally used by the Wapishana people of the Brazilian/Guyanese border, and which have been the subject of recent patent applications have caused much recent controversy. The first comes from the seeds of the **greenheart tree** (*Chlorocardium rodiei*), used for generations by the Wapishana to prevent infections and stop haemorrhages, as a contraceptive and to induce abortion. A second substance is a compound (cunaniol), extracted from the leaves of the **barbasco bush** (*Clibadium sylvestre*), which acts as a neuro-muscular stimulant.

and people of Samoa (which has already taken the form of financial assistance to one particular community), 20% of all profits from the development of the compound by the Aids Research Alliance is to be returned to the country. No one has yet come up, however, with a reliable way of deciding what a **fair share** of the profits from a plant-based drug might be. Unethical collecting by some people has led to a widespread suspicion of all bio-prospecting, even where carried out under agreements **in compliance** with the CBD.

Patenting

Despite the fact that patent law requires an invention to be 'novel', patents are **routinely granted** on medicinal (and many other) properties known for generations. The Asian neem tree (*Azadirachta indica*) is a case in point. Used for thousands of years by local people in medicine and agriculture, 182 patents currently exist in the USA that relate to it, several for medical applications (including infertility), to the dismay of Indian scientists and others.

Although it is a written tradition, many now fear that the whole **Ayurvedic traditional base** of 7,000 or so plants is being systematically patented. The US National Cancer Institute has collected and screened some 50,000 plant and animal samples, many of them from species already widely used by indigenous peoples in the 30 countries involved. Most of this collection was made, however, before the CBD, but the USA is the only major nation **not to agree** to the convention. The US army now holds patents on tropical disease treatments, as well as micro-organisms that could be used in biological warfare.

Patent laws are being employed to privatize any gene or medicinal property that may make a profit.

The 1992 UN Convention on Biological Diversity (CBD) was intended to encourage the free flow of **scientific information** and the dissemination of genetic material, alongside the equitable sharing of any financial rewards that came from the development of products based on biological diversity.

In practice, however, problems of legal interpretation have created a complex situation, which has made the practical application of the convention **open to debate**. In the meantime, led by America, patent laws are being rapidly employed to privatize any gene or medicinal (or other) property in any plant, animal or other organism that has potential to make a profit. Many people in developing countries fear they will lose the right to use medicinal plants that have been part of their society for thousands of years.

After centuries of exploitation and the often genocidal policies imposed by many countries, such patent applications are **fiercely opposed** by South America's indigenous peoples.

The CBD (which legally defines the biological diversity of any country as its own property) allows specific agreements to be made about benefit sharing and the rights to the biological and intellectual property of **local people**. Despite the problems, a number of agreements have been made, in which governments, research institutes, reserves of some kind and, increasingly, village communities are to 'benefit' from such deals. One well-publicized case is that of the development of the anti-viral compound **prostratin**, isolated from the Samoan tree *Homalanthus nutans*. As part of a carefully negotiated settlement involving the government

Hospital in London on the anti-viral effects of plant compounds. In 1987, the sugar mimics were tested for anti-viral activity. Although they had little effect on many common viruses, such as herpes and flu, castanospermine in particular had a dramatic effect on the AIDS virus HIV. The surface of HIV is covered with proteins to which many sugar chains are attached. Castanospermine alters the properties of these chains and leaves the virus more susceptible to attack by the infected person's immune system and less effective at attacking these cells.

Natural castanospermine, however, was found not to be effective enough to use as a drug, so chemists have formulated a derivative, Bucast, based on its unique structure, which is several times more powerful. It has been used in human clinical trials in San Francisco since 1994 and more recently in Japan. Similar castanospermine derivatives are being tested elsewhere.

A wide range of plant-derived compounds have now been shown to possess anti-HIV activity in various ways, including gossypol (from cotton seeds), papaverine (from poppy seeds), glycyrrhizinic acid (from liquorice), hypericin (from *Hypericum* spp.) and prostratin (from the Samoan tree, *Homalanthus nutans*).

Anti-AIDS drugs have greatly improved the prognosis for people with HIV/AIDS in industrialized countries. But most people with the infection live in the developing world where these treatments are still prohibitively expensive and access to them remains virtually non-existent.

Medicinal plants and the law

If plants are the history of medicine, they are also part of its future. But it is a future in which complex legal and ethical questions will increasingly rule our use of plant-based medicines, whether they be traditional herbal remedies or synthetic drugs modelled on natural ingredients.

Regulators of herbal medicines face a difficult task in implementing safety measures without overburdening the complementary medicines sector with legislation. Many countries give a special status to traditional medicines, recognizing that a long history of use is a form of long-term safety testing. For example, legislation proposed by the EU in 2002 will allow straightforward registration and over-the-counter sales of the less potent herbal remedies, so long as these can be shown to have been used within the EU for at least 30 years. More potent medicinal plants will still be available, as at present in the UK, when prepared by a herbal practitioner. Increasingly, such practitioners are members of a professional body, such as the National Institute of Medical Herbalists.

The safety of herbal remedies is an important issue – traditional Chinese medicines have, for example, been found on sale in Europe containing poisonous ingredients because the plant sources have been accidentally misidentified. The Royal Botanic Gardens at Kew runs an identification service to help importers of Chinese traditional medicines ensure that their herbal medicines contain the correct ingredients.

239

Plants that **transport** us

Unless you go barefoot, whichever way you travel and whatever your destination, whether the office, a foreign country or the moon, plants will be helping you get there — and back.

● Cork tiles cut from the bark of the cork oak have been regularly used in the construction of the heat shields and fins of spacecraft, helping to insulate its astronauts against the intense heat generated when re-entering the atmosphere after orbiting the earth.

● Cork was used as an insulating material on the external fuel tanks of the rockets that propelled the most recent Space Shuttle into orbit.

● Each time a jet aircraft makes a flight, its engines make use of plant oils such as coconut and castor bean.

Most of us use plants in less spectacular ways to get us from A to B, but without them, and one in particular, we would find it hard to move at all: rubber.

The source of the natural rubber indispensable to tyre manufacturers is the rubber tree *Hevea brasiliensis*. Although native to a wide area of the southern half of Amazonia, most of the world's rubber supply now comes from trees grown in commercial plantations and small-holdings in Thailand, Indonesia, India and Malaysia.

When the bark is cut – to a depth of about 1 mm (1/10 in) – an interconnecting system of tubular vessels just beneath its surface exudes rubber latex, a thick milky liquid that is, in fact, a suspension of rubber particles in a watery fluid. Tapping is a highly skilled operation since the best latex-yielding vessels occur closest to the cambium growth layer. If damaged, regeneration of the bark is hindered and the tree's life shortened.

A remarkable feature of the rubber tree, however, is its ability to renew quickly the supply of latex in the area of bark just tapped, after the initial flow (which lasts for 1–3 hours) has stopped. On commercial plantations and small-holdings tapping is most often done by cutting a groove in a 'half spiral' round part of the trunk. By removing a thin sliver of bark from its lower edge, this same groove can be repeatedly tapped. In conventional systems, tappers work successive panels of the tree, first using easily accessible virgin bark and then returning to cover renewed bark six or seven years later. The tapping of a single tree is usually carried out for some 10–20 years, but if skilfully done, may continue for over 25 years. The development of chemical stimulants, which are applied to the tree, delaying the natural coagulation of the rubber from the cut, has enabled yields to be maintained whilst reducing tapping frequency. Chemical stimulants are also used to treat areas of bark subject to the more recent method of 'puncture' tapping, which involves making one puncture a week on a scraped area of bark. Unlike the conventional method in which the latex is caught in and emptied from small open cups attached to the trees on the day of tapping, puncture-tapped latex is collected after two or three days – still liquid because the receptacles that catch it are closed.

The latex that will be used for tyre manufacture is converted to dry rubber by the addition of acid, which coagulates it, clumping the rubber particles together. As it dries, it is then formed principally into blocks and sheets.

Previous page: Rubber tyres.
Opposite: A Nenets man in Siberia bends larch planks over the embers of a fire to make sled runners.

Rubber tyres

Tyre making is a complex process since both radial and cross-ply tyres comprise many layers that are built up step by step. Production begins by masticating the raw rubber: working it to soften it and to prepare it for mixing with other chemicals. Early on in the production of rubber goods at the beginning of the 19th century, it was found that raw rubber was an awkward material to process and shape satisfactorily, and that a means of softening it was needed. Though mastication (invented by Thomas Hancock in 1820) solved this problem, and enabled rubber to be used to make such innovative products as the waterproof raincoat, it had the inconvenient attribute of making the rubber mixture stiff when cold, and soft and sticky when hot.

By adding sulphur to the rubber and heating the two together, the chemical linkages between the rubber molecules were stabilized and the problem was solved. A compound was thus made that was strong, elastic and relatively temperature insensitive. Invented by Charles Goodyear in 1839, this process, known as vulcanization (after Vulcan, the Roman god of fire), is essential for the production of all modern rubber products, including tyres. Though the process is now far more

The past price of rubber

Like many other plant products we take for granted today, the potential of rubber was **first discovered** by Amerindian peoples of South and Central America. A famous game once played by Mayan Indians revolved around the hitting of a heavy rubber ball – using only the head, legs or shoulders – through a stone ring fixed to the side walls of a special court. **Columbus** found smaller rubber balls being used as toys by children in Haiti and on his return to Spain presented several to Queen Isabella. A little later, in the early 16th century, **Cortés**' men in Mexico used rubber latex to waterproof their clothing, having observed the use made of it by local people. The **Omagua Indians**, who populated the banks of the central Amazon in large numbers until

the arrival of Europeans, had long made pouches, bags, footwear and even syringes from natural rubber.

Though the substance had intrigued those **travellers** who encountered it, the scientific world was not really interested until Charles Marie de la Condamine brought samples to Europe in 1744. Commercial development, however, did not begin until 1823, when Mackintosh took out his patent for various kinds of rubber-proofed fabrics. But the most significant impetus was **Dunlop's patent** for 'a hollow tyre or tube made of India-rubber and cloth, or other suitable materials', which precipitated what had been a moderate but profitable industry into a wild-commodity boom.

From about 1880 to 1911, with the spiralling demand for rubber for the

motor tyre industry, a barbarous greed gripped certain individuals who set themselves up to 'manage' for their companies the tapping of rubber latex from Amazonian *Hevea brasiliensis* trees. Thousands of Amerindians were forced, using horrific cruelty, to gather the **precious latex**. This systematic persecution resulted in the deaths of vast numbers of people: some 40,000 Indians from the Putumayo region alone were killed in just six years.

By 1914 the great fortunes that had been made at the expense of the native people of the Amazon had collapsed. The Brazilian seeds taken to Southeast Asia in the late 1870s had prospered and their yield of latex had at last exceeded that from wild South American trees. The Amazon rubber boom was **mercifully at an end**.

sophisticated, sulphur is still the main agent used. To make tyres, however, many other chemicals are added to the rubber base before vulcanization, both to assist this process and to protect the rubber from the effects of heat, light and ozone in the air. Synthetic rubber of different kinds may also be added, as well as pigments and special fillers mixed in to give strength and stiffness. The most common pigment, carbon black, turns the pale or dark brown rubber to the familiar black of tyres. Complex blending processes produce rubber of many different kinds with specialized properties, which are used alongside synthetic rubber for the different components of the tyre, including its tread, side walls, carcass and other smaller items.

Cloned commodity or way of life

As well as the introduction of chemicals, which stimulate the flow of latex, careful selection and breeding techniques have greatly increased the output of rubber trees over the years. Research continues to focus on improvements in productivity and quality, and on developing plants resistant to South American leaf blight. The search for genes suitable for modification, to produce highly productive or disease-resistant transgenic clones, has been a significant feature of this work.

Despite their enormous importance in providing genetic material that has boosted cultivated rubber trees in other parts of the world, wild rubber trees native to the forests of the Amazon continue to be destroyed at an alarming rate. Many of Amazonia's rubber-tapping communities, however, whose way of life has ensured the protection of wild trees, have fought hard – despite years of intimidation and death threats – to maintain their traditional livelihoods. Although their rubber has been considered 'uneconomic' for the supply of major industry, many now supply latex for smaller ventures, such as the production of 'vegetable leather', used for many products including bags and shoes.

The seeds of world rubber

Eager to have their own rubber plantations in Southeast Asia and to break the **Brazilian monopoly** on this lucrative raw material, the British tried for some time to grow *Hevea brasiliensis* seeds and plants in England and abroad.

All these attempts met with failure until a batch of 70,000 seeds, carefully packed in **banana skins**, were shipped from Brazil by Henry Wickham, arriving at the Royal Botanic Gardens, Kew, in 1876. Only 2,397 of these seeds germinated in Kew's orchid greenhouses, and most were sent to Sri Lanka, for planting. It was in Malaysia, however, that the Kew seedlings were to assume greater economic importance, following the development of **rubber tapping** techniques in Singapore. By 1990, some 2 million ha (4.9 million acres) of rubber trees were under cultivation in Malaysia. By 2000, however, with the shift away from rubber to other crops – especially **oil palm** – Malaysia had dropped to fourth place in global rubber production. Thailand is now the world's biggest supplier with nearly 2.5 million tons in 2001, followed by Indonesia (1.6 million tons) and India (622,000 tons). Total world production is in excess of 7 million tons.

243

Rubber for roads, rails and runways

Latex from the rubber tree is present in the vast majority of the world's motor vehicle, aeroplane and hard bicycle tyres, and makes travel more comfortable for the passengers of many our newest, most efficient trains. This single tree species made possible the development of vehicles that could reach their destinations much more quickly, safely and comfortably than before; in fact, it revolutionized the potential of every form of transport using wheels.

Motor vehicles and aeroplanes

Although solid rubber tyres for horse-drawn carriages came first, the pneumatic tyre made its successful debut on the wheels of a bicycle. In 1881, John Boyd Dunlop launched his own version of the inflatable tyre and the painfully bumpy bike ride was no more. With the development of the motor industry at the turn of the century, the tyre had its future assured and became the major product of the rubber manufacturer. This has been the case for many years, with tyres accounting for around 70% of all the uses to which natural rubber is put. Demand for motor vehicles, and therefore tyres, is, however, greatly affected by economic growth. The recent slowdown in the global economy resulted in only half of all natural rubber being used for tyres in 2001.

Whilst the first tyres and all rubber goods until the Second World War were made entirely from a carefully processed base of natural rubber, the Japanese takeover of Malaysian, Indonesian and Thai rubber plantations in 1941 prompted the USA to develop rapidly synthetic

Other uses for rubber in vehicles

Tyres are not the only parts of motor vehicles that depend on this versatile, long-lasting material.
- Natural rubber plays a part in interior **suspension**.
- Engine **mountings** are made primarily from natural rubber compounds to cut down vibration and noise.
- Many **small components**, such as bushes and gaskets hidden under the bonnet or in the chassis of the car, as well as some mats for our feet, use natural rubber.

Rubber from other plants

Hevea brasiliensis is not the only plant that yields a useful rubber latex. Many other species, mainly native to the tropics, and even some fungi, have been found to contain it, and until the end of the 19th century several were in use. These include:
- The **Ceará rubber tree**.
- Several species of the **Castilla tree** in tropical America.
- The India rubber tree in Asia.
- A **dandelion**, *Taraxacum bicorne*, which provided Russians with latex

during the Second World War. Although *Hevea brasiliensis* is the major source of commercial rubber used today, **guayule** (*Parthenium argentatum*), a shrub native to northern Mexico, Texas and southern California, supplied 10% of the world's natural rubber in 1910. When supplies of Malaysian and Southeast Asian rubber were **cut** during the Second World War, large investments were made in guayule cultivation in the USA, a major advantage being its

tolerance of very arid conditions. Today, guayule is the focus of commercial interest because its latex does not contain the allergens found in hevea latex (which can cause severe allergic reactions), making it suitable for the production of items such as **rubber gloves**. Whilst work is being carried out to increase latex production using genetic modification, there is also interest in guayule's anti-insect and anti-fungal properties as a wood preservative.

substitutes. The increased demand for car tyres after the war and cheap and plentiful supplies of oil encouraged worldwide expansion of the use of these new kinds of rubbers, which account today for about 70% of all rubber used. Synthetic polymers such as styrene butadiene (SBR), now the most common synthetic rubber, made from coal and oil by-products, are now present in most vehicle tyres in varying proportions, but natural rubber is still a vital component.

Another plant product, rosin (extracted from the trunks of living pine trees, decayed pine stumps and pulped pine wood), is used in the processing of natural and synthetic polymers to improve tack. Alginates from seaweeds have been used to help cream and stabilize liquid latex.

Different vehicles, of course, require different kinds and thicknesses of tyre according to the job they must perform. In general, the greater the load carried and the higher the stress, the more natural rubber will be present. This is largely due to two important attributes, which make natural rubber superior to other man-made materials in these respects: its good building tack (adhesion) and strength, and its low heat generation when flexed continually. These properties are especially valuable for large-tyred vehicles carrying heavy loads.

For large aircraft, including Concorde, where tyres are subjected to tremendous pressures during take-off and landing, natural rubber is likely to account for 100% of the polymer used. In smaller aircraft, the proportion is over 90% of the total polymer. The great strength of natural rubber, its unique ability to dissipate the heat generated in landing and take-off and ability to remain flexible even after long exposure to the sub-zero temperatures encountered in high altitude flying, have made it the pre-eminent material for aircraft tyres of all kinds. Natural rubber also enables aircraft tyres to be re-treaded up to eight times.

Heavy commercial road vehicles such as trucks and buses also have a high proportion of natural rubber in the tread of their tyres as well as in the inner layers, where it supplements man-made polymers and steel and textile reinforcements. In trucks (which account for one-third of all natural rubber used) natural rubber may account for 80% of the total polymer. Many off-road vehicles too, including tractors, military and logging vehicles, and earthmovers, depend on a very high proportion of natural rubber. Since the problem of heat generation is not so severe with cars, the external tread of their tyres is usually made almost entirely of synthetic rubber, such as styrene butadiene, which wears well and gives good grip on roads.

Tyres are complex, composite structures, made up of different components that include – besides the tread – the carcass, side walls and inner liner. The most common type of tyre for passenger cars today is the radial, in which the tread is braced by steel and layers of synthetic fabric. Tyres of this kind, especially all-weather

Messieurs Michelin

It was the French who **pioneered** rail transport on rubber tyres. After André Michelin had spent a far from restful night on an express train thundering its way to the Côte d'Azur in 1929, he asked his brother Edouard to try to find a way of improving this means of travel. Edouard's answer to the problem was the **pneumatic** tyre. By 1931, the first Micheline, a commercial rail car prototype, had been unveiled. The new vehicle, powered by a petrol engine, looked much like a lorry but had tyres adapted to run along thin rails. As designs progressed, vehicles came to look more like **modern trains**.

Running on 24 pneumatic tyres – and with rubber used to mount the engine and improve the suspension – the rail cars were a **great success**. They were able to travel fast and quietly to please the passengers and they also cut down railway maintenance.

In 1951, the Paris metro became the first exponent of a brand new design by Michelin, which enabled ordinary tyres to run on a simple concrete track, with **steel wheels** and rails to provide guidance and ensure safety in the event of tyre failure. The great success of the Paris metro led to the building of the Montreal and Mexico City systems, which now carry millions of passengers each year.

245

radials, may contain about 30% natural rubber, chiefly in the inner casing components and side walls. Natural rubber is vital to their manufacture as its properties of strength and tack enable the shape of the tyre to be maintained while it is being built up. Cross-ply tyres, which have generally been replaced worldwide by radials, use considerably less natural rubber in their construction.

Rubber on and under roads

Since the 1930s, rubber has been mixed with bitumen in some areas, with impressive results. Continued research has shown that chief among these are an approximate doubling of the life of the road surface, the enhancement of water-drainage and a reduction of road noise. Natural and synthetic rubber can be added in the form of lattices or as powders or crumbs to the bitumen, or (though research suggests that it is less effective) in the form of discarded car tyres. Despite the clear advantages of using rubber compounds to produce safer, quieter and more durable roads, their relatively high cost and the reported dominance of bitumen suppliers has meant that there has been a relatively slow up-take of this beneficial practice.

But if not in the surface of the road, there is often rubber underneath it. While France, Germany and the USA tend to use synthetic rubber, in the UK (with its largely cooler climate) the bearings supporting road (and rail) bridges are almost always based on natural rubber, allowing the structures to expand and contract as temperatures go up and down.

High speed trains on rubber

Astronauts, road users and plane passengers are not the only travellers to benefit from rubber trees. The tyres of many of the trains that speed along the Paris, Lyons and Mexico City metros, and all of those using the underground systems of Montreal, Marseilles and Santiago de Chile, are also made in part from natural rubber.

Although the trains are not the fastest in the world, their rubber tyres enable them to accelerate and decelerate quickly and to tackle gradients as steep as 1 in 7, reduce noise levels considerably and lessen vibrations, making them among the quietest and most comfortable to travel in. Sapporo, the largest city on the northern Japanese island of Hokkaido, and several other major cities in Japan now have subway trains that run on rubber tyres. Some of the country's commuter trains and monorails, like those of Tokyo and Kyushu that run above the ground, are similarly equipped.

Like most other 'over-ground' trains, the high-speed Eurostar depends on different kinds of rubber in many different areas, including its shock absorbers, couplings and secondary suspension system.

London's underground

'Tube' trains do not run on rubber tyres, but **natural latex** still plays an important part in the railway's functioning.

■ The **escalator handrails** that usher us up and down are made in part from it.

■ **Rubber blocks** help support the track.

■ **Rubber suspension units** cushion the bodywork of the carriages.

Trees for transport

Wood has been indispensable to the development of rail, road, air and sea travel all around the world. Though its use is now reduced considerably in many forms of transport, it is still vital for a number of construction purposes.

Rail travel

Until the early 1990s, the flooring of London Underground's trains was made of rock maple (see overleaf) imported from Canada. Though man-made materials have now replaced it in newer trains, many still display it in their distinctive grooved flooring. Beneath the train, shoe beams supporting the collector shoes, which pick up the current, may still be made of wood: commonly beech wood impregnated with synthetic resin and made dense (and therefore extremely strong and resilient) by heat and pressure.

The Paris metro ran until recent years on a track made of the West African rainforest timber ekki (see overleaf).

Ekki has also been used in Europe alongside several other tropical hardwoods such as afzelia from Africa to build the underframes and bottom cladding of railway vehicles. Other species used for railway sleepers (although not in Britain) include kempas from Southeast Asia and mora from South America. Though many have now been replaced with concrete, oak was in the past the principal timber for sleepers used by British Rail, and a number are still in use today.

Inside the passenger carriages on British railways, decorative veneers of teak, sapele, walnut, cherry, box and ash were once very common. But these have now mostly been replaced by man-made composite materials.

Wood on the road

Road vehicles still rely on wood to varying degrees. Although the frames and bodywork of modern cars are now generally made of steel, aluminium or reinforced plastic glass, the British Morris Minor and Morris Mini Travellers and many of the older European and American 'station wagon' vehicles have wooden frames. They are joined in some developing countries by a large number of trucks and buses, whose bodywork develops a particular charm with age.

The last new Morris Traveller was built in 1971, but some 5,000–6,000 are still on the road. Along with almost any other renovation work required, these cars can still have their wooden frames and interior fittings repaired at special centres. The wood used for the Traveller's rear frame and roof rails is ash (see right).

Ash trees provide a particularly resilient wood.

Ash

■ Ash (*Fraxinus excelsior*) is one of Britain's toughest native hardwoods and is well suited to frame making, since it bends easily when steamed and absorbs shock very well. The creamy golden colour of the timber, with a surface grain that resembles the contour markings on a map, is a distinguishing feature.

247

Rock maple

■ Known in Canada as the sugar maple (the source of maple syrup) (see page 110), in the USA as white maple and classified as 'hard maple' in the timber trade, *Acer saccharum* is a native of northern temperate regions and produces pale, normally straight-grained timber, which sometimes carries a distinctive 'bird's eye' figure.

■ The wood is slightly heavier than beech and of high density. It is very resistant to abrasion and the surface wears smoothly without disintegrating. For these reasons, maple timber has been much used to make the flooring of industrial buildings, dance halls, squash courts, bowling alleys and gymnasiums, as well as underground trains. Rock maple was also formerly used by London Underground for escalator treads.

Ekki

■ This outstanding wood, shipped mainly from Cameroon, comes from one of the largest African trees, *Lophira alata*, which can grow to 60 m (197 ft) in height. Its clear, straight bole makes it suitable for cutting into very long timber lengths.

■ Dark red-brown or purple-brown in appearance, with fine whitish flecks, ekki is 50% heavier than oak, and extremely strong. Its exceptional density (similar to greenheart, see page 258) means that it cannot be nailed or fastened unless the wood has first been drilled, but this makes it extremely durable, and resistant to fungi and wood-boring insects.

Caravan frames were once made extensively from tropical hardwoods, such as keruing (*Dipterocarpus* spp.) and mersawa (*Anisoptera* spp.) from Southeast Asia. Today, however, most caravans are made externally of aluminium, although interior panelling makes use of hardboard, chipboard and plywood. Small pieces of solid timber – often ash – may be used to frame windows and doors.

Until the early 1990s, the floors of all London buses were made of strong plywood (usually faced with birch wood) overlaid with man-made coverings. It had been found that this material was simpler and cheaper to install than aluminium or similar alternatives. Until the late 1990s, the seat backs of London's buses were also made from plywood panels.

Wood-based boards are still used in the construction of car interiors. Plywood, for instance, lines the boots and door panels of many modern cars.

Wooden luxury

Perhaps the most familiar item to be made of plant material in modern cars – generally the more expensive ones – is the dashboard or fascia and, in certain models, the door-cappings (or waistrails), picnic tables (fitted to the rear of the front passenger seats) and centre consoles.

In most cases, thin veneers of woods selected for the beauty of their grain and markings overlie a framework made of plywood or solid timber, carefully cut and assembled by hand. Although other woods may also be used, the pre-eminent material for decorative veneers is black walnut (*Juglans nigra*) (see opposite). The sapwood is a pale straw colour and clearly defined from the heartwood, which is generally greyish-brown, but walnut timber varies considerably in appearance, according to local conditions. Infiltrations of colouring matter – usually an attractive smoky-brown or reddish-brown – produce darker streaky patterns in the heartwood, and this decorative effect is often accentuated by the naturally wavy grain.

A particular kind of walnut wood is especially favoured for veneers. An irritation or injury to the tree causes it to form a burr on its branches or on the side of the trunk, sometimes extending beneath the ground. When cut or 'peeled' carefully by machine, a beautiful streaked and mottled pattern is revealed, varying with every burr – a figuring that has been highly prized for centuries by cabinet makers and all those working with decorative woods.

Bentley cars make particularly important use of burr walnut today. Each year more than 6,500 sq m (69,900 sq ft) of this fine wood are fitted to their interiors, adding significantly to the style and elegance of each car. The decorative veneers are cut, then opened out to give a mirror image or matching quarters of the natural patterning. They are

Black walnut, a North American native, is highly sought after for decorative veneers.

so carefully aligned that they give the appearance of one solid piece of wood. The veneer for each car is cut from the same log and is applied to the upper surfaces of all the wooden parts, harmonizing not only the grain and figuring of the wood, but the car interior as a whole. Solid, straight-grained walnut is used for the fascias and some other parts. Other woods also in evidence include the lighter 'Australian striped walnut', not in fact a walnut species but *Endiandra palmerstonii*, used for decorative cross-banding, sometimes with a box or ebony inlay. Oak, maple, madrona, amboyana and cherry have also been used as veneer.

All the wood used for these sumptuous cars is carefully checked for quality, colour and grain before assembly. Minor flaws in the natural figuring are corrected with a paintbrush and matching pigment.

Different woods are involved again in the structures underlying the veneers. Most interior panels are made as pressed laminates and strength is provided by African mahogany or tulip poplar. Fascia and console panels are made from birch-faced plywood, overlaid with a backing veneer of makoré and straight grain walnut on the front, to prevent twisting. Valued for its great strength and resistance to warping and twisting, birch-faced plywood is also used elsewhere in the cars.

Walnut

■ The black walnut (*Juglans nigra*), a stately, wide-spreading tree when mature, is one of 25 walnut species. Its dark, rugged bark and numerous toothed leaflets help distinguish it from the common walnut (*Juglans regia*). Native to North America, it is found throughout most of the eastern half of the USA, naturally occurring in 30 states, including Kansas, Missouri, Kentucky and West Virginia. America's most important hardwood timber, with a heartwood that is very durable, black walnut is highly prized for many different uses, notably the supply of veneers for furniture and all kinds of high-quality woodwork.

■ The high commercial value of black walnut timber has promoted much research into its genetics, growth and cultivation, and, with the development of clones selected for the particular straightness of their stems, it became the first tree species to receive a US patent.

■ Various unrelated timbers are sometimes described as walnut, including satin walnut (*Liquidambar styraciflua*) from the USA, African walnut (*Lovoa trichilioides* or *Plukenetia conophora*), East Indian walnut or koko (*Albizia lebbeck*) from the Andaman Islands, and Queensland walnut (*Endiandra palmerstonii*) from Australia. However, only *Juglans* species are true walnuts.

249

Finishing touches

Leather car upholstery also owes part of its distinctive quality to plants. Though, like most soft leathers used today for clothing or shoe uppers (see page 89), the cattle hides made into seats are tanned with chromium to give them suppleness, they are also immersed in plant tannin solutions to give them strength and to stop the leather from stretching with constant use. The main sources of plant tannins used in Europe, North America and Japan today are black wattle trees, grown chiefly in southern Africa, quebracho from Paraguay and Argentina (see page 89), and sweet chestnut trees from Italy (see page 90).

To leave the car in sparkling condition, plants can help polish up the paintwork. Carnauba wax, which forms on the young leaflets of the Brazilian wax palm (see page 40), is a component of many car polishes today. Only small amounts are needed, as the wax is very hard, but it is still one of the best materials for the job, supplementing the larger quantities of oil-based paraffin wax and small amounts of beeswax which, together with various solvents, make up the polish base.

Trees for air travel

As a light but strong material, easily worked and in plentiful supply, wood was the natural choice for the builders of the world's early aircraft, and is still used today for the components of many light

Plastic panelling from plants

Many modern car interiors benefit in different ways from plants. As long ago as 1941, Henry Ford used resin-stiffened **hemp and flax** fibres to build the bodywork of one of his cars. Today, hemp along with other 'hard' fibres such as kenaf, jute and flax is being used to make composite panels for use inside a range of vehicles.

Consisting of around 85% cellulose, hemp fibres are extremely strong in relation to their weight, and have outstanding **mechanical** properties. Ford discovered that his hemp car was able to withstand ten times the impact of those made of metal. Now, new composite materials that

incorporate hemp have been developed, which improve on the **strength** and reduce the weight of glass fibre and petroleum-derived plastics by as much as 30% and which are less prone to splintering. Part of the move towards producing 'greener', recyclable car parts, natural fibres are also cheaper to process than traditional materials. With momentum led by **Germany**, which accounted for two-thirds of the European production of natural fibre composites in 2001, car manufacturers such as Mercedes (Daimler/Chrysler), BMW and Audi/Volkswagen have led the way. Following the earlier introduction of

jute-based door panels in Germany, fibre usage in European cars increased from 4,000 tons in 1996 to over 21,000 in 1999 and is expected to continue to rise substantially.

It is estimated that each of the vehicles made world-wide each year could use some 20 kg (44 lb) of **natural fibres** to make parts such as door panels, parcel shelves and boot linings. About 4 billion tons of hemp, flax, jute and kenaf are currently grown each year. A typical door panel may comprise 25% hemp, 25% kenaf and 50% polypropylene, specially formulated not only for strength and lightness, but for fire-resistance.

aeroplanes, including gliders and microlights. Leonardo da Vinci's man-powered Ornithopters, for which several designs were sketched in the late 15th century, were to have flown by means of bird-like wings (ingeniously controlled by human hands and feet) made of a wooden frame (probably beech) covered with a lightweight fabric. Unfortunately, Leonardo, who spent many years perfecting his delightful 'flapping machines', misunderstood, like the few who had preceded him and most of those to follow for the next 300 years, the basic principles of flying.

By the early 19th century, however, as aerodynamic principles were beginning to be understood, experiments with kites and gliders made of lightweight wood and fabric brought the reality of aviation within reach. In 1853, the first successful glider, designed by Sir George Cayley, made its maiden flight. Further experimentation by many individuals and the perseverance of the Wright Brothers at last brought the first powered take-off, by the Wright Flyer, in 1903.

The materials used to make the framework and vertical and transverse struts as well as the wing spars of many of these early aircraft – chiefly bi-planes up to 1935 – were ash or spruce, or sometimes bamboo lengths. The whole structure was assembled by gluing, pinning or bolting the component parts together, and was braced with numerous tension wires. The wings were covered with strong fabric, such as varnished silk, unbleached muslin or rubberized, waterproof linen, and had ribs of plywood, generally birch-faced, a very versatile material that was also used to panel the fuselage where necessary. Steel tubing and small amounts of aluminium were also used to help strengthen the aircraft. For lightness, seats were made of wickerwork.

In 1912–13, the French-built Deperdussin racers introduced the monocoque design for aeroplanes. For the first time the stresses were carried by a single outer shell, obviating the need for any internal skeleton. Their streamlined body was made from three layers of tulip wood veneer glued together over a strong hickory frame. Though these materials proved too expensive to be commonly used, the strength and flexibility of other woods and wood veneers, which could be curved when heated, was essential for the early monocoque designs.

From the 1920s, military aircraft in particular began to make more use of metal for their construction, using an all-metal 'stressed skin' in place of wood. But the Second World War brought a dramatic revival of this now somewhat maligned material, mainly because of the need to conserve scarce and valuable aluminium resources (see right).

The famous British combat aircraft the Mosquito – made almost entirely of wood – turned out to be one of the most successful aircraft of all time. A mix of hard and softwoods, the Mosquito's streamlined shell was made of balsa wood sandwiched between two birch veneers.

Stinging nettles for aeroplanes

With aluminium in short supply during the Second World War, some surprising plant materials came to be investigated, including fibres from the common **stinging nettle** (*Urtica dioica*). Britain's Ministry of Aircraft Production experimented with the use of a very strong, high-grade paper made from nettle fibre for **reinforcing plastic** aircraft panels as well as gear-wheels and other machine parts. With its considerable tensile strength – greater than anything else tested at the time – nettle-reinforced **gear-wheels** were reported to have been almost as strong as steel. By 1943, however, in view of persistent problems with supply (enormous quantities would have been required to produce paper on a large scale) and processing, and the availability of superior materials, experimentation with nettles came to an end.

251

Today, steel- or aluminium-framed planes predominate, many of which have a bodywork of composite materials such as glass-reinforced plastic. But the bodywork of some light aircraft and occasionally the frame is still made of wood.

● For larger components, such as spars, Sitka spruce from North America has been the wood of choice, but is now often replaced with white pine, also from North America.

● Wing ribs and occasionally wing edges, as well as seat bases, floor pans and instrument panels, are commonly made of plywood. The strongest of these is birch – commonly from trees grown in America and Finland – but other woods, such as Douglas fir, balsa, okoume or gaboon mahogany, basswood and African mahogany, are also used.

● Mahogany (*Swietenia* spp.) played an important part in early aviation for its superior weight to strength properties. Propellers were often made entirely from this wood. Central American mahogany was chiefly used, cut into a number of laminates that were glued together and shaped by hand to form the blades. After 1935, 'densified' wood was generally adopted, however, using other timbers. This was made by compressing multiple laminates together to double their density.

● Wooden propellers are still used on lower-performance sports aircraft, as wood is far better than more modern materials at absorbing vibration. Current timbers used include spruce and sugar maple.

Plants for water transport

About 500 km (312 miles) from Lima at Huanchaco on the northern coast of Peru, local fishermen still ride out to sea on *caballitos de totora*, their 'little horses' made from reeds. Identical craft, distinguished by their graceful curving prows, were being used by the Moche people in pre-Inca times. Later, in the 16th century, Spanish chroniclers marvelled at the fishermen's prowess: 'everyone set on horse-back, cutting the waves of the sea, which in their place of fishing are great and furious'.

To make the boats, totora reeds (*Schoenoplectus californicus*, also recognized by botanists as *S. riparius*) grown in the vicinity are cut, dried and lashed together in two thick bundles, about 3.5 m (11 ft) long. These are then tied side-on to form the raft-like, buoyant base, with the rear end blunt and the front end curving to a sharply tapered point. The fishermen kneel or sit astride these *caballitos*, and use a paddle to help manoeuvre them out to sea

Some 1,200 km (750 miles) to the south, a different species or possibly a subspecies of the more northerly totora reeds (*S. tatora* or

S. californicus subsp. *tatora*) have been used by the Uros and Aymará Indians of Lake Titicaca for centuries to make elegant boats for travelling, fishing, hunting ducks and carrying supplies. The reeds, which grow at the edges of the lake and are also cut by the Uros to form the 'floating' islands on which they live, are stacked up to dry. They are then lashed together with ichu (*Jarava ichu*), a tough bushy grass that characterizes the altiplano (the surrounding treeless, windswept plain). Four long bundles, which taper at both ends, are constructed and tied firmly together to make the distinctive shape of the boats.

Totora reed boats in Peru. Tough fibres grouped around the outside of the long, straight, lightweight stems of reeds and rushes make them durable and easy to work.

Sails, too, are sometimes made from totora reeds, which have been split and sewn together lengthwise, but the Indians generally propel and steer their boats with the help of long poles. As the reeds begin to rot and disintegrate after about six months in the water, new ones are cut to replace them.

Reeds and reed-like plants have been used by various peoples throughout history, especially where trees are scarce, to make rafts and simple vessels. But much more durable raw materials for water transport have come almost everywhere from trees. From the bark canoes of North American Indians to the sturdy hulls and framework of Medieval sailing ships; from Phoenician merchantmen and Roman quinqueremes to Britain's 'clippers' and from balsa rafts to modern luxury yachts – throughout history, trees have provided essential natural materials for sea and river-going craft.

Naturally, a great variety of different timbers – and one exceptional grass, bamboo (see page 256) – have been used by different peoples for boat construction, depending on the type of craft and its function. In northern Europe, from at least the time of the Vikings until the 19th century, the great merchant ships and naval vessels depended largely on oak for their strength and durability. For many of the larger ships, whose often extraordinary voyages made possible the expansion of Western civilization, huge single timbers, such as those that formed the stern posts, were much in demand to take the ships' tremendous loads and stresses, though small, overlapped and interlocking pieces were often substituted. Curved members such as frames were similarly often formed from individual sections fitted together lengthwise, though single pieces taken from the naturally occurring curves in certain trees were preferred. Such timbers could only be cut from enormous ancient oaks, and as the size of fleets increased, the difficulty in finding suitable trees increased.

Wooden warships

The building of a Tudor warship like the *Mary Rose*, which sank in 1545, was a major undertaking both in terms of **manpower** and **materials** used. While a figure for the amount of wood needed for this ship has not been calculated by those involved in its restoration, for an 18th-century gunship like the *Victory*, it has been calculated that **6,000 trees** (90% of them oaks) were used. Frames, deck beams, flooring, side bracing and supports as well as out-board planking were all frequently made of this resilient timber. Although the **British Navy** had a guaranteed supply from the royal forests, most dockyards still had to obtain much of their timber from contractors who monopolized the supply and sometimes provided wood of inferior quality.

253

By the beginning of the 19th century, iron and steel were being introduced into the hulls of merchant ships, but teak and other tropical hardwoods were highly valued. The famous British clipper, *Cutty Sark*, was planked with teak and greenheart timbers, which were bolted to wrought-iron frames. Mahoganies and cedar from South and Central America and agba from West Africa, as well as other, relatively light, decay-resistant timbers, were also widely used around this time. But by the end of the 19th century, most deep-sea cargo ships had hulls, decks, spars and rigging made of steel, and the last major wooden passenger ship, the *Torrens*, set sail for Adelaide in 1903.

Today the framework of container ships and most large naval vessels is made of steel or aluminium. Smaller pleasure boats, including cruisers, yachts and sailing boats, are often made of fibre- or glass-reinforced plastic. But wood, despite high labour costs, is still in use.

Modern preservative treatments, special epoxy resin glues and resilient paints have made marine plywood (often faced with tropical hardwoods such as gaboon mahogany and sapele) durable, strong and versatile. The use of this plywood and wood veneers (moulded and laminated to almost any shape) has reduced the amount of timber required for boat building as well as the degree of waste. As new standards for compression strength and absorption rates are set, however, new composite materials, such as that made from recycled cellulose and glass fibre (double the strength of marine plywood), are being rapidly developed. Balsa-cored, reinforced plastic panels have also come to be widely used for the construction of boats of many different kinds.

A South American hardwood distinguished by its extraordinary softness and lightness, balsa (*Ochroma pyramidale*) has given us a timber much used in the past for raft-making by South and Central American Indians and still in use today in certain areas by fishermen. Thor Heyerdahl's historic 8,000 km (5,000 mile) voyage from Peru to the Tuamato Islands near Tahiti in 1947 was also made on a balsa raft, the *Kon Tiki*. Since the mid-1950s, balsa wood has found a different

254

Tree trunks for boats

It is easy to imagine that the first successful means of water transport was developed simply by observing the ability of **trees to float** and their imperviousness to water. Today, the trunks of many large forest trees, which have been **skilfully hollowed** out and further shaped and hardened

with the help of fire and steel instruments, serve thousands of communities of sea-, lake- or river-dwelling peoples.

In parts of West Africa, for example, the trunk of the large forest tree **iroko** (*Milicia excelsa*) is shaped and hollowed out to make a sturdy dugout canoe. In Polynesia, **breadfruit**

trees (*Artocarpus altilis*) provide good boat-building timber that is easily worked. Palm species and mango trees are also useful for making hulls, whilst coconut fibres are sometimes utilized for lashing **outrigger poles** (often sections of bamboo) into place.

but nonetheless important use in modern shipping. As the lightest of all commercial timbers – noted for its low thermal conductivity and high sound absorption – it is used in large quantities as an insulation material in ships carrying liquid gas and other goods that need similar cold storage. Balsa wood has also been widely adopted as the core material in reinforced plastic panels used in the construction of boats of all kinds, including dinghies, fishing vessels, military craft such as mine-sweepers and large luxury yachts, and for shipping containers.

The extreme buoyancy of this wood made it suitable in the past for the manufacture of life-saving equipment such as floats and buoys. Cork was also formerly used in lifejackets.

A traditional jangada. A raft made from balsa wood cut from the Atlantic forest of northern Brazil. These boats have been used for centuries by local fishermen.

For racing boats and dinghies in particular, marine plywood can offer a much better strength to weight ratio than some of its competitors. Most modern wooden boats are built using the 'shell first' method, in which frames are laid inside the completed hull. For a small cruiser, for example, the shell may be built up of separate 'skins', individually contoured and shaped, then glued in alternate directions and heated on a mould to cure for maximum strength. Woods used might include Honduran mahogany and obeche, and utile, iroko and opepe from West Africa, with the outermost ply sometimes laid horizontally to give the impression of traditional carvel planking.

The shell of the cruiser may also be made by a process of 'cold moulding' – that is, by laying long, thin strips of wood diagonally over one another. The advantage of this method is that a variety of softwood timbers can be used, such as Douglas fir or different pine species. Interior frames, positioned to strengthen the hull, are also often made of laminates. Whilst decking timbers might be made of larch, Scots pine or cedar, interior joinery woods have included teak, oak or mahogany.

For lifeboats and fishing vessels, built for the roughest seas, wood was once pre-eminent. Some internal furniture is still made of plywood, but all new British lifeboats are now made from fibre-reinforced plastic. In Britain, the building of wooden trawlers is now carried out by only one company, in northeast Scotland. The bulk of these boats is made from solid oak, but some sections, such as the keel, are made from British larch, while Douglas fir or iroko may be used for the decking.

Despite their impressive size and use of metals for construction, modern passenger liners such as the *Voyager of the Seas* – which became the world's largest cruise ship in 1999 – rely on wood extensively for interior joinery and decking. The decks and handrails of this grand ship are in fact all made of teak.

255

Manila hemp

■ Native to the Philippine Islands, *Musa textilis* is closely related to the banana. Its fibres (known also as abaca) are extracted mainly from the edges of the leaf bases or petioles that form the 'stems' of these giant plants, by crushing and then scraping away the pulp. The strength, lightness and resistance of manila hemp to sea-water made it the premier material for marine ropes, and for many years it competed with synthetic fibres.

■ The Philippines currently produces most of the world's manila hemp, with over 77,000 tons in 2001, while Ecuador produced over 24,000 tons.

■ Today, this fibre is mainly used in the paper-making industry and is valued for a range of specialized papers including tea bags, sausage casings, cigarette filters and currency. In the Philippines, superior grades are used for textiles, as part of a thriving handicraft industry.

Knowing the ropes

From the earliest times, different peoples have made rope and cords from locally available materials to help them build, sail, load and moor their sea- and river-going craft.

Until relatively recently, with the exception of wire cables developed at the start of the 19th century, plant fibres were the only practical source of cordage for the world's shipping, both on a large and small scale. The development of man-made materials and machinery to maximize production has greatly diminished the modern application of natural fibre ropes at sea.

Today's ocean-going ships and tankers as well as smaller pleasure boats of almost every kind, including speed boats, yachts and cruisers, rely chiefly on a variety of synthetic ropes made of nylon, polyester, polypropylene, polyethylene, aramid and other 'advanced' fibres. The superior strength, immunity to rot, mildew and marine decay, and low water absorption of these ropes have all favoured their adoption.

However, several plant fibres of great historical importance are still being used on a minor scale by rope makers in Britain for some nautical and various other applications.

● Manila hemp from the leaves of *Musa textilis*, a relative of the banana (see left).

● Coir from coconut husks.

● Sisal from the leaves of two *Agave* species (see opposite).

● Flax stem fibres from *Linum usitatissimum* (see page 72).

● Hemp from the stems of *Cannabis sativa* (see page 75).

With their hard-wearing, low-stretch characteristics, plus their resistance to heat and sunlight, ropes made of manila hemp and sisal are still widely used for the lashing down and handling of cargo on board ships, while some of the barges on the River Thames have also

Bamboo for boats

Bamboo has been of great importance to the **boat-builders** of China. The sectioned culms of the bamboo plant are said to have given the Chinese the idea of making watertight bulkheads, and these were indeed fitted to their **junks** 2,000 years before they appeared in the West. Today, the huge variety of junks, sampans and rafts that navigate the waterways of China

carrying every sort of cargo are made from a range of different materials, but **bamboo** once featured prominently in many. Besides being woven into screens, roofing mats and deck-housing, split lengths were often peeled and their fibres made into ropes and cables of **great strength**, resistant to rot and stretching and very light to handle. Bamboo was also used to

make rods and laths that helped hold sails to the masts of certain boats.

In Taiwan, the stems of the giant bamboos *Dendrocalamus giganteus* and *D. strictus* were once commonly fashioned into light, sea-going trapezoidal **rafts**. Today, these bamboos have largely been replaced with plastic tubing, but bamboo is still used on some rafts for bracing purposes.

Mayan women sorting and grading sisal fibres in a processing plant near Mérida in Mexico.

returned to using sisal (which can be cheaper than synthetics). Distinguished by its great strength and flexibility, home-grown hemp was for many centuries the pre-eminent fibre for ships' rope and rigging in northern Europe.

Tarred anchor cables were also often made from this resilient fibre until the early 19th century. Whilst steel wire came to fulfil many of the hemp rope's former functions, other fibres from the tropics were also being introduced. Coir ropes, for example, of very large sizes and made from coconut fibre, were used for mooring because of their great elasticity and ability to float.

Although sisal fibres (from *Agave* spp.) are not exceptionally durable, their length (often around 1 m/39 in) and great strength made them very important cordage materials, for many years supplying approximately half of all the 'hard' plant fibres in the world. Today, with the general substitution of synthetic fibres for ropes and the large-scale abandonment of agricultural twine, due to modern harvesting techniques, global demand for sisal has greatly decreased. The fibre has, however, increased in popularity for carpets and matting, and significant quantities are pulped for paper making.

Many thousands of years ago, the properties of several other plant fibres were utilized to great effect in southern Europe and North Africa. The Egyptians twisted alfalfa grass, papyrus and date-palm

Agave

■ Two species of desert-loving *Agave*, both native to Central America (*A. sisalana* and *A. fourcroydes*), produce very similar fibres, both often known as sisal. *A. sisalana*, however, is the source of 'true' sisal whilst *A. fourcroydes* produces henequen.

■ When Spanish conquistadores arrived in Central America in the early 16th century, they found fibre from these plants in common use for making cordage and clothing. Fine clothing in the form of scarves, shirts and dresses made of agave fibre is available in Mexico today but its main commercial applications have traditionally been agricultural twine and rope.

■ Brazil currently produces the most sisal in the world: over 180,000 tons in 2001. This was followed by China (36,000 tons), and Mexico (27,000 tons). Kenya and Tanzania were the largest exporting countries after Brazil.

Greenheart

■ One of the world's heaviest, hardest, strongest and most durable hardwoods, also resistant to marine borers, the unique properties of greenheart (*Chlorocardium rodiei*) have made it the ideal timber for dock and harbour work. Such large volumes have now been cut, however, that the tree, which occurs only in northern Brazil, Venezuela and the Guyanas, is currently facing extinction.

■ Traded in the past in lengths of up to 17 m (56 ft), and with a high commercial value, greenheart, despite its rarity, is still available as solid timber from suppliers in the USA, who export mainly to Europe.

■ Some 12,000 cubic m (423,600 cubic ft) of greenheart from Guyana (where much of the total population remains) was recently used to rebuild the groynes along the seafront at Eastbourne in the UK. The production of plywood is now reported to be a major end-use of greenheart timber.

fibre (as well as camel hair) to make ropes for an impressive range of barges, boats and warships. Some Egyptian galleys relied on ropes not just to set their sails but to give their vessels rigidity and to reinforce the sides. They also used flax, as did the Greeks and Romans, who added hemp and esparto grass to their range of raw materials.

Sea and river travel are not the only forms of modern transport to make use of natural fibre ropes. The steel wire cables that raise and lower modern elevators need a fibre core for lubrication purposes and to help give them their vital flexibility. Though this may be made of synthetics, sisal of high quality is generally the preferred material, particularly in high-rise offices in the USA. The fibre core acts essentially as an oil sump, lubricating the cable and thereby reducing the frictional forces exerted on it as it travels to and fro over pulleys and other machinery. Despite new developments in elevator design (such as the use of magnetics), the use of natural fibre cores is expected to continue for some time.

And to assist us to negotiate our stairs by foot, ropes made of flax (widely used in gymnasiums too) are now replacing wooden hand rails in many homes and offices.

Dock construction

The strong, durable timbers of a number of temperate and tropical hardwoods and some preservative-treated softwood trees continue to play an important part in general marine and fresh-water construction work.

Dock and lock gates, piling, beams and decking planks for wharves, jetties and piers as well as rubbing strakes and fenders, which absorb impact and give protection, still rely to a large extent on the special natural attributes of certain timbers.

Amongst the tropical hardwoods, greenheart (see left), opepe, iroko and ekki (see page 248) have been widely used. The great density of these particular timbers, their resistance to marine borers and to abrasion, and their availability in large sizes has made them especially suitable for marine work, but extensive use has now made these trees vulnerable to extinction in the wild.

Amongst the temperate hardwoods, oak has exceptional qualities (see page 180). After preservative treatment, however, softwoods such as pine and fir species (see page 163) are also useful in dock construction. The docks at Port Newark in New York are set on piles made of treated Douglas fir, with protective fenders made of decay-resistant oak.

Dock fenders as well as those used on boats (which are sometimes made from coir) are also widely manufactured from a mixture of natural and synthetic rubber.

Plants for engines

The liquid still fuelling most of our cars, indeed most internal combustion engines, and the oils that lubricate them are derived in part from plants. The petrol, diesel and liquid petroleum gas (LPG) fuel and the engine oil used by hundreds of millions of vehicles each day are the result of the compression over millions of years of minute sea-dwelling animals and plants, which once lived on earth, and the mud and sediments that covered them as they died. Subjected to intense pressure and heat, this 'organic soup' underwent chemical and physical changes to end up as droplets of oil. The migration of these droplets through porous rocks and fissures led to the formation of underground reservoirs contained by layers of impervious rock, which are the source of most modern fuel oils, lubricants and petrol. The earth's crude oil has also given us, of course, power, heat and light and the petrochemicals from which thousands of items that we take for granted daily – from pharmaceuticals to plastic teaspoons – are made.

It is hard to equate the tiny sea plants that existed in prehistory with the petrol and oil in our cars today, but some much more familiar plants are helping us to extract crude oil from the earth. As the hollow, rotating drill cuts through impacted rock towards the oil deposit, a specially prepared mud or sand is pumped down the drill pipe under pressure. This lubricates and cools the bit and forces the rock debris dislodged by the drill up to the surface around the outside of the pipe. The mud is prepared with the addition of plant-derived materials. Potato starch, palm oil and a number of plant gums make the mud stable, viscous and relatively fluid by absorbing water. The plant gums include gum ghatti, exuded from *Anogeissus latifolia* (a tree native to the dry deciduous forests of India and Sri Lanka), and guar gum, from the seeds of *Cyamopsis tetragonolobus*.

The castor oil plant is the source of an oil whose special properties have made it ideal as a lubricant for some of our fastest engines. The plant has been cultivated for thousands of years.

259

Dynamite from plants

The dynamite used by geophysicists in their **search for oil** by means of seismic surveys almost certainly incorporates material from plants.

The various kinds of modern dynamite are complex mixtures which do not always include nitroglycerine (the traditional explosive base), though most still do. Liquid nitroglycerine by itself is very unstable and liable to explode easily. To stabilize it and make it safe to handle, it was formerly mixed with kieselguhr, a fine, earthy material composed of the fossilized silica coats of minute **brown algae**, which lived in vast numbers in fresh and marine waters millions of years ago. Today, kieselguhr has been replaced by **wood pulp** as the all-important absorbing medium.

Other plant-derived materials that have been used for this purpose include '**gun cotton**' (nitrated cellulose used to make gelatinous dynamite) and **charcoal** from cork which can absorb about 90% of its weight of nitroglycerine.

Castor oil tree

■ The castor oil tree (*Ricinus communis*) is a native of the tropics, where it may grow up to 4 m (13 ft) high. The pale yellow oil extracted from its highly poisonous seeds or 'beans' is used in the production of jet aircraft and racing car engine lubricants, plastics, paints and inks, and is the source of many industrial chemicals. It has pharmaceutical, cosmetic and medicinal uses too, and is well known for its laxative properties. Castor oil beans contain several poisonous compounds, including ricin, a highly toxic protein. Today, most castor oil is produced in India, Brazil, the Netherlands, the UK and Indonesia.

■ The seeds produced by a member of the mustard family, *Lesquerella fendleri*, native to the southern USA and Mexico, have been found to contain an oil that may become a replacement for castor oil in many applications. Three of its fatty acids are similar to rinoleic acid, the main fatty acid in castor oil. The focus of biotechnological development in America, lesquerella has excited researchers because of the potential market for its oil, and for the gum and meal, also derived from the seeds. While oil containing less natural pigmentation (making it more acceptable to cosmetic manufacturers) has already been developed, the gum (which has been patented) is being investigated for use as a texturizer in frozen, processed foods and as a thickener in products such as paints and drilling fluids.

Ground coconut and walnut shells are also used in some oil well drilling operations. A fluid containing walnut shells, along with sand and aluminium pellets, for example, may be used to make small fractures in sandstone rock, allowing oil to flow into the well. The starch from cereal grains, meanwhile, is now used to make some sealants used during bore drilling.

Lubricants from plants

Lubricating oils for the majority of motor vehicles in use today are made from refined petroleum, but engines that must work at very high temperatures to power our fastest forms of travel rely to a large extent on the oils produced by living plants.

All modern turbine engine oils (used, for example, in aeroplanes and space rockets) are based on polyol esters, which are synthesized from natural sources such as coconut or castor oils (see left) and beef tallow. While some petroleum-derived components may also be used, ordinary mineral oils cannot survive the intense heat generated by today's most powerful engines.

Before the development of the gas turbine engine, aviation piston engines had, in fact, used castor oil-based lubricants. Their formulation, however, meant that they were not suitable for high-performance engines and the search began for an alternative which had the properties of natural oils whilst eliminating their tendency to gel at low temperatures, and to form gums and lacquers.

Mineral oils, which were used to lubricate most early jet aircraft engines, seemed to solve the problem for a while, but their use was unsatisfactory as it led to problems of deposition and degradation, frequent oil changes, increased oil consumption and higher mechanical failure rates. It was the simultaneous development of the new gas turbine engine and its ester-based lubricants, made in part from fractionated plant oils, that provided the answer – improving thermal stability and general engine performance, and lowering maintenance costs.

From about 1960, aircraft powered by gas turbine engines progressively replaced their post-war, piston-engined counterparts in airline operation. Today these engines power most military and civil aircraft and are lubricated by the new 'synthetic' aviation oils, comprising mixtures of organic polyol esters (made by reacting organic acids, wear additives, corrosion inhibitors, dispersants and anti-foam agents.

Meanwhile, some of our fastest combustion engines that power racing cars and racing motorbikes rely on the special properties of the plant that has lent its name to a well-known brand of engine oils. Castor oil has been described as the most effective natural lubricant yet discovered, and particular grades of 'Castrol' have been used to help break world speed and distance records in almost every category of motor sport. Castrol R30, for example, comprises around 66% castor oil. Brake fluids also incorporate a high proportion of castor oil.

A range of 'bio-lubricants', which are rapidly biodegradable, non-toxic and made from rape- and sunflower seed oils, are now available for use in Europe in two-stroke engines, such as lawn mowers, strimmers and chain-saws. They were developed for use where there is a risk of damage to the immediate environment through leakage of oil into the soil or water. In the USA, these lubricants (based on soya oil) are now being used by some cars that have short service intervals (for oil change). Following co-operation between vehicle manufacturers and oil companies aiming at sustainable development, the use of bio-lubricants could become commonplace. In 2002, the Alternative Crops Technology Interaction Network in the UK predicted that with regulatory support, from a technical point of view, over 90% of all lubricants could be made from biodegradable plant oils in the foreseeable future.

Millions of vehicles, like this pick-up truck in Campo Grande, Brazil, are now powered either partially or totally by alcohol, derived from **sugar cane**.

Plants for future fuel

As concerns about our depleting stocks of fossil fuels have mounted, in tandem with alarming evidence of the serious air pollution caused by vehicle exhausts, various alternatives to the internal combustion engine, which runs on fossil fuels, have been suggested or tried, alongside 'greener' and cleaner fuels.

The term 'bio-fuel' refers to any plant or animal product that can be used (as an alternative to oil and gas) to produce power or heat, ranging from wood – perhaps the oldest fuel of all – to chicken or pig manure. Unlike fossil fuels, not only are bio-fuels made from renewable resources, but their use produces less of the main greenhouse gas, carbon dioxide. In many parts of the world, renewable energy sources are being adopted as a more responsible alternative to fossil fuels.

Alcohol fuels

By pouring neat alcohol – that is, ethanol (ethyl alcohol) (see right) – into the converted engines of our cars, we can partially or totally replace the petrol needed to run them. Several countries are now doing just this.

Recent support from America's maize-growing states and approval at government level have promoted the modern development of this fuel, and the number of cars using ethanol (in a blend with petrol) in the USA is said to be increasing. Brazil, however, has long been the world leader in the production of ethanol fuel. First adopted for

Ethanol

■ Industrial ethanol is made by the fermentation of plant matter with a high sugar content, such as sugar cane, or which is rich in starch (a more complex sugar), such as maize, barley and wheat. As long ago as the 1880s, it was used by Henry Ford to power one of his earliest vehicles, the 'quadricycle', and some subsequent Model T's. It is reported that Ford's vision was to build a vehicle that was affordable to working families and powered by a fuel that would boost the rural farm economy.

261

Bioethanol

■ After Brazil and the USA, China is the world's third largest ethanol producer, but many other countries, including Australia, India, Thailand and Mexico, are now working to convert their farm surpluses into fuel. In the UK, experts have concluded that bioethanol could be home-produced on significant scale from wheat and root crops such as sugar beet. These crops can be grown with much lower inputs of pesticides than currently used for food production.

Methanol

■ Almost all methanol is now made from fossil fuel sources, mostly natural gas, but it could once more be made from biomass (energy-rich organic material) such as waste wood. Using the blend of 85% methanol/15% petrol is said to produce half the smog-forming emissions of the same vehicle using just petrol, though its potential to affect global warming has been assessed by Friends of the Earth to be the same.

vehicles in 1903, its use was compulsory in some regions during the First World War. The result of a government programme begun in 1975 – intended to reduce the country's dependence on imported oil – means that around 4 million cars are now reported to run on ethanol. Non-diesel vehicles currently use either hydrated ethanol (95% ethanol: 5% water) or 'gasohol' (22% ethanol: 78% petrol). Though cassava is also a source, the juice crushed from sugar cane stems has been the main raw material for ethanol production in Brazil.

One negative aspect of this industry is that large tracts of Amazon forest were destroyed during the 1980s to grow much of the sugar cane needed to produce the fuel, and the main by-product, a very rich organic sludge, was dumped in rivers causing serious pollution. On the more positive side, using a blend of 85% ethanol to 15% petrol is said to reduce greenhouse gases by as much 37 per cent.

Bioethanol (see left) can be used in blends up to 15% without engine modification. Although converted petrol engines running on neat ethanol consume more fuel than those running on petrol, the performance and fuel consumption of pure-ethanol cars is expected to be roughly equivalent to petrol cars in the future. Ethanol is sometimes used as an alternative to lead in the production of high-octane, lead-free fuels. As an additive, ethanol enhances the performance of petrol-based fuels.

Methanol (methyl alcohol) (see left) – originally produced from the distillation of wood – has fuel characteristics that are similar to ethanol and it is widely used as a fuel additive. It has been available in the USA (as a blend of 85% methanol and 15% petrol) as an alternative fuel for racing cars since the 1960s, and has been promoted or demon-strated for use in 'flexible-fuel vehicles' (including ordinary cars and speedway bikes) for some years. No new methanol-powered cars were made after 1998, but this liquid has been assessed as an excellent base fuel for the fuel-cell vehicles of the future, which have traditionally run on hydrogen or hydrogen-rich fuels such as methane.

Vegetable diesel

Crop plants are now helping to solve our energy problems in yet another way. As long ago as 1911, Rudolph Diesel wrote: 'The diesel engine can be fed with vegetable oils and [this] would help considerably in the development of agriculture of the countries which will use it.' While Diesel's engine was designed to run on peanut oil, a 3 to 1 mixture of sunflower oil and diesel is, in fact, very similar in efficiency to ordinary diesel. Various plant oils are now being used to make a fuel that is helping to reduce our dependency on fossil oil. Providing that they are effectively filtered, raw vegetable oils can be used directly to power diesel engines, but their calorific value is much lower than conventional diesel. By adding methanol, however, the oil can be up-graded to become 'bio-diesel' (or RME) for use in unconverted vehicles.

A large number of countries (including Austria, France, Germany, Sweden, Italy, Malaysia, Nicaragua, the UK and the USA) are now making or using bio-diesel. It has developed rapidly in recent years to full-scale industrial production, and it is finding increasing acceptance by the diesel vehicle industry. Europe produces and markets the largest quantities, Austria having developed the first sophisticated standards for bio-diesel quality. A significant advantage of this fuel is that it can be used either alone or in blends with fossil diesel, in modern diesel engines, without modification.

Oil seed rape growing in Hunan state, in China. It forms part of an integrated system that includes rice production and fish-farming. Many countries are using rapeseed oil as a way of reducing their dependence on petroleum.

263

Copaiba

■ Almost all copaiba balsam (*Copaifera* spp.) has been obtained by tapping wild trees which may reach 30 m (98 ft) in height – often from very remote sites, mostly in the Brazilian state of Amazonas, but also in Venezuela, the Guianas and Colombia. Whilst some trees have been grown on experimental plantations, the economics of such ventures and of the large-scale use of copaiba for fuel have not been considered worthwhile. For aromatherapy use in Japan, however, copaiba oil currently fetches a retail price of over $4,000 per litre ($880 per gallon).

Plants currently used vary from country to country, but the main ones that are used are:

● Oil seed rape, the source of the first bio-diesel to be commercially produced (in 1988), now accounting for around 80% of raw materials.
● Sunflower seeds.
● Soya beans (mainly in the USA).
● Yellow mustard seed (in the USA).

Other raw materials being converted into bio-diesel include palm oil, linseed oil, beef tallow and recovered frying oils (of plant or animal origin). It was a surplus of soya beans in the USA and falling prices that were a major impetus to the development of their use for fuel there. About 3.3 kg (7.3 lb) of soya bean oil are needed to produce 4.5 litres (1 gallon) of bio-diesel.

Mustard seed oil is regarded as a low-value waste product because it is inedible. But as it comprises 90% mono-saturated fatty acids, it is considered perfect for bio-diesel. It is estimated that the future use of this oil will add many billion of litres to the bio-diesel supply. The US postal service and vehicles run by many transport authorities, national parks and recycling companies are amongst bio-diesel's many users.

In Europe, as in various countries worldwide, conversion to bio-diesel is being viewed by many as a way of making good use of agricultural over-production and waste materials. In the UK, 'Global Diesel' (a mixture of ultra-low sulphur diesel and bio-diesel made from rapeseed oil, and said to emit 5% less carbon dioxide and up to 28% fewer particulates than ULSD) is now commercially available. Long-term strategies for growing non-food crops (primarily oil seeds, which, many feel, do not need to be genetically modified) for bio-fuel have been suggested as part of a blueprint for sustainable farming in the UK.

Until the 20th century, oats and hay – as food for horses – provided the 'fuel' for road transport in much of Europe. A century ago, a fifth of all farmland in England was devoted to the production of these crops. Those campaigning for bio-fuels today claim that 10% of Britain's transport fuel, in the form of ethanol or bio-diesel, could be home-grown, on 10% of the nation's farmland, by 2010.

Diesel from trees

In the Amazon rainforest, a number of trees have been found to produce oils that can be used directly as fuel. The liquid balsam or oleoresin stored within the copaiba tree (see left) was reported in the 1970s to be almost identical to diesel fuel and capable of powering a diesel engine simply by being poured into its fuel tank.

The traditional commercial use of this precious liquid – distilled to provide an essential oil – has for years been as a raw material for the perfumery industries of North America and Europe, and, to a lesser extent, for pharmaceuticals. In Brazil, where most originates, it is used as an antiseptic and anti-inflammatory, and products containing copaiba oil are promoted for their value in treating skin and other disorders.

Copaiba balsam has been used directly by local people in Amazonia for lighting and as an engine lubricant, but its potential as a source of fuel has not been realized. Estimates have varied considerably regarding the quantity of balsam the tree can produce – determined in part by the species and age of the tree, the length of time since previous tapping, and by the season. In addition, there is great variability between individual trees growing in the same conditions.

The oleoresin is present in thin capillaries within the tree, and accumulates in a network of connected cavities, forming reservoirs of

Those campaigning for bio-fuels today claim that 10% of Britain's transport fuel, in the form of ethanol or bio-diesel, could be home-grown, on 10% of the nation's farmland, by 2010.

clear liquid. Tapping is done by cutting a hole inward and downward into the centre of the trunk about 1 m (39 in) off the ground, which serves as a reservoir. The oleoresin gradually drains into this hollowed-out well. A wooden tube is then inserted, through which the liquid flows. When the flow has stopped the hole is plugged and the tree may not be visited again for many months.

While yields of 15–20 litres (3⅓–4½ gallons) every six months and 53 litres (nearly 12 gallons) from a single tree have been reported, other studies have shown much lower and variable yields. One report stated, however, that 18 litres (4 gallons) of 'fuel' could be produced in two hours.

As its name suggests, the petroleum nut tree (*Pittosporum resiniferum*), found in the Philippines and Borneo, also produces a high octane oil that can be used directly as a fuel. During the Second World War, the Japanese used it to power their tanks, though tribal people of the Philippines have long appreciated it as a fuel for lamps. The oil is pressed from the nuts or seeds of the tree. Just six trees have been reported to be able to produce some 320 litres (over 70 gallons) of oil in one year. Recent experimental plantings in the Philippines for fuel production are said to be promising and have shown that a harvest of 18 kg (40 lb) of fruit, per tree, per year, can produce about 3,000 litres (660 gallons) of oil per hectare (½ acre).

Plants that **entertain** us

Whether doing a crossword or reading a book, playing cricket or a double bass, painting a picture or taking one, plants are there again, giving us the raw materials, providing countless means of entertainment and communication.

Plants are essential for a tremendous range of recreational activities, especially in the fields of sport and music, where different woods, in particular, have played a vital role. Often the fact that we are using substances derived from plants is not at all obvious. Photographic film, for example, is made from a base of plant cellulose, yet this is the last thing we are likely to think of as we load the camera. Perhaps the most commonly used material of all, for recreation of any kind, and of course for communication, is paper. This is also made from cellulose. The artist, bookworm and globetrotter armed with paper currency or traveller's cheques would all be lost without it. Even the most sophisticated computer needs something to print its binary message on. In the main, they all use trees.

In 2000, over 323 million tons of paper and cardboard were produced worldwide, made from over 187.5 million tons of pulp. This pulp was made from de-barked hard and softwood trees grown in plantations and cut from natural forests. Though some claim that the true figure is nearer 50%, official estimates state that some 14% of the timber commercially harvested worldwide is destined for paper products, including books, newspapers and magazines, computer and fax paper, cardboard boxes, tissue paper, and much more.

Wood has not always been the chief source of paper, though, and as basic pulp-making technology has changed, so have the raw materials. At least 5,000 years ago, the Egyptians were using thin strips of papyrus (*Cyperus papyrus*) peeled from the stems of this reed-like plant to write on. The strips were laid side by side and over one another at right angles and, after moistening, were pressed flat and left to dry. The dried material was hammered to make it more compact and rubbed with a hard implement, perhaps of bone, shell or ivory, to produce a smooth surface. To make a durable and flexible scroll, several of these flat sheets would be stuck together – the strength supplied by the hard fibres in the plant's stems. Though the use of papyrus was certainly very significant, having left us not just with the word 'paper' but with innumerable writings from Egypt and the Middle East, paper as we know it had its origins in a different use of plants.

By at least 200 BC the Chinese had begun to make fine sheets of paper, influenced, it is said, by the nest-building activities of wasps. The brittle 'paper' made by wasps is essentially the result of redistributing the cellulose fibres contained in wood or woody matter by breaking down the sticky substances that hold them together and leaving the fibrous mass to dry. Though paper making is now a highly sophisticated operation relying largely on chemicals to produce the basic pulp and a variety of additives to give the customary finish to the sheet, these underlying principles have hardly changed.

Previous page: Guitar made from FSC-certified Amazonian hardwood.
Opposite: Taiga forest in Siberia.

The first papers

The Chinese began their operations using the inner bark of the paper mulberry (still used in Japan to make strong paper), hemp waste and that most versatile giant grass, bamboo. Expanding on the work of the Chinese, the Arabs developed the practice that was to last until the 19th century of using hemp, cotton and flax in the form of rags as their chief raw materials. As they came to dominate different lands, their technology of paper making went with them, replacing the use of papyrus in Egypt and the Mediterranean countries. Indeed, it was the Arab invasion of Spain in the 11th century and the dissemination of their skills and learning that brought paper making to Europe.

It was a further observation of the work of wasps, this time by the Frenchman René de Réaumur in the early 18th century, which helped to turn the 'industrial eye' to wood for paper making. However, the commercial substitution of logs for rags, which were becoming increasingly scarce as the demand for paper rose, did not take place until the early 19th century. This major change was to be preceded by another of the greatest importance – the mechanization of the paper-making process. In 1803, the Fourdrinier brothers successfully developed and patented a machine that could deliver a continuous sheet of paper to a pair of rollers. The paper-making process was now revolutionized in terms of the size and quantity of sheets that could be made, and this development helped considerably to meet demand. But

Mexican bark paper

In parts of Mexico, *papel amate* – bark paper – was being made over 1,400 years ago by many **indigenous peoples**. The Aztecs, Toltecs and Mixtecs all produced fine paper sheets and large quantities were sent by way of annual tribute to Montezuma II, the Aztec ruler.

Bark paper is still made today by the Otomi people of San Pablito in the state of Puebla, but theirs is the only paper-making centre left in Mexico. Until the recent popularization and commercialization of their craft, the Otomi used the bark of various fig tree species (chiefly *Ficus tecolutensis*) and mulberry (mainly *Morus*

celtidifolia) to make paper of different kinds and thicknesses. Although these trees are still used, their over-exploitation, plus changes in land use from forestry to commercial agriculture, has necessitated the substitution of **other kinds of bark**. A number of trees including *Ulmus mexicana*, *Brosimum alicastrum*, *Sapium pedicellatum*, *Urera caracasana* and *Myriocarpa cordifolia*, whose barks have different qualities and properties, may now be used. However, since the 1980s the fast-growing *Trema micranthum* has been the most important resource for bark paper. To make the paper, the fibrous inner phloem fibres are

separated from the outer bark in strips. They are then **soaked** overnight and **boiled** for several hours in water containing lime. This softens the fibres and makes them separate more easily. After rinsing, the strips are arranged in a grid-like pattern on a smooth board and then **beaten** with a special flattened stone until the fibres mesh. Once they are sufficiently intermeshed, the newly formed sheet (still on the board) is left to **dry** in the sun.

The object of a flourishing tourist trade, Mexican bark paper is instantly recognizable by its often fluorescent illustrations.

the paper mills were plagued by shortages of rags and a new material needed to be found.

With the German development in 1840 of a means of reducing logs to pulp by holding them against a revolving grindstone beneath a jet of water, the earliest mechanized technique of making paper pulp from wood was born. Shortly after this, chemicals were introduced to do the job and both the chemical and mechanical processes are used today. The main advantage of chemicals was that they dissolved the lignin and other softer components of the wood without damaging the cellulose fibres, as mechanical grinding had done, leaving a pulp of far superior strength.

Pulping processes

Processing plants now produce enormous quantities of chemical pulp of various kinds, depending on the end use. The most common method is that known as sulphate or 'Kraft' (meaning 'strength' in German), also invented in Germany. In this process, wood chips from almost any tree species can be reduced to pulp by first steaming and then saturating them with hot sodium hydroxide and sodium hydrosulphide at very high temperatures, under pressure in a digester. The subsequent pressure 'cooking' separates the fibres from the other unwanted ingredients, most of which are recovered and reused. It is the residual lignin that gives strong brown paper such as that used for grocery bags its distinctive brown colour, but for conversion into all kinds of writing, printing and drawing papers, the fluffy, crumb-like particles that emerge after washing and 'flash' drying must be bleached. The standard bleach used for many years has been chlorine or chlorine dioxide but many mills are now using oxygen and hydrogen peroxide in their bleaching processes.

Mechanical means of reducing logs to pulp, by grinding them on giant wheels or pulp stones (sometimes under pressure and with the addition of chemicals), are generally used for the production of coarser papers such as newsprint or tissue. Since this pulp will contain almost all the wood's components and not just its fibres, it is not suitable (used on its own) for high-quality paper. As can be seen from old newspapers and paperback books, exposure to heat and light for any length of time turns paper made from this pulp yellow. Various kinds of chemical and mechanical pulp, however, may be used alone or in combination, with or without the addition of various fillers and sizing substances (see page 278), to make a great variety of papers. The pulp is first fed onto a fast-moving screen of wire mesh on which it drains, leaving the fibres interlocked. It then passes through a series of heavy rollers and heated cylinders which remove most of the remaining moisture and press the paper absolutely flat, to produce a continuous roll.

Wood chips for paper from FSC-certified Norway spruce grown in Sweden. Young trees of about ten years old are harvested as forest-thinnings, and must be chipped as soon as possible for the best pulp.

THE REAL PRICE of a piece of paper

It is often claimed that rainforests are not being turned into paper. Sadly, two recent reports * show only too clearly that the Indonesian pulp and paper industry is responsible for causing massive destruction of its native forests and that it is likely to

prominent British high street bank. Such financial institutions have been keen to invest in Indonesian paper and pulp, giving them access to cheap raw resources designed to guarantee a competitive advantage. According to Friends of the Earth, 'The APP is

Indonesia is regarded as the second richest country in the world in terms of its biodiversity. Though only occupying 1.3% of the world's land area, its forests contain about 10% of the world's flowering plants and 17% of all birds, reptiles and amphibians. The forests are also home to endangered species such as the Sumatran tiger and rhino, the Asian elephant and the orang utan.

Indonesia is being deforested at the astonishing rate of 2 million ha (4.9 million acres) a year.

continue to do so until there is little left to destroy.

The reports provide evidence that, to date, some 835,000 ha (2.1 million acres) of highly diverse rainforest has been destroyed to supply the pulp and paper industry. In 2000, one of the largest pulp mills, which supplies Asia Pulp & Paper (APP) (Indonesia's biggest producer), acquired 75% of its logs from clear-cutting natural forests.

This activity has been funded by the international financial community for the last ten years. Over $15 billion has been supplied by a wide range of investors, including at least one

flooding the UK market with un-brand-ed and re-branded paper products. APP has covered up the environmental impacts of its operations ... obscuring the fact that most of the raw material is sourced by clear cutting natural rainforest.'

Inevitably, local peoples, such as the Sakai, have suffered the most, losing ancestral lands, livelihood and way of life, and finding their protests met with violence. The rights and welfare of millions of forest-dependent peoples have been ignored. It is estimated that nearly 75% of all logging in Indonesia is probably from illegal sources.

The World Bank have estimated that Indonesia is being deforested at the astonishing rate of 2 million ha (4.9 million acres) a year (an area the size of Belgium), while Global Forest Watch has calculated that 72% of its original forest cover has already been destroyed.

*Barr, Christopher, Profits on paper; the Political-Economy of Fiber, Finance and Debt in Indonesia's Pulp and Paper Industries, Center for International Forestry Research (CIFOR) and WWF-International's Macroeconomics Program Office (2000)
*Matthew, Ed, & Willem Van Gelder, J., Paper Tiger, Hidden Dragons, Friends of the Earth & Profundo (2001)

272

Trees for pulp

Naturally enough, the modern production of wood pulp and paper has tended to be concentrated in highly industrialized, well-forested countries. For many years, Canada, the USA and Scandinavia produced over 80% of the world's pulp, chiefly using softwoods. But as different countries have begun to exploit their natural resources, and huge plantations have been set up, the sources of supply have also changed.

In 2000, the USA and Canada produced over 84 million tons of paper pulp – nearly half the world's supply – followed by the EU (mainly Finland, Sweden, Germany, France and Austria), together producing around 20 per cent. The third biggest individual producer was China, which accounted for nearly 18 million tons. Despite the age of electronic communication, more paper is being used than ever before. Between 1999 and 2000, the UK's consumption of paper and board rose from 9.4 to 12.9 million tons; the use of email alone is estimated to have led to a 40% rise in paper use.

The huge and ever-growing demand for paper worldwide has led to the development of an industry dominated by multinational companies that obtain and make pulp from a great variety of sources. These, however, vary considerably according to market forces and are constantly changing and, since pulps are often mixed, it is extremely difficult to identify the tree species involved. Statistics concerning the composition of various pulps also vary greatly, making the precise picture far from clear. According to some estimates, the UK's bleached chemical pulp is made up of 80–90% hardwoods, whilst others give a figure of about 57 per cent. Of this, the greatest proportion (appearing to range from nearly 30% to over 80%) is comprised of eucalyptus species and their hybrids, trees bred to give a high yield of fibre under intensive cultivation (see right).

Though impossible to identify the individual trees involved, hardwoods from tropical and temperate forests of many different kinds are also being converted to pulp for paper. The destructive logging of the world's remaining old-growth forests, in regions as far removed as Siberia and Indonesia, has been widely condemned. Estimated to account for 16% of the world's wood pulp, in boreal regions they remain a significant source of pulp. In recent years, Indonesia has felled nearly three-quarters of its species-rich rainforest, much of it illegally for paper pulp and timber, destined for use in countries around the world, including Britain. Chile, meanwhile, has been systematically felling its unique forests of southern beech (*Nothofagus* spp.) for pulping by Japanese companies, much of it for conversion to fax paper.

The hardwood pulp imported into the UK is also likely to include a proportion of familiar temperate species grown in northern Europe and in North America, such as oak, birch, beech, maple, poplar and aspen. In North America, black gum and sweet gum are also used. A number of other hardwoods have also come to be cultivated on a large scale for pulp production. These include:

● *Leucaena leucocephala*, a legume native to Central America with a great range of uses and plantation-grown in Southeast Asia and other tropical regions.
● The fast-growing *Gmelina arborea*, widely planted in the tropics.
● Balsa, also widely planted in the tropics.

Softwood pulp, meanwhile – likely to comprise some 30–40% of the bleached paper pulp imported into the UK – has tended to be made

Eucalypts

■ A large proportion of the world's plantation-grown hardwood pulp is reported to comprise currently two eucalypts and their hybrids:

■ *E. globulus* (said to be the preferred species for 'fine papers') is widely grown in Spain, Portugal and various other countries, including Chile.

■ *E. grandis*, crossed with *E. urophylla* or *E. saligna* (widely used for tissue paper manufacture), is extensively grown in South Africa and Brazil.

■ From a total of some 2 million tons per year, 95% of which is destined for export, Brazil's largest pulp company, Aracruz Cellulose, is projected to produce some 870,000 tons of eucalyptus pulp in 2003. This is being produced from fast-maturing trees, which can be harvested after 5–7 years, now growing on some 285,000 ha (704,000 acres) in the states of Espirito Santo and Bahia.

Eucalyptus logs – grown on a plantation – at a sawmill in Bahia, Brazil.

Save trees – recycle paper

The destruction of long-established forests – as is well known – is generally as disastrous for **the local people** who have used and, in many cases, maintained them as it is for the flora and fauna that make them up. The spread of inappropriate monoculture plantations often further damages habitats and threatens species (including people) in a number of ways.

According to WWF in 1998, the American paper industry alone was consuming 32,200 sq km (12,430 sq miles) of forest **each year**. Though the Americans make up only 5% of the global population, it has been calculated that they use 27% of all the wood commercially harvested worldwide, and that the average American uses 331 kg (730 lb) of paper each year (about the equivalent of **nine mature pine** trees). The top four paper-consuming nations of the developed world, the USA, Japan, Germany and the UK, use half of all paper produced.

Equally alarming is the fact that paper makes up nearly 40% of all the **rubbish** Americans throw away, and in Britain around 5 million tons of paper ends up in landfill sites or is incinerated every year. While the use of non-wood fibres becomes increasingly important in lowering our consumption of

from spruce and pine species grown chiefly on plantations in northern Europe and North America. Other trees include western hemlock, redwood, western red cedar and eastern red cedar, harvested from old-growth forests.

As with hardwood pulp, what is bought depends to a large extent on price and availability and the position of the dollar. In the USA, the percentages of hardwood to softwood used are a rough reversal of those in Britain, with most pulp coming from native softwood trees, many of which are known collectively as 'southern pine'. These trees, chiefly loblolly pine (*Pinus taeda*), longleaf pine, shortleaf pine and slash pine (*P. elliottii*), have been genetically 'improved' to have a higher wood density, to be highly resistant to disease and insects, and to be harvestable several years earlier than ordinary pines.

Similarly 'improved' *P. patula*, *P. taeda* and *P. elliottii* trees are being grown for pulp production on a massive scale in southern Africa. In Swaziland, some 70 million pine trees now cover 65,000 ha (161,000 acres) of what was once open grassland. On average, the trees are harvested at 15–20 years of age, having reached a height of around 25 m (82 ft). Large stands are clear-felled at a time and some 4.5 million seedlings are planted out each year.

Plantations of another very fast-growing softwood – Monterey pine (*P. radiata*) – have also been established in Chile, New Zealand and Australia.

Different plants for different paper

All plant fibres have their own special characteristics, and these contribute directly to the quality and type of paper that can be made from them. Softwood fibres are generally longer than hardwood fibres, and since they intermesh over a greater area, they make a stronger paper. Whilst the shorter hardwood fibres make a weaker pulp – often forming, for example, the majority of the pulp for disposable handkerchiefs and bathroom tissue – they give a better surface for printing on. Hardwood and softwood pulps, then, are likely to be mixed to produce a typical sheet of writing paper.

It is not just trees, however, that make the basis of the paper pulp in use today. A number of plants provide useful raw materials that can be added to it to alter or enhance its performance, and these are being used around the world to make paper that is completely wood-free. In Britain, speciality papers of very high quality may comprise a proportion of esparto grass fibres (*Stipa tenacissima*), exported from North Africa. Though they are very short and of small diameter, they give a bulk and opacity to paper, plus a closeness of texture and smoothness of surface, which are unique. Paper made with esparto also watermarks very clearly and expands less than other papers when wetted.

trees, the recycling of our used waste paper remains a **critical challenge** in the most wasteful industrialized countries.

In the UK, 'waste paper' provided about 65% of the source materials for all paper made in 1999, but much of this was 'pre-consumer waste' – that is, unused, unprinted paper such as printers' off-cuts. We need to **recycle** much more of our used paper. The de-inking of pre-used paper involves chemicals and water, but the process is **less harmful** to the environment than that of manufacturing brand new paper. Every ton of paper that is recycled is estimated to save 17 trees and 31,900 litres (7,000 gallons) of water, while 60% less energy is needed to manufacture paper from recycled stock.

275

Plantation-grown parana pine (*Araucaria angustifolia*) (see page 170) being transported to a pulping plant near Curitiba, Brazil, for South America's largest producer of cardboard and paper. In the wild, parana pine is now seriously endangered.

For paper that must be very strong yet flexible, such as bank note paper and some legal documents, mixtures of hemp, flax and cotton fibres are used, but mostly in the form of rags. Almost the whole of the world's production of raw cotton fibre – the purest form of natural cellulose – and linen is used by textile and yarn manufacturers so paper makers are obliged to use them in this form or as the waste from processing. Some cotton linters, the shorter seed fibres, too short for spinning, are also incorporated. Paper using organic cotton is also now being made.

The environmental impact of paper production based on wood pulp is a major concern to many people. Quite apart from the vast consumption of trees that it entails (a significant proportion of them from natural ecosystems), and land (much of it inappropriate – on social or environmental grounds – for conversion to plantations), are the problems caused by processing. Though other methods may also be used, the bleaching of wood pulp with chlorine and chlorine dioxide, for example, produces the carcinogenic chemical dioxin. This is one of the most toxic environmental poisons (with government safety limits measured in parts per billion), and is known to accumulate in meat, dairy foods and human breast milk. In many parts of the world, toxic waste is still being discharged by paper mills into rivers and streams, destroying aquatic life for miles. In recent years, the UK's paper mills have become much more environmentally friendly in terms of emissions, waste, energy and water use, but the use of non-wood

Inside the Ortviken paper mill in Sweden. Wood pulp is converted to paper in a high-tech, computer-monitored operation involving many different stages.

Hemp and banana papers

Widely acclaimed as a much more environmentally sound alternative to the wood-based paper currently consumed in Europe is paper made from **hemp**. Already used in various speciality papers, hemp fibres don't require bleaching and use 90% fewer chemicals than wood-based paper as they are low in lignin and naturally a creamy colour. Hemp paper is also far **more durable**, is said to last 50–100 times longer than that made from timber, and, as an annual crop, produces more pulp per ha (acre). A mixture of 50% hemp and 50% straw produces a paper which, it is said,

can be recycled 30–40 times, in contrast with ordinary paper pulp, which can only be reused (estimates vary according to type) **2–8 times**. Hemp has a long history of use for paper, and until the late 19th century much of the world's supply was made from it. **Hemp canvas** was used by many of the world's greatest artists, including Van Gogh and Gainsborough.

Another non-wood fibre, which would also otherwise be wasted, is provided by **banana stems**. A by-product of the Costa Rican banana harvest, discarded banana stems have become an environmental threat

in the region, often clogging the rivers into which they have been dumped. Currently some **230,000 tons** of banana fibre are produced in Costa Rica each year. But a smooth-textured paper is now being produced containing some 5% reclaimed banana fibre and 95% unbleached waste paper.

Other more unusual paper-making materials being used in combination with waste paper are the waste from Central American tobacco plantations and, in El Salvador, **brewed coffee** and **coffee bean skins**. Coffee plantation by-products have created serious solid waste problems.

fibres for paper is generally not only much less polluting but advantageous to people and the environment in a variety of ways.

Local plants for local paper

Locally available plants have largely determined what different nations make their paper from. The UN estimates that non-wood fibres account for a third of paper production in developing countries. One of the world's most important non-wood sources of paper pulp is bamboo.

China, Thailand, Brazil, Vietnam and Bangladesh all convert huge quantities of bamboo to pulp, but the single biggest producer is India, which uses bamboo for around 65% of its paper production, manufacturing some 711,000 tons of pulp each year.

The fibres have a high cellulose content and, since they are more slender than wood fibres, can produce a smooth, flexible paper. However, disadvantages have included the high processing cost (the pulp has more impurities than wood pulp) and the unpredictable flowering habit of some species: after flowering the plants will die. In Thailand, the utilization of a range of selected bamboo species, however, plus earlier and improved harvesting techniques, have overcome these problems.

Plants that save waste

Other important sources of non-wood fibres for paper, particularly in tropical regions, are manila hemp or abaca, sisal and kenaf. Straw, however, from various cereal crops such as wheat, maize or rice, is the most commonly used raw material worldwide, including China and India where 80% of the world's tree-free paper is made. This is followed by bagasse, the waste from sugar cane, which has been an important source of fibre in Brazil and Mexico. Other sources of paper include water hyacinth – a prolific tropical aquatic plant – pineapple and jute waste, coconut fibre, mulberry bark, scraps (including blue jeans) from clothing manufacturers and recycled currency. Some European and North American manufacturers are now using combinations of these fibres, such as 15% hemp and 85% sugar cane, and wheat or rice straw mixes, making use of raw materials that would otherwise be destroyed. Californian farmers alone, for example, are said to burn 1.5 million tons of rice straw each year after the rice harvest.

Fast-growing bamboo species are a major source of pulp in various countries, including China, where it was one of the first plants to be used for making paper.

277

The use and transportation of commodities – such as wood pulp – over great distances, from all around the world, is done at high environmental and economic cost. The logical alternative, and a basic principle of sustainable development, is 'local production for local needs'.

Local paper for local needs

The use of annual fibre crops, which can be locally grown and which are well suited to smaller-scale processing, using new technologies to become effluent-free, must be the way forward. This would help make decentralized paper production an economic reality. Tree-free and chlorine-free papers are gradually carving a larger niche of the paper market. Pulped hemp fibres, for example (see box, page 276), are being blended with post-consumer waste paper or pulp made from local agricultural residues such as wheat or rice straw or sugar cane. Hemp fibres are extremely strong in relation to their weight, and newspaper made from them is just as strong as that made from wood pulp but only a third as heavy. Already widely used in China, hemp is viewed by many as the perfect crop for small-scale paper production using non-toxic, sustainable technology.

More plants for paper

Before we come to lay our hands on finished books or magazines and the paper money that we use to buy them, plants are used again in two important ways. First, to make sizing agents, which improve the paper and prevent it from becoming too absorbent, and second as components of printing inks.

Sizing agents

Though synthetic sizes have now been developed that are effective in strongly alkaline paper pulps, several plant-derived size materials are still in use – chiefly natural starches, gums and resins, and cellulose. The most widely used include modified maize, potato and cassava starch, gum karaya, guar gum, cellulose ethers, alginates and rosin.

When mixed into the pulp, the selected size is usually just one of various additives such as dyes, brighteners and fillers (chiefly china clay and chalk) which give the desired characteristics of colour, smoothness and opacity to the paper. In this form, or as an external coating, the size helps to bind all the ingredients and determine the moisture resistance of the paper to be printed. If it is left out, too much ink will penetrate the paper and leave it smeared. Plant sizes each have different attributes, which will naturally give differing results.

● Cellulose ethers – derived from cotton linters or wood pulp – have oil- and grease-resistance properties and help to give an even finish.

● Sodium alginates improve the paper's surface and give added smoothness as well as better ink acceptance.

● Starch derivatives, available in different formulations as an additive for pulp or as a surface size, also generally improve the paper's moisture retention, filling up the pores in the paper sheet.

The main use of guar gum (see page 33) in paper processing is as an additive to pulp. The chemical pulping process removes not just the lignin from the fibres but a large proportion of the hemi-celluloses present in the wood. Guar gum, and to a lesser extent locust bean gum, replaces or supplements the effects of natural hemi-celluloses in bonding the fibres, distributing them more evenly and improving folding, bursting and tensile strengths. The gums also make the paper less prone to curling at the edges.

Pine trees also help achieve the qualities of many of the paper products that we use. Rosin – a natural resin with the turpentine removed – is obtained from a variety of pines. Added as a size to the paper pulp, it helps improve the finished surface of the sheet and enables it to be printed on without the ink smearing or 'feathering'.

Norway spruce stacked ready for pulping.

Sustained destruction?

Many paper products now sport statements or logos declaring that they have been made using raw materials obtained from **sustainably managed forests**, or from forests in which the number of trees felled is offset by the number being replaced. Such claims, however, are in many cases misleading and disguise poor environmental management. Some labels, however, consider themselves to be much more rigorous than others.

The Forest Stewardship Council (FSC), for example (see page 166), has drawn up independently certifiable criteria for **'well-managed'** forests, for pulp and timber use, including 'ecological' management, the involvement of local people and good employment practice. It includes members of the timber trade and forestry professions, community forest groups, indigenous people's organizations and forest product certification organizations. Today, over 35.5 million ha (88 million acres) worldwide (about 3% of production forests) are FSC certified. The FSC aims to have certified **200 million ha** (494 million acres) of forest by 2005.

Critics of the scheme, however, have pointed out that 'certification' may merely provide a survival option for an increasing number of large logging companies, the overall impact of whose activities does not provide **the necessary protection** for our beleaguered forests. They are concerned that much damage is still done by logging roads, for example (which provide easy access for other destructive activities, encouraged by settlers), and that the **underlying causes** of forest destruction remain intact. Critics argue that instead of boycotting all tropical timber in favour of locally produced woods, responsibility for halting destructive, predatory logging practices has been transferred from governments into the hands of consumers, presenting them with a **confusing array** of eco labels. We should not, they believe, think that the cause of our social and ecological malaise can become the cure – and that forests can be saved simply by 'ethical' shopping. However, many people agree that the certification scheme offered by the FSC is an important step in the right direction.

Gum arabic

■ The term 'gum arabic' is sometimes used for gum taken from any *Acacia* species – including *A. seyal* and *A. karroo* – and occasionally that taken from trees of another genus. But most of the gum arabic in international trade is obtained from just one of some 1,200 species of *Acacia* tree distributed throughout the tropics and subtropics: *A. senegal*.

■ Gum is collected by hand from wild trees, native to the sub-Saharan and Sahel regions of Africa, and those planted (increasingly as a means of combating desertification). The bark of the spindly, spiny trees is cut and a 'tear' of amber gum forms at the scar. After harvesting, the gum is cleaned and sorted into grades for export. Processing – into milled or spray-dried powder – is usually done by the importer. Sudan is the world's biggest producer of gum arabic, followed by Nigeria, but many other countries are also suppliers, including Chad, Egypt, Kenya and Senegal. The world market for the gum is estimated to be in the region of $100 million per year.

■ Gum arabic is used in a number of important industries. It thickens many convenience foods as well as pharma-ceuticals and cosmetics, and may appear in textile sizes, watercolour paints, printing inks, ceramics and adhesives. Once used to make the glue for postage stamps, it is still added to some beers as a foam stabilizer.

Plants for printing inks

For many years, petroleum-based inks were those most commonly employed for the industrial printing of books, newspapers and magazines, although they also made use of various plant materials. Today, printing inks based on vegetable oils have become widely adopted in lithographic printing processes. Though many still contain varying percentages of petroleum, soya oil, in particular, has become a standard ingredient of high-quality inks for much colour printing. Among other things it gives a vibrant colour, is resistant to 'rub-off', provides greater coverage per pound of ink, and does not have the environmental problems associated with petroleum-based inks. Vegetable oils emit fewer volatile organic compounds (VOCs) – carbon-based compounds that contribute to air pollution – than inks made from mineral distillates, which are also, of course, non-renewable. Over 90% of American daily newspapers now use soya-based ink when printing in colour.

The distinctive properties of other plant products, however, continue to make them very valuable to the printer. Classed as 'industrial' oils, linseed and tung oils have been especially useful for lithographic printing since they dry relatively quickly and give hardness and 'rub resistance' to the finished product. Tung oil, which comes from the seeds of several species of deciduous *Aleurites* trees native to China, was an original ingredient of 'India ink' still widely used by China's calligraphers. Both oils dry faster and are harder when dry than soya bean oil.

Although hard resins are also available that are derived from mineral oil, the printing industry currently consumes large quantities of rosin and rosin derivatives, from pine trees, for many different kinds of ink. These additives provide adhesion and give gloss and hardness to the product as well as helping it flow freely.

Until the last 15 years or so gum arabic (see left) was added as a binder (fixing the solid pigment particles to the surface to be printed) and suspending agent, controlling the consistency of the ink and the settling of the other ingredients. Soluble in water without becoming highly viscous, it is still highly valued for its thickening, suspending, stabilizing and emulsifying properties. Early industrial inks were simply dispersions of lamp black in water, but gum arabic was added to improve their texture. This gum, which is non-toxic, odourless and almost colourless, was also to play an important part in the development of lithographic printing.

Today, Britain's leading environmental printer uses light-sensitive silicone-coated printing plates as part of a sophisticated process of 'waterless printing', which has eliminated the need for solvents (such as isopropynol), which are major polluters in conventional printing.

Plants for budding artists

Linseed oil, already mentioned for its use in printing inks, is still an important ingredient of many artists' oil paints. Forming the base that carries the pigments (see overleaf), it acts as the binder, fixing the pigments to the surface to be painted as it dries. Linseed has long been regarded as the best all-round choice for formulating oil colours, but many other plant oils that also dry by absorbing oxygen may be used, including walnut, poppy seed, safflower and sunflower. These lighter-coloured oils help minimize the natural yellowing that occurs with age.

While other materials are available, linseed oil is also an important additive that alters or enhances the characteristics of the colour (for example by changing the rate of drying, increasing gloss or improving flow). Alkyd paints also often incorporate linseed oil to increase flexibility. To reduce ageing and cracking and assist the refraction of light penetrating the paint film, some oil paints also include natural resins, resulting in a slightly more brilliant appearance.

Today's oil paint varnishes may be based on synthetic resin formulations, but linseed oil, which has been heated to 298°C (570°F) in an oxygen-free atmosphere, causing it to thicken to a honey-like consistency, is sometimes used, as well as mastic and dammar resins. Applied in liquid form, these resins 'hold' the paint when dry and give a special lustre and transparency. Mastic is the resin exuded from the stems of the Mediterranean shrub *Pistacia lentiscus* (see right).

Dammar resin is obtained primarily from trees of the *Dipterocarpaceae* family (tropical Asia's most important timber trees), notably *Shorea* and *Hopea* species from Indonesia and Malaysia. It is available in the form of lumps or crystals, or ready-dissolved in gum turpentine. The luminous depth of many oil paintings has also been achieved with the help of copals. These are fossilized resins produced by species of *Copaifera* (mined from West Africa), and the related genus *Hymenaea*, from tropical America and Africa, and by *Agathis* species from New Zealand and the Asia–Pacific region. While good quality copal is no longer available for artists' use, dammar resin is practically confined to oil paint varnishes.

As an oil paint thinner and solvent for brush cleaning, turpentine from pine trees has long been indispensable and is still very popular with painters. Today, however, petroleum-based solvents (such as 'white spirit') and those based on 'd-limonene' from citrus oil (mostly extracted from the rind of oranges, lemons and grapefruits) and now widely used in cleaning products, are available as alternatives. 'Venice turpentine', meanwhile, which has the consistency of honey, is used not as a thinner but occasionally in artists' oil paints, where, together

281

Pistacia lentiscus
■ Produced in the form of pale yellow, glassy 'tears' after the bark has been cut, mastic resin has an interesting range of traditional uses, chiefly medicinal and culinary. Though no longer common because its long-term performance is considered poor, it is still cultivated in Greece for high-grade artists' varnish.

Madder

■ *Rubia tinctorum* is one of some 60 species in the genus *Rubia* and is the largest of the bedstraw family. Native to Eurasia, it has trailing, angular stems with narrow, evergreen leaves that are dark and shiny, arranged around the stem in whorls. Small yellow flowers are followed by blackish, berry-like fruits. The thick, fleshy root, which produces madder dye, has a dark outer skin with a ruddy-coloured inner part. Wild madder, *R. peregrina*, grows in France and southwest Britain, but produces a weaker red dye than *R. tinctorum*. Madder plants were once widely cultivated throughout Europe and the Middle East. Cloth remnants dyed with madder, dating to around 1400 BC, and wall paintings found in ancient tombs also prove its long history of use in Egypt.

with stand oil or sun-thickened linseed oil, it gives better brushing quality and makes a tougher surface.

Pigments

The pigments used in watercolours, oil paints and most other artists' materials, whether coloured pencils, chalks or wax crayons, are chiefly derived from chemical compounds containing different metals such as cobalt, cadmium and iron, from petroleum or from earths dug from the ground, such as 'umbers' and 'ochres'. Two exceptions do remain in artists' paints, however: madder, for distinctive pinks and reds, and gamboge for yellow. Madder is extracted from the crushed roots of the herbaceous perennial *Rubia tinctorum* (see left), whilst gamboge is prepared from a natural exudate of the tropical tree *Garcinia hanburyi* (see below), native to Cambodia, Vietnam, Thailand and Laos.

The madder grown commercially today comes chiefly from Turkey and the Middle East. The fleshy roots of the plants are dried and their outer bark removed, before the inner portion is dried again and milled to a very fine powder. Both bark and inner root produce the colouring matter, which is extracted from the powder by fermentation and hydrolysis with dilute sulphuric acid, but the bark gives an inferior colour. Madder Lake and Rose Madder for artists' pigments are prepared from the extract by adding alum and mixing this with an alkali to form a precipitate. Natural madder contains the colouring matter alizarin, together with the closely allied purpurin and certain others in small quantities. After the invention of synthetic alizarin from chemicals in 1868, the plant was largely ignored as a source of dye and artists' paint. But the preference of traditionalists for this colour and the fact that it is one of the most stable of the natural organic colouring materials has extended the use of madder as an economic plant.

The name 'gamboge' is a corruption of 'Cambodia' (now Kampuchea), one of the chief countries from which the raw material has been exported. The pigment is contained within the resin produced by various species of tropical *Garcinia* trees – mainly *G. morella* and *G. hanburyi*, native to parts of tropical Asia. Only *G. hanburyi*, however, has been exploited commercially. Once these trees have reached at least ten years old, a spiral cut is made in the bark during the rainy season and the resin exuded is collected in bamboo lengths placed at the base of the cut. The resin takes about a month to harden, after which the bamboo containers are heated and the gamboge taken out as cylindrical sticks. Alternatively, the gamboge is collected as 'tears' or fragments that form near the incision. Though the raw resin is a reddish- or brownish-orange, it turns brilliant yellow when powdered and gives a deep yellow emulsion when mixed with water. It fades rapidly in sunlight as a watercolour, but it is fairly permanent when ground in oil.

Apart from its modern use in the best watercolour paints, it is the principal colouring ingredient in certain golden spirit varnishes, and in Thailand in a golden yellow ink for writing on black paper.

Pencils and wax crayons

Wax crayons and the cores of coloured pencils use plant-derived materials chiefly to bind all their ingredients together. Cellulose derivatives made from wood pulp or cotton linters may make up 15–20% of a crayon core, helping to produce an even flow of colour onto the paper, while various plant waxes are also likely to be present. Carnauba wax (see page 40) is mostly used as well as Japan wax, not a true wax, but a vegetable fat found between the kernel and outer skin of the berries of two species of sumac tree: *Toxicodendron vernicifluum* and *T. succedaneum*, long cultivated in Japan and China for the lacquer exuded by their bark. Since Japan wax has now become a very expensive material, it is often replaced with a substitute made from vegetable oils, as well as small quantities of natural resins.

The outside of a coloured pencil is usually coated with lacquers made in part from plant materials. Nitrocellulose or cellulose acetate resins made from a base of wood pulp or cotton linters treated with acids are common ingredients, though shellac (see page 185) is also sometimes used.

As for the supporting body of the pencil itself, wood, of course, is the traditional material. One wood in particular became so much associated with pencil production in the past that it is still commonly referred to as the pencil cedar. It is a juniper species native to North America, *Juniperus virginiana*. Its wood, once dry, does not distort or splinter, and its straight grain and softness allow it to be easily whittled or sharpened to a point. This raw material was used extensively for many years – indeed so much so that by the end of the First World War it had become extremely scarce. In the 1920s, however, another conifer with similar suitability for pencil manufacture was selected to replace it and was (until the late 1980s) the main supplier of wood for pencils: the incense cedar (see overleaf), also native to North America. The straight grain, uniform texture and relative softness of this wood have made it very suitable for precision sawing, staining and waxing, and, for the pencil user, for sharpening into a satisfactory point.

The cedar logs are first cut into squares which are seasoned before the pencil-making process can begin. Each square is then sawn into blocks approximately 185 mm (5½in) long, and these are sawn into the slats from which the pencil body will be made. The slats are then graded and treated under pressure with wax and a stain. The stain gives the finished pencil body a uniform colour, and the wax lubricates it during manufacture. Grooves cut along the length of the slat hold the

Gamboge

■ Gamboge has been used as a pigment for centuries in the Far East. It was the traditional dye for the silk robes of Buddhist monks and priests and the unique brilliance of its colour made it the prime ingredient of certain yellow paints, as well as special varnishes and inks. The Dutch introduced gamboge to Europe during the mid-17th century and it was subsequently used by early Flemish oil painters.

Watercolours

Whilst watercolour paints do not require plant oils or turpentine for their manufacture, they do use other plant materials. **Gum arabic** (see page 280) is an important ingredient of most smudge-proof watercolours, since in acting primarily as a binder, it keeps the particles of solid pigment fixed to the paper. As it **absorbs water** to form a paste, this gum also holds the pigments in suspension so they can be evenly applied, increasing transparency and gloss to give greater brilliance of colour.

Dextrins made from maize and potato starch have also been used in the past, for lower-quality children's colours, to bind and viscosify the paints' ingredients. Like gum arabic, as yet unsurpassed in quality by modern chemical equivalents, these water-soluble materials are **especially valuable** because of their non-toxicity.

283

Incense cedar

■ Native to the forests of the Oregon Cascades and the Sierra Nevada ranges of California, the incense cedar (*Calocedrus decurrens*) is not one of the four 'true' cedars (*Cedrus* spp.), which belong to the *Pinaceae* family, but a member of the *Cupressaceae*.

■ Very slim when young, with dense, bright green foliage, it does, however, develop an open crown with level branches where its growth is unimpeded. Harvested annually, the trees are grown commercially in forests, where they are native. Their soft, light wood is easily sharpened without splintering, and is finely textured, making it perfect for pencil production.

Box

■ The box (*Buxus sempervirens*), native to the chalky soils of southern England, is often clipped to form a dense evergreen hedge, but it can reach a height of over 10 m (33 ft) as a fully grown tree. It produces a hard, even-textured yellow wood, so heavy that it will sink in water when green. Its great stability, and the fact that it does not warp, made it very suitable for the manufacture of rulers and set squares, which needed to be strong and which could be precisely marked.

■ In the 18th century, box became popular for the blocks from which artists made their wood engravings. Other uses included chess pieces. The great demand for box has led to the best and largest trees being felled, leaving only the less vigorous specimens in their natural habitat.

pencil 'leads' in place, and a sandwich made of two slats stuck together forms a solid block from which the pencil lengths are cut.

About 20% of the world's pencils are now made from incense cedar grown chiefly in the forests of the Oregon Cascades, northern California and the Californian Sierra Nevada mountain ranges. A considerable proportion of the rest are made from basswood, grown in China. Others are made from jelutong, Brazilian pine or local tree species, according to the country in which they are made.

In an effort to help save trees, and promote recycling, a number of pencils are now being made from disposable polystyrene coffee cups – some 4 billion of which are used in the UK alone each year.

Many woods can be used to make artists' charcoal. In Britain, almost all artist's charcoal is made from willow – chiefly varieties of *Salix triandra* – grown and prepared in one location in Somerset.

Erasers

When it comes to erasing pencil marks, we have traditionally turned once more to plants – specifically to that most important economic product made by several of them – rubber.

It was Dr Joseph Priestley's observation in 1770 of 'a substance excellently adapted to the purpose of wiping from paper the marks of a black lead pencil' that led him to coin the name 'rubber' for this revolutionary substance. At this time, lumps of raw rubber were being sold in London for use by artists and all pencil users. Whilst this practice continued long after other uses for natural latex had been found, the development of new synthetic rubbers from coal tar and petrochemicals (from the 1930s onwards) brought a change in the components of the household eraser.

Today many rubbers are made from polyvinyl chloride (PVC) or from synthetic rubber. But a proportion of those used by artists are still made from the rubber latex tapped from *Hevea brasiliensis* trees (with added pumice stone powder which enhances the ability to erase), or a mixture of natural and synthetic rubbers.

Drawing instruments

Like rubbers, rulers are most likely to be made of man-made substances – generally transparent or opaque plastics – but for many years boxwood was the chief material used (see left).

Two tropical timbers, unrelated to box but with similar properties and available in larger sizes, have been used more recently for mathematical instruments, including T-squares and rulers. These are ramin from Malaysia, Indonesia and the Philippines, and maracaibo boxwood, exported from Colombia and Venezuela. Beech wood from European forests is also used by some manufacturers of rulers.

A green light for photography

Photography, one of our most popular leisure activities, depends almost entirely on plants. Photographic film is made from a base of cellulose acetate, manufactured by reacting the cellulose from a range of soft and hardwood trees and cotton linters with acetic anhydride and other chemicals. To make film, the syrupy substance that is formed in this way is first moulded into pellets and these are subsequently mixed with solvents to produce a clear, honey-like liquid. To form the solid transparent sheet that is the basis of the film, a constant flow of this liquid is spread in a uniform layer onto a turning wheel. As the solvents evaporate and the film dries, a fine continuous sheet is produced.

This base layer is then coated with a very fine emulsion of gelatin in which light-sensitive silver halide microcrystals are suspended. Purified saponins extracted from the Chilean tree *Quillaja saponaria* have been used for many years to help disperse these silver halide salts evenly, and they continue to be used today. Colour films need several layers of different emulsions and additional colour-forming chemicals to capture the image in view.

It is estimated that in 1998 Americans spent about $1.6 billion on around 500 million rolls of (mostly 35 mm) film. Though chemicals control the developing processes, the images we capture with our cameras could not be reproduced as slides or snapshots without trees.

Sporting plants

Despite the advances of modern technology, which have provided more resilient materials to match the new demands and greater physical strength of today's sportsmen and women, traditional materials are still a vital part of many sports.

Cricket bats and balls

The distinctive sound made by a leather-covered ball hitting a cricket bat is certainly a heavenly one to many of the game's most ardent devotees. Its distinctiveness is caused, of course, by the materials from which the bat and ball are made.

As almost every serious cricketer and fan will know, the most important piece of cricket equipment starts its life growing on a river bank or piece of marshy ground. Willow trees have given countless generations of cricketers the chief material for making bats and their wood is still in great demand today.

Yew wood is still considered the best for making longbows.

Cricket bat willow

■ A useful natural attribute of the cricket bat willow (*Salix alba* subsp. *caerulea*) is its ability to grow exceptionally fast. Traditionally planted along low-lying fields and river banks in Essex, Suffolk and Norfolk, the trees are usually felled at 12–15 years old, by which time they may have grown to 20 m (66 ft) or more in height.

■ One of several varieties of white willow, the cricket bat willow can be distinguished by its purple twigs and blue-grey, finely tapering leaves. Its bark is dark grey and develops a network of thick ridges.

■ Though important for their timber, these trees, like many other willows, are of great value to the countryside since their roots help bind the soil along river banks and reduce erosion.

Of the 18 willow species native to Britain only a variety of white willow (see left), has come to be considered as the best material for bats, and has thus assumed the name 'cricket bat willow'. Though the wood of the Kashmir willow is also used today as an alternative (this is the same variety of white willow as the cricket bat willow, but grown in southern India), the toughness, resilience and comparative lightness of the English cricket bat willow are unique. Due to the tropical climate, Kashmir willow puts on two bursts of growth in one season. It therefore tends to be harder and more brittle than cricket bat willow, and is less likely to give such a good performance over the same period of time.

A bat's 'blade' has to withstand tremendous punishment during the course of a professional match and absorb the impact of a ball travelling at anything up to 160 km/h (100 mph). It is cut from a triangular debarked wedge or 'cleft' of solid willow wood, which has been seasoned for 8–12 months. After its initial rough shaping using a circular saw, the embryonic blade is left for another period to mature, during which the first stage of grading is carried out. The blade will then be carefully cut and shaped by hand, a critical process requiring a high degree of skill to ensure that the finished article is perfectly balanced.

To increase its strength and to avoid scarring when the ball hits the bat, the blade is compressed in a special machine, which also reveals any hidden faults or weaknesses that may be present. The 'V'-shaped slot is then made in the neck of the blade ready for the handle to be inserted. After sanding and polishing and once the handle is in place, some blades may be finished with a fine cloth covering, whilst others are coated with a white or clear polyurethane film. Traditional 'natural' blades, however, receive no other treatment than a coat of raw linseed oil (see page 191).

The handle of the cricket bat is also made of specially adapted plant materials – but not of solid wood. The need for an exceptional capacity to absorb shock and diminish the impact of the ball led to the development of handles made from up to 16 separate pieces of rattan cane (the sinewy stems of climbing tropical palms from Southeast Asia, see page 186) cushioned with rubber. The canes are split, shaped and glued to make slabs, and these have rubber layers inserted between them. After drying and further shaping, the blade end of the handle is spliced into the body of the bat whilst the protruding part is bound with twine and fitted with a rubber grip. Together the shock-absorbing layers of rubber and cane form a springy handle that will greatly help protect the batman's hands.

Willows, rubber trees and rattans are not the only plants essential to the game of cricket. White ash, also known as American or Canadian ash, is used to make the stumps and bails. Furthermore, the inside of the leather-covered ball is made up of a central core of cork (cut from the cork oak, see page 180), surrounded by layers of oiled wool, soft cotton 'quilting' and more layers of cork.

Other bats and rackets

Many of the woods that first became established for use in various sports, such as ash (see right) and hickory, are still in use because of special properties such as strength and lightness, the ability to resist repeated impact without splitting and to absorb shock. Though modern tennis racket frames are made of man-made composites, in the early days of the game these were made of solid ash. By the 1930s lighter laminates had come to be used, reducing not just the weight of the racket but the tendency of the frame to distort. Until the larger headed rackets began appearing on the courts, these laminate frames, which comprised seven or eight layers of ash and/or other woods were used, such as beech, or various tropical timbers. Decorative wedges of walnut, sycamore or mahogany sometimes filled the area between the handle and the frame. But the need for a much stronger, lighter frame able to take the higher string tension of the larger head necessitated the move away from wood in all but the cheapest non-professional rackets.

Traditional squash racket frames were also once manufactured from various woods, including ash, hickory, mahogany, beech, sycamore and obeche, but they too are now made of composite man-made materials. Conversely, whilst the cheaper baseball bats are now manufactured from aluminium or plastic, those preferred by professionals are made of solid white ash or rock maple.

Polo mallet 'heads' were first made of bamboo, and other woods including ash, birch, willow, maple and sycamore have also been tried. Today, the most widely used timber is *tipuana blanca* (*Tipuana tipu*), exported from Argentina, the best coming only from the province of Salta. Some companies also use kahikatea (*Dacrycarpus dacrydioides*) from New Zealand, in its whole or laminated form. Mallet handles, meanwhile, have traditionally been made of rattan canes, using sections taken from the root end only of individual plants: the stiffest end (nearest the root joint) forming the handle end, with the 'whippier' end attached to the mallet head. Over-harvesting of rattans, however, and the difficulty of finding canes of the correct, very particular specifications, has resulted in the adoption by some manufacturers of composite shafts made of graphite, and these now account for a large proportion of the handles made.

Lacrosse sticks

Hickory is, like ash, extremely strong and can be steam bent into various shapes whilst still retaining its toughness. Many lacrosse sticks

Ash

■ The toughness, resilience, straight grain and good bending properties of ash (*Fraxinus* spp.) have made it exceptionally useful to manufacturers of sports equipment. The common or European ash (*F. excelsior*) is one of some 65 species all native to the northern hemisphere, many of which yield a pale yellow wood of commercial importance. The tall, graceful tree reaches a height of 30–42 m (98–138 ft) on good soil and its greenish-grey bark develops deep fissures with age.

■ In pre-Christian times, the ash was worshipped in Scandinavia as a symbol of life, and was believed to have medicinal as well as mystical properties. According to Norse mythology, the most powerful god, Odin, carved the first human being from a piece of ash wood.

287

Shagbark hickory

■ One of the most important hickories for commercial use, the shagbark hickory (*Carya ovata*) is the source of very strong, impact-resistant timber. It is used, amongst other things, for lacrosse stick manufacture and, when burned, for smoking foods.

■ Like its famous relative the pecan (*C. illinoinensis*), grown for its delicious nuts, the shagbark is native to North America, but has a wider natural distribution – from Quebec to Texas.

Persimmon

■ Although its own timber is very pale, the persimmon (*Diospyros virginiana*) belongs to the same genus as trees whose heartwood yields the famous black ebony (see page 293). There are about 500 different *Diospyros* species – mostly tropical and evergreen – and several, like persimmon, produce an edible fruit. *D. virginiana* is native to eastern and central North America.

■ Its off-white sapwood timber has a fine, even texture and straight grain and is only used for specialist purposes. Since the wood is very dense – about 15% heavier than beech – it is exceptionally hard, strong and impact-resistant.

288

are still made from North American hickory species, chiefly *Carya laciniosa* and *C. cordiformis*, as well as shagbark hickory (see left). To withstand the strain of play, a single length of wood is used and carefully bent into shape with the aid of steam. The wood is stiff and very dense – some 15% heavier than ash and superior to it in its ability to withstand sudden impact.

Skis

Most ski-makers have abandoned wood in favour of the modern, very tough but lightweight composite materials. In Norway, however, some skis are still being made using laminates (40–50 per ski) of birch and/or hickory wood, some of which are edged with beech that has been boiled in oil and then compressed to half its volume, to produce a substance harder than aluminium. Able to withstand flexing and abrasion without splitting, these wooden skis are both extremely strong and flexible.

Golf clubs and balls

The shafts of most golf clubs are now made of steel, but hickory was a favourite candidate for many years and is still used for some. Titanium and stainless steel have generally replaced wood for golf-club heads but some manufacturers still use solid persimmon (see left), well suited to the job it must perform, for those traditionalists who prefer the feel of wood.

The outer covering of most top-quality golf balls was once formed from balata rubber, a natural latex exuded from a number of trees belonging to the genus *Manilkara*, but chiefly *M. bidentata* (see opposite), native to Trinidad and parts of South America. The main advantage of balata, which has distinctive qualities and feel, is that it has much more back-spin than its alternatives, meaning essentially that once the ball hits the ground and begins to roll forward, this movement is counteracted, giving the player much more control. Most of the world's professionals have preferred to play with golf balls covered with a thin layer of balata, which also spin a lot faster than those coated with other materials, but are more easily damaged.

By 2002, fewer than 5% of all golf balls were being coated with balata. This, however, is not natural balata but a synthetic reproduction, with the same chemical composition as the latex tapped from the tree. The rest of the world's golf balls are also covered with synthetic materials currently manufactured from petro-chemicals.

Billiard cues

The green baize of the billiard table is graced by wood in the form of the cues wielded by amateurs and professionals alike. A cue comprises

A young lignum vitae tree in Mexico. Large trees have been seriously over-harvested in the wild.

two main pieces, the butt, or handle, and the shaft. A large number of different woods have been and still are involved, many of them now increasingly rare tropical hardwoods, but favoured for their beautiful grain, mottling or other features.

Butts, which may comprise several different woods spliced together, and shafts may include:

● Purple heart from Amazonia.

● Padauk from the Andaman Islands.

● Brazilian rosewood, kingwood, tulip wood or cocobolo (all from various *Dalbergia* spp.) (see overleaf).

● Various ebony species (see page 293), including macassar ebony from Sulawesi, and Gabon ebony, *gonçalo alves* (also known as zebra wood).

● African blackwood or *mpingo*.

● Bocote.

● Rock or bird's-eye maple (see page 248).

Shafts consisting of a single piece of wood are often made from rock maple or ramin from Southeast Asia.

In between the butt and the shaft of the billiard cue, a dyed piece of veneer, often of sycamore or maple, about 2.5 mm ($\frac{1}{10}$ in) thick, is inserted to create an attractive figuring as the cue is turned. Satin wood (from a variety of tropical trees), tulip wood and blood wood, pink ivory and holly are also sometimes used.

Flat green and crown green bowls

A tropical wood well known for its association with one of the oldest British games is lignum vitae (see overleaf). Flat (or lawn) green and crown green bowls were once made almost exclusively from this extremely hard and heavy wood – probably the most dense timber in commercial use. A severe shortage of lignum vitae of the right size due to over-cutting has led to its widespread substitution with other materials (such as polyester, urethane and synthetic resins), but this precious timber is still used for crown green bowls. Lignum vitae logs are chosen for straightness, with their inner pith always forming the horizontal axis of the bowl. Traditionally, only seasoned heartwood has been used, with any cracks carefully filled before the bowl is turned and accurately shaped. Though a limited number of virgin lignum vitae bowls are still made, most crown green bowls are now remodelled from old lawn green bowls. As an alternative to lignum vitae, supplies of cumaru wood – also from tropical America and which have FSC certification – are described as an excellent substitute.

Balata

■ The latex tapped from wild balata trees (*Manilkara bidentata*) has long been valued for its special non-elastic properties when reacted with other compounds and set hard.

■ Native to most of tropical America, balata grows to a height of around 40 m (131 ft), and is usually tapped three times a year. As well as providing a coating for golf balls, the latex has sometimes been substituted for chicle gum (*M. zapota*), the chief ingredient of chewing gum (see page 138).

Billiard balls from cotton

The first successful plastic ever made took the form of a **billiard ball**! In 1868, an American scientist, J. W. Hyatt, won a $10,000 prize for successfully making a plastic billiard ball from a mixture of camphor and nitro-cellulose – a substance made by reacting **cotton linters**, themselves almost pure cellulose, with acids.

Plants instrumental to good music

Lignum vitae

■ Six species of *Guaiacum* are native to the warmer tropical regions of South America. These small, evergreen trees and shrubs were given the name 'lignum vitae' (meaning 'wood of life') because at the time of their discovery by Europeans in the 16th century, their resin was thought to be a cure for several diseases, particularly venereal disease. In 16th-century Europe the wood of *G. officinale* in particular was used with mercury for the treatment of syphilis, and almost exterminated as a consequence through over-harvesting. The remarkable timber of this species and of *G. sanctum* – a distinctive greenish-black in colour and very rich in resin – became and has remained a very valuable commodity for its tremendous hardness, strength and density. The fine-textured, closely interlocked grain, which makes lignum vitae 70–80% heavier than oak, combines with its resin content to make the wood waterproof and exceptionally resistant to rot and abrasion.

■ Much commercial lignum vitae has come from the West Indies, Central America, regions of Colombia and Venezuela, and recently from Mexico.

■ The continued exploitation of these slow-growing trees has led to their inclusion on Appendix II of CITES (the Convention on International Trade in Endangered Species). This is intended to ensure that harvest is not detrimental to their survival and that exports have a government permit. It remains to be seen whether these measures will ensure their survival.

We will never know whether the earliest musical instrument was made when someone hit or blew air into a naturally resonant material and was fascinated by its sound. And was it bone or wood or maybe animal skin or sinews stretched taut that gave the first 'musical' note? Certainly all sorts of animal products have been used for making music around the world, from jaguar bone flutes and conch shell trumpets to armadillo-backed guitars. But plants have provided us with a vast pool of natural resources for music making.

When it comes to the different tree species used to make the world's musical instruments, the picture becomes very complex. Locally grown or native trees have naturally provided the basic raw materials for most of the instruments played by different peoples, but where more suitable foreign timbers have become available these have often been adopted. Bagpipes, for example, which were introduced to Scotland in the 13th century from central and southern Europe, would have made use of temperate trees. Today, they are largely made of African blackwood, a rosewood from tropical Africa (see page 292) considered by some to be unequalled by any other wood for the purpose. For the last 150 years or so, bagpipe makers have also made use of mopane or ebony from various tropical countries. Bagpipes of a lower tonal quality are made in Pakistan from shee-sham wood, another rosewood (*Dalbergia sissoo*).

Shee-sham is the material used to make the neck and pegs of the *jowzé*, a traditional Iraqi bow-stringed instrument, which has a resonator made from half a coconut, and a sound box made from the amniotic sack of a goat. This 'spike fiddle', as it is also known, is currently played in classical ensembles in Iraq.

Rosewoods of various kinds have been the traditional choice for the manufacture of a wide range of musical instruments and, despite the rarity of many of these timbers, are still much in evidence today. The devotion of many musicians to these outstanding materials has contributed to their very serious decline in recent years. Making the situation more complex is the fact that they are often traded under a confusing set of names.

● 'Palisander', for example, is used for woods from various species of *Dalbergia*, *Jacaranda* and *Machaerium*, though it is often Brazilian rosewood (*D. nigra*).

● The wood traded as 'grenadillo' comes technically from *D. granadillo*, while the term 'grenadilla' actually refers to African blackwood, which, as discussed on page 292, is actually *D. melanoxylon*.

A mature pear tree, near Chepstow, UK. One of the hardest and most finely grained of the fruit woods, pear wood has a long history of use for woodwind instruments. A renewed interest in perry pears has led to the planting of many old pear varieties.

Woodwind instruments

It would be impossible to list all the kinds of wood that have been hollowed out or fashioned into drums (and drumsticks) worldwide, and almost as difficult to name the plants that have been used for making simple woodwind instruments. Plants with naturally hollow stems such as reeds and bamboo, however, have been the natural choice for many peoples where these materials have been available, and some of our loveliest and simplest music is produced by them.

The giant reed *Arundo donax*, native to the Mediterranean, has been of particular importance to Western music. European panpipes made in France and the Balkan countries have traditionally made use of its bamboo-like stems – gathered, in the latter region, from the Danube delta area – and the plant continues to provide the best 'reeds' for clarinet, oboe and saxophone mouthpieces. While cane is grown for local use in various Mediterranean countries, the cultivation of reeds for woodwind instruments has largely been limited to a small area in southeastern France, though commercial supplies have also been grown in Texas and California. A large proportion of the clarinet and saxophone reeds currently supplied worldwide are cultivated in southern France and central Argentina.

A whole range of flutes and panpipes are made by Andean people of Peru, Bolivia and Ecuador from native bamboo species such as those belonging to the genus *Chusquea*, and from introduced reeds such as *A. donax*. The light airy sounds made by the tiniest *sikus* or *zampoñas*, and the extraordinary deep and husky bass notes of the *toyos* from Bolivia, are produced from different lengths and widths of hollow stem.

For wind instruments of the East, bamboos have been especially significant since the earliest times. The traditional Japanese *shakuhachi*, is made from the lower end of small-culmed bamboos, while the Chinese *xiao*, a notched flute played in modern orchestras, is made from the bamboo *Phyllostachys nigra*.

Bamboo has many other musical uses too. In western Java, more than 20 percussion, wind and string instruments are made from various species but mostly from the genus *Gigantochloa*. In Bali, the *fungklih* xylophone has bamboo keys, whilst others have resonators made of this material.

Recorders may utilize box, considered by some to be the best wood

THE REAL PRICE of good music

The devotion of many musicians to instruments made only from the best, tried and tested raw materials is, of course, understandable. Such preferences, however, have sadly contributed to the demise or threat of extinction now facing a number of the tree species that have provided much-loved woods. One of those at risk is African blackwood, also known as mpingo (*Dalbergia melanoxylon*), a member of the coveted rosewood family. With superior turning qualities, and as one of the more stable woods with respect to changes in humidity and temperature, African blackwood has been considered hard to equal by the makers of wind instruments such as oboes, clarinets and bagpipes.

A fairly small, slow-growing tree, its timber often contains 'pitch pockets' and burls, which mean that as much as 60% is likely to be wasted from a

typical log when it is used, for example, for bagpipe manufacture. Although it is found in a range of woodland habitats, in some 26 sub-Saharan countries, over-exploitation has greatly reduced populations of mature trees, and the species is reported to be facing commercial extinction. Other rosewoods are similarly threatened, as well as various ebonies and mahoganies, much prized by musicians.

'SoundWood', a project set up as part of the Global Trees Campaign and developed in association with the World Conservation Monitoring Centre, has been designed to try to help safeguard the future of trees used to make musical instruments. This campaign aims to protect the world's most threatened tree species and their habitats through conservation, education and wise use. Rather than advocate a ban on the use of timber from threatened trees,

SoundWood is aiming to promote the sustainable management of timber trees through field projects run in partnership with local people, education programmes, the use of certified woods and other initiatives. The project has established strong links with the music industry and is endorsed by many manufacturers and musicians.

In Tanzania and Mozambique projects aimed at conserving African blackwood are now under way, based on the implementation of regional management plans. A variety of other projects are helping similarly endangered trees in other parts of Africa and in South and Central America. These include pau brasil, or pernambuco (*Caesalpinia echinata*), used for violin bows, but now seriously threatened in Brazil, and several rosewood and mahogany species, as well as lignum vitae in Central America.

for the purpose and also valued for its fine golden-yellow finish, as well as satin wood, pearwood, ebony, kingwood, olive wood and maple. Bassoons are still commonly made from sycamore or fiddle-back, also known as bird's eye maple – cut from the burrs of *Acer saccharum*.

African blackwood is still the wood of choice for clarinets and oboes and, increasingly, for wooden flutes and piccolos. Because of the increasing scarcity of traditionally-used woods, some modern clarinets are being made from a blend of 95% 'grenadilla' (African blackwood) powder with 5% carbon filler and epoxy resin, and are said to have the same tonal qualities and projection as their solid wood counterparts.

African blackwood or mpingo trees in coastal Tanzania. Over-exploitation of these slow-growing trees has greatly reduced wild populations. The trees shown here are in a forest area that is regenerating.

Pianos

As is true for the manufacture of any specialist equipment, the production of musical instruments to a consistent standard depends not just on how they are put together, but also on the particular properties of the materials used.

A piano, for example, is made up of a large range of different woods, each one selected for the special function it will perform:
● The rim of a concert grand piano is likely to be made of 18 or 19 layers of hard rock maple, a resilient, stable timber that resists distortion.
● The soundboard and its ribs require a lightweight wood with a high degree of elasticity, enabling it to vibrate and give resonance to the strings. Spruce has been an ideal timber for this purpose, though other light woods are sometimes substituted.
● Hammer shanks and mouldings are often made of maple.
● Lime wood grown in America (also referred to as basswood) (*Tilia americana*) has often formed the wood under the ivory veneers of white piano keys, since it will not twist or warp.
● Black keys have often been made of ebony (see right).

Stringed instruments

Violins, violas, cellos and other instruments of the violin family likewise comprise various kinds of wood, chosen for their different characteristics. Since their top-side or belly has similar requirements to a piano soundboard, it is invariably made of a light coniferous

Ebony

■ Distinctive black ebony wood (*Diospyros* spp.), valued for centuries for cabinet work and decorative inlays, is still used today to make components of a variety of musical instruments and some sports equipment. Organ stops and black piano keys, fittings for violins, guitars and bagpipes, as well as billiard cues and castanets have utilized this hard, heavy wood. The most important ebony species for commercial use have been *D. ebenum* from India and Sri Lanka and *D. reticulata* from Mauritius.

■ Over-exploitation of *D. ebenum*, still regarded as the best commercial black ebony, has led to India and Sri Lanka banning exports of this species. Supplies of the related *D. crassiflora*, however, despite its threatened status, are being exported from West Africa.

293

wood – usually Norway spruce (*Picea abies*), favoured for its strength, lightness and resilience. The narrow grain produced by such conifers, grown on poor soils in harsh climates (today likely to be harvested in Scandinavia, the Balkans or Romania), has been found to produce the best sound. The backs, sides, necks and scrolls, meanwhile, are traditionally made of sycamore or other maple species. The distinctive rippled grain of *Acer saccharum* burrs, and the 'curl', 'figure' or 'flame' produced by a waviness in the grain of other *Acer* spp., has become so much associated with violin making that it is commonly referred to as 'fiddle-back'.

The master violin maker Antonio Stradivarius (1644–1737) was the first to use a bridge of maple to support the strings of violins. The quality of his instruments, however, is said to have been determined to a large degree by the way in which the wood was seasoned and the composition of the finishing varnish. In 2000, a Stradivarius fetched $1.3 million at auction in New York. Today, some 300 years after such masterpieces were made, modern manufacturers are still choosing the same woods and following the same basic designs for their instruments.

Some of the best old Italian and modern cellos are backed with poplar or willow, though beech or pear wood is sometimes used. Ebony is still used for the fingerboard and tailpiece, and two smaller parts – the top nut and the saddle – of many violins. Tuning pegs are often cut from ebony or Honduras rosewood (*Dalbergia stevensonii*), and it also sometimes used for the chin rest. This is a very high-density timber also commonly selected for high-quality xylophones, though boxwood has been considered smarter.

That now ubiquitous Moorish invention, the guitar, similarly still relies on wood to a great extent. To achieve good resonance, harmonic stability and durability, the makers of Spanish or acoustic guitars have long chosen wood types with different densities best suited to the different functions the various components must perform. With its particular vibrancy, providing an ideal medium for the transmission of sound, by far the most popular wood used for the soundboard of modern acoustic guitars is Sitka spruce. This is followed by Engleman spruce and cedar of various kinds. Backs and sides are very likely to be made of mahogany from Brazil, rosewood – Brazilian, Honduran or East Indian – ebony, African blackwood, cherry koa, native to Hawaii, or maple. Amongst other attributes, these woods are all favoured for their stability. Fingerboards, necks and bridges also make use of these woods, as well as others such as pear, walnut and purple heart.

Perhaps surprisingly, 99% of all electric guitars are also made mostly of wood. The body of many may comprise a single block of any of a great range of woods (from many different countries), features

that will be reflected in the price. The body of others, however, might consist of a back piece of ash, alder or poplar, or (for the most expensive) mahogany – a heavy wood, which gives a warm, deep resonance to the sound – with a decorative front of maple, a lighter wood which gives a 'brightness' to the strings' vibration. Fingerboards are often cut from pieces of ebony or rosewood, or from bubinga, a heavy timber from West Africa that is similar to rosewood in appearance, with a distinctive, highly decorative colouring and grain.

Whatever the instrument we play or listen to, musicians and music lovers everywhere will surely agree that plants have played a vital part in helping us create the sounds that identify our different cultures and unite us all.

This boy, in Manaus, Brazil, is learning how to make high-quality guitars using a variety of sustainably harvested Amazonian hardwoods. He is part of a project helping street children to learn skills, to try to bring a brighter future.

295

Further reading

A very large number of books, articles, reports and websites were consulted during the research and writing of *Plants for People*. It is impossible to name them all, but the following list presents a sample of books that may be useful for further or background reading. No articles or reports are listed other than those specifically mentioned in the text.

General
Anderson, E. S., *Plants, Man, and Life,* Missouri Botanical Garden Press (1997)
Balick, M. J. & Cox, P. A., *Plants, People, and Culture: the Science of Ethnobotany,* Scientific American Library (New York, 1996)
Brown, L., *Eco-Economy: Building an Economy for the Earth,* Earthscan (London, 2001)
Cotton, C., *Ethnobotany: Principles and Applications,* John Wiley (Chichester, 1996)
Dickson, C. & J., *Plants and People in Ancient Scotland,* Tempus (Stroud, 2000)
Hobhouse, H., *Seeds of Change: Six Plants that Transformed Mankind,* Papermac (London, 1999)
Khor, M., *Rethinking Globalization: Critical Issues & Policy Choices,* Zed Books
Koziell, I., *Diversity not Adversity: Sustaining Livelihoods with Biodiversity,* IIED & DfID (2001)
Mabberley, D. J., *The Plant Book. A Portable Dictionary of the Vascular Plants* (2nd ed.), Cambridge University Press (Cambridge, 1997)
Madeley, J., *Hungry for Trade: How the Poor Pay for Free Trade,* Zed Books (London, 2001)
Musgrave, T. & W., *An Empire of Plants: People and Plants that Changed the World,* Cassell & Co. (London, 2000)
Sauer, J. D., *Historical Geography of Crop Plants. A Selective Roster,* CRC Press (Boca Raton, Florida, 1993)
Simpson, B. B. & Conner-Ogorzaly, M., *Economic Botany: Plants in our World* (3rd ed.), McGraw-Hill (Boston, 2001)
Tokar, B. (ed.), *Redesigning Life: the Worldwide Challenge to Genetic Engineering,* Zed Books (London, 2001)
Wickens, G. E., *Economic Botany: Principles and Practices,* Kluwer Academic (London, 2001)
ten Kate, K. & Laird, S. A., *The Commercial Use of Biodiversity: Access to Genetic Resources and Benefit-sharing,* Earthscan (London, 1999)

Part 1: Plants that care for us
Bauer, K., Garbe, D. & Surburg, H., *Common Fragrances and Flavor Materials: Preparation, Properties and Uses* (3rd ed.), Wiley-VCH (Weinham, 1997)
The Bitter Fruit of Oil Palm, World Rainforest Movement (2001)
Lawless, J., *Illustrated Encyclopaedia of Essential Oils: the Complete Guide to the Use of Oils in Aromatherapy and Herbalism,* Element Books (Shaftesbury, 1995)
Manniche, L., *Sacred Luxuries: Fragrance, Aromatherapy and Cosmetics in Ancient Egypt,*

Opus Publishing (London, 1999)
Nicholson, P. T. & Shaw, I. (eds), *Ancient Egyptian Materials and Technology,* Cambridge University Press (Cambridge, 2000)
'The Politics of Extinction', Environmental Investigation Agency (1998)
Price, S., *Aromatherapy Workbook: a Complete Guide to Understanding and Using Essential Oils,* Thorsons (London, 1998)
Weiss, E. A., *Essential Oil Crops,* CABI International (Wallingford, 1997)
White, E. C., *Soap Recipes: Seventy Tried-and-true Ways to make Modern Soap with Herbs, Beeswax and Vegetable Oils,* Valley Hills Press (Starkville, Massachusetts, 1995)

Part 2: Plants that clothe us
Balfour-Paul, J., *Indigo,* British Museum Press (London, 1998)
Buchanan, R., *A Weaver's Garden: Growing Plants for Natural Dyes and Fibers,* Dover Publications (New York, 1999)
Cannon, J. & M., *Dye Plants and Dyeing,* Herbert Press in Association with Royal Botanic Gardens, Kew (London, 1994)
Dean, J., *Wild Color. The Complete Guide to Making and Using Natural Dye,* Watson-Guptill (New York, 1999)
Harris, J. (ed.), *5000 Years of Textiles,* British Museum Press in association with the Whitworth Art Gallery and the Victoria and Albert Museum (London, 1993)
Hinrichsen, D., 'Requiem for a Dying Sea', *People and the Planet Magazine* (Vol. 4, No. 2) (1995)
Hollen, N., Saddler, J. & Langford, A. L. *Textiles* (6th ed.), Macmillan (New York, 1988)
Nicholson, P. T. & Shaw, I. (eds), *Ancient Egyptian Materials and Technology,* Cambridge University Press (Cambridge, 2000)
Robbins, N. & Humphrey, L., *Sustaining the Rag Trade,* IIED (London, 2000)
Smith, C. W. & Cothren, J. T. *Cotton: Origin, History, Technology and Production,* John Wiley & Sons Ltd (New York, 1999)

Part 3: Plants that feed us
Brody, H., *The Other Side of Eden: Hunter-Gatherers, Farmers and the Shaping of the World,* North Point Press (New York, 2001)
Davidson, A., *The Oxford Companion to Food,* Oxford University Press (Oxford, 1999)
The Great EU Sugar Scam: How Europe's Sugar Regime is Devastating Livelihoods in the Developing World, Oxfam Briefing Paper (2002)
Harlan, J. R., *The Living Fields: Our Agricultural Heritage,* Cambridge University Press (Cambridge, 1995)
Heiser, C. B., *Seed to Civilisation: the Story of Food,* Harvard University Press (Cambridge, Massachusetts, 1990)
Kiple, K. F. & Ornelas, K. C. (eds), *Cambridge World History of Food,* Cambridge University Press (Cambridge, 2000)

Madeley, J., *Food for All: the Need for a New Agriculture,* Zed Books (London, 2002)
'Mugged: Poverty in your Coffee Cup', Oxfam International (2002)
Norberg-Hodge, H., Merrifield, T. & Gorelick, S., *Bringing the Food Economy Home: Local Alternatives to Agribusiness,* Zed Books (London, 2002)
Roberts, J., *Cabbages and Kings: the Origins of Fruit and Vegetables,* HarperCollins (London, 2001)
'Seeds of Doubt', The Soil Association (2002)
Shiva, V., *Stolen Harvest: the Hijacking of the Global Food Supply,* Zed Books (London, 1999)
Shiva, V., *The Violence of the Green Revolution,* Zed Books (London, 1992)
Smartt, J. & N. W. Simmonds (ed.), *Evolution of Crop Plants* (2nd ed.), Longman (Harlow, 1995)
'Unpeeling the Banana Trade', The Fairtrade Foundation (August 2000)
Vaughan, J. G. & Geissler, C. A., *The New Oxford Book of Food Plants,* Oxford University Press (Oxford, 1997)
Zarb, J., 'Small Holding Up' (the *Ecologist,* Vol. 30, No. 9, Dec 2000)
Zohary, D. and Hopf, M., *Domestication of Plants in the Old World* (3rd ed.), Clarendon Press (Oxford, 2000)

Part 4: Plants that house us
Borer, P. & Harris, C., *The Whole House Book: Ecological Building Design and Materials,* Centre for Alternative Technology (Machynlleth, 2001)
Crook, G., *Basket Making,* Crowood Press (Ramsbury, 2000)
Farrelly, D., *The Book of Bamboo,* Thames & Hudson (London, 1996)
Janssen, J. J. A., *Building with Bamboo: a Handbook* (2nd ed.), Intermediate Technology Publications (London, 1995)
Magin, G., *Good Wood Guide,* Friends of the Earth/Flora & Fauna International (2002)
'Partners in Mahogany Crime: Amazon at the Mercy of "Gentlemen's Agreements"', Greenpeace (October 2001)
Sullivan, F., Dudley, N. and Jeanrenaud, J.-P., *Bad Harvest: the Timber Trade and the Degradation of World's Forests,* Earthscan (London, 1996)
Velez, S., Kries, M., Dethier, J., Steffens, K. (eds), *Grow your own House: Simone Velez and Bamboo Architecture,* Vitra Design Museum (Germany, 2000)
Woolley, T., Kimmins, S., Harrison, P. & Harrison, R., *Green Building Handbook* (Vols 1 & 2), Centre for Alternative Technology (Machynlleth, 1999 & 2000)

Part 5: Plants that cure us
Barnes, J., Anderson, L. & Phillipson, J. D., *Herbal Medicines: a Guide for Healthcare*

Professionals, Pharmaceutical Press (London, 2002)

Bartram, T., *Encyclopaedia of Herbal Medicine*, Grace Publishers (Christchurch, Dorset, 1995)

Bown, D., *The Royal Horticultural Society New Encyclopaedia of Herbs and their Uses*, Dorling Kindersley (London, 2002)

Hatfield, G., *Memory, Wisdom and Healing: the History of Domestic Plant Medicine*, Sutton Publishing (Stroud, 1999)

Hoffmann, D., *The Complete Illustrated Holistic Herbal: a Safe and Practical Guide to Making and Using Herbal Remedies*, Element Books (Shaftesbury, 1996)

Lange, D., *Europe's Medicinal and Aromatic Plants: their Use, Trade and Conservation*, TRAFFIC International (Cambridge, 1998)

Mills, S. & Bone, K., *Principles and Practice of Phytotherapy: Modern Herbal Medicine*, Churchill Livingstone (Edinburgh, 2000)

Minter, S., *The Healing Garden*, Headline Book Publishing (London, 1993)

Ody, P., *The Herb Society's Complete Medicinal Herbal*, Dorling Kindersley (London, 1993)

Sumner, J., *The Natural History of Medicinal Plants*, Timber Press (Portland, Oregon, 2000)

Swerdlow, J. L., *Nature's Medicine: Plants that Heal*, National Geographic (Washington DC, 2000)

Part 6: Plants that transport us

Coppen, J. J. W., *Non-wood forest products: 6* ('Gums, Resins and Latexes of Plant Origin') , Food and Agriculture Organization of the UN (Rome, 1995)

Dean, W., *Brazil and the Struggle for Rubber: a Study in Environmental History*, Cambridge University Press (Cambridge, 1987)

Horne, B., *Power Plants: Biofuels Made Simple*, Centre for Alternative Technology (Machynlleth, 1996)

Lewington, A., *Antonio's Rainforest*, Wayland (Hove, 1992)

Sethuraj, M. R. & Mathew, N. M. (eds), *Natural Rubber: Biology, Cultivation and Technology*, Elsevier (Amsterdam, 1992)

Tickell, J. & K., *From the Fryer to the Fuel Tank: the Complete Guide to Using Vegetable Oil as an Alternative Fuel*, Veggie Van Publications (1998)

Webster, C. C. & Baulkwill, W. J., *Rubber*, Longman Scientific & Technical (Harlow, 1989)

Yung Johann, J. C. (ed. Prance, G. T.), *White Gold: the Diary of a Rubber Cutter in the Amazon 1906–1916*, Synergetic Press Inc. (Oracle, 1989)

Part 7: Plants that entertain us

Barr, C., *Profits on Paper: the Political-Economy of Fiber, Finance and Debt in Indonesia's Pulp and Paper Industries*, Centre for International Forestry Research and WWF–International's Macroeconomics Program Office (2000)

Bell, L. A., *Papyrus, Tapa, Amate and Rice Paper: Papermaking in Africa, the Pacific, Latin America and Southeast Asia*, Liliaceae Press (McMinnville, Oregon, 1985)

Carrere, R. & Lohmann, L., *Pulping the South: Industrial Tree Plantations and the World Paper Economy*, WRM & Zed Books (London, 1996)

Dawson, S. *The Art and Craft of Papermaking*, Aurum Press (London, 1993)

Hunter, D., *Papermaking: the History and Technique of an Ancient Craft*, Dover Publications Inc. (New York, 1978)

Ilvessalo-Pfaffli, M.-S., *Fiber Atlas: Identification of Papermaking Fibers*, Springer-Verlag (Berlin, 1995)

Matthew, E. & Willem van Gelder, J., *Paper Tiger, Hidden Dragons*, Friends of the Earth & Profundo (2001)

Miles, A., *Silva*, Ebury Press (London, 1999)

Rault, L., *Musical Instruments: a Worldwide Survey of Traditional Music-making*, Thames & Hudson (London, 2000)

Waring, D., *Making Wood Folk Instruments*, Sterling Publishing (New York, 1991)

Useful addresses

297

ActionAid, Hamlyn House, Macdonald Road, London N19 5PG
Tel: 020 7651 7651 Website: www.actionaid.org

Alternative Crops Technology Interaction Network, Pira House, Randalls Road, Leatherhead, Surrey KT22 7RU
Tel: 01372 802054 Website: www.actin.co.uk

BioRegional Development Group, BedZED Centre, Helios Road, Wallington, Surrey SM6 7BZ
Tel: 020 8404 4880 Website: www.bioregional.com

Centre for Alternative Technology, Machynlleth, Powys SY20 9AZ
Tel: 01654 705950 Website: www.cat.org.uk

Centre for Economic Botany, Royal Botanic Gardens, Kew, Richmond, Surrey TW9 3AB
Fax: 020 8332 5768 Website: www.rbgkew.org.uk/scihort/eblinks

Environmental Investigation Agency, 62/63 Upper Street, London N1 0NY
Tel: 020 7354 7960 Website: www.eia-international.org

Ethnomedica, Centre for Economic Botany, Royal Botanic Gardens, Kew Richmond, Surrey TW9 3AB
Website: www.rbgkew.org.uk/ethnomedica

The Fairtrade Foundation, Suite 204, 16 Baldwin's Gardens, London EC1N 7RJ
Tel: 020 7405 5942 Website: www.fairtrade.org.uk

Fauna & Flora International, Great Eastern House, Tenison Road, Cambridge CB1 2TT
Website: www.fauna-flora.org

Food and Agriculture Organization of the United Nations, Viale della Terme di Caracalla, 00100 Rome, Italy
Tel: +39 06 57051 Website: www.fao.org

Forest Stewardship Council, Unit D, Station Building, Llandidnoes SY18 6EB
Tel: 01686 413916 Website: www.fsc-uk.info

Friends of the Earth, 26–28 Underwood St, London N1 7JQ
Tel: 020 7490 1555 Website: www.foe.co.uk

Greenpeace International, Keizersgracht 176, 1016 DW Amsterdam, The Netherlands
Tel: +31 20 523 62 22 Website: www.greenpeace.org

Henry Doubleday Research Organisation, Ryton Organic Gardens, Coventry, Warwickshire CV8 3LG
Tel: 024 7630 3517 Website: www.hdra.org.uk

International Institute for Environment and Development, 3 Endsleigh Street, London WC1H ODD
Tel: 020 7388 2117 Website: www.iied.org

People and Plants Initiative, WWF–UK, Panda House, Weyside Park, Catteshall Lane, Godalming, Surrey GU7 1XR
Website: www.rbgkew.org.uk/peopleplants

Society for Economic Botany, PO Box 1897, Lawrence, KS 66044-8897, USA
Website: www.econbot.org

The Soil Association, Bristol House, 40–56 Victoria Street, Bristol BS1 6BY
Tel: 0117 929 0661 Website: www.soilassociation.org

Survival International, 6 Charterhouse Buildings, London EC1M 7ET
Tel: 020 7687 8700 Website: www.survival-international.org

TRAFFIC International, 219a Huntingdon Rd, Cambridge CB3 ODL
Tel: 01223 277427 Website: www.traffic.org

UNEP World Conservation Monitoring Centre, 219 Huntingdon Road, Cambridge CB3 0DL
Tel: 01223 277314 Website: www.unep-wcmc.org

World Wide Fund for Nature, Panda House, Weyside Park, Catteshall Lane, Godalming, Surrey GU7 1XR
Tel: 01483 426444 Website: www.wwf.org.uk

■ Further images relating to plant use can be viewed at:
www.plantsforpeoplephotos.com

Index of plant names

301

Index of subjects

Acknowledgements

This book, which has taken nearly two years to bring to fruition, has involved the labours of many people, to all of whom I am indebted. First of all, I must thank Mike Petty at the Eden Project, a tireless optimist, who has overseen the project from beginning to end, surmounting obstacles and giving encouragement and advice throughout. I am also deeply grateful to Dr Mark Nesbitt, of the Royal Botanic Gardens, Kew, for his most valuable advice, his suggestions regarding research help and his outstanding thoroughness in checking facts used in the text. I should add here, however, that some of the personal opinions I express in the book are not necessarily shared by Dr Nesbitt. To my editor, Emma Callery, I give my heartfelt thanks. Her professionalism, patience and sensitive editing of a manuscript that threatened to spill into another book, have been truly wonderful: a constant source of encouragement, during what at times seemed like a never-ending task. I am extremely grateful, Emma. And I must warmly thank Sue Miller, who has worked so hard to create such a beautiful design for every page. Thank you, Sue.

I am also especially grateful to Dr Sasha Barrow, Georgina Pearman, Helen Sanderson and Caroline Servaes, who carried out much of the research for the new edition, from the Royal Botanic Gardens, Kew. Consulting a vast number of publications, of many different kinds,

wading through websites and checking with experts, they played a vital supporting role in the creation of this book. I must also thank the many other members of staff at Kew, who supplied information or checked details. Special thanks are also due to Sue Minter, of the Eden Project, who read through the first draft and made many helpful suggestions, and Andrew Ormerod (also from Eden) for his help with information.

In the endeavour to make *Plants for People* as up to date and accurate as possible, a very large number of experts and individuals (listed below) from industry, academic institutions and non-governmental organizations have been consulted and kindly given advice. A number of these have taken the time, too, to check and update sections of text, and for this I am extremely grateful. I must also thank Spence Gunn for his hard work on Part 5: Plants that cure us.

A special thank you is due to Francis Sullivan and Perdita Hunt of WWF–UK for the generous sponsorship of this book at a critical time. This help is greatly appreciated.

Last, but not least, I should like to thank my family: Edward Parker, my husband (and the book's photographer), who has supported and encouraged me through thick and thin, and my dear children, Eppie and Aaron, to whom I have promised never to do anything quite like this again!

303

Part 1: Plants that care for us
Catherine Cartwright Jones; Dr Stephen Case-Green, The British Wax Refining Co. Ltd; Dr Simon Charlesworth, Downderry Nursery; Alan Connock, A. & E. Connock (Perfumery & Cosmetics) Ltd; Matthew Crisp, Sappi Saiccor; Alan Critchley; Rita Donoghue, Department of Trade and Industry; Siegfried Falk, ISTA Mielke GmbH (Oil World); Gene Hale, International Aloe Science Council; John Hancock, Federation of Oil Seeds and Fats Association Ltd; Joan Head, Norfolk Lavender; Alex Morgan, The Henna Page; Celsi Richfield, Organic Botanics; Harald Sauthoff, Cognis Deutschland; Jens Schaller, JenaBio Science; Julie Towle; Hans de Vries, Central Soya European Lecithins; Jane Webb, Lush; Antonia Wheatley, Estée Lauder; Dr Peter Wilde, The Oil Factory.

Part 2: Plants that clothe us
Graham Agg, Rayonier; William Baker, Royal Botanic Gardens, Kew; Jenny Balfour-Paul; Dr Elizabeth Barber; Ian Bartle, Alternative Crops Technology Interaction Network; Campbell Bland, Acordis UK Ltd; Dr B. Braddock, Acordis Acetate Products; Martin Brink; Edwin Datschefski; Alun Davies, Acordis Cellulosic Polymers; Pauline Delli-Carpini, Masters of

Linen; Isabelle DuJardin, Levi Strauss, Brussels; Gillian Edom; Mary Eve, M. & R. Dyes; Exposure Ltd; Simon Ferrigno, Pesticide Action Network; Tom Fox, International Institute for Environment and Development; Polly Ghazi; Dr Kerry Gilbert, Bristol Dye Research Group; Brian Glover; Nigel Goodburn, Avebe UK Ltd.; Tony Gower, Tanning Extract Producers Federation; Dr Pat Griggs, Royal Botanic Gardens, Kew; Professor Mike Guiry, University of Ireland; John Hobson, Hemcore Ltd; Michael Holt, Acordis Cellulosic Polymers; Daniel Jenkins, Proctor & Gamble; Professor Philip John, University of Reading/ Spindigo Project; Kirsty Lyall, Hort Research; Gordon Meredith, Levi Strauss (UK) Ltd; Sabine Modde, Levi Strauss, Brussels; Charles Myring, Clariant UK Ltd; Dr Richard Percy, Maricopa Agricultural Centre, Arizona; Thomas Petit, Vericott Ltd; Dora Radwick, NPD Group, New York; Maria Luisa A. Ramos, Embassy of the Philippines, London; Steve Rose, Shoe and Allied Trade Research Association; Shawn Rossiter, Levi Strauss, USA; Naomi Rumball, Royal Botanic Gardens, Kew; Roy Russell, M. & R. Dyes; Damien Sanders; Sue Scheele, Landcare; Dr Kelvin Tapley, University of Leeds; Dr Neville Slabbert, Tanning Extract Producers Federation; Jim

Taylor, Tencel Ltd; Rebecca Unsworth, The Textile Institute; Derek Weightman, Sappi Saiccor; Women's Environment Network; Susanna Wood.

Part 3: Plants that feed us
Melanie Adams, The Tea Council; Bryan and Cherry Alexander; Ted Batkin, Citrus Research Board, California; Brogdale Horticultural Trust; Dr Tom Cope, Royal Botanic Gardens, Kew; Dr Aaron Davis, Royal Botanic Gardens, Kew; Steve Davis, Royal Botanic Gardens, Kew; Professor James Fox, The Australian National University; Diana Gayle, The Fairtrade Foundation; John Higginbotham, Tate & Lyle; Tadesse Woldemariam Gole; Dr Jai Gopal, Central Potato Research Institute, India; Harriet Lamb, The Fairtrade Foundation; Jane Pettigrew, The Tea Council; Joe Simrany, The Tea Association of the USA; Dr Mike Witty, University of Cambridge; Helen Wolfson, ActionAid.

Part 4: Plants that house us
Dr P. J. Bienz, Timber Trade Federation; Ximena Buitron, TRAFFIC; Belinda Bush, the *Ecologist*; Ralph Carpenter, Modece Architects; James Chamberlain; Dr Osman Chowdhury, Bangladesh Jute Diversification Centre; Jenny

Coleback, Wigglesworth & Co. Ltd; Dr John Dransfield, Royal Botanic Gardens, Kew; Dundee Heritage Trust; John Excell, The Cane Workshop; Norman Falla, Paint Research Association; Forest Peoples' Programme; Roger Hunt; Duncan Johnstone, Corus Building Systems; Amanda Kent, Interface Europe Ltd; Ingrid Lewis; Paul Long, Lawrence Long Ltd; Virginia Lulling, Survival International; Leonard K. Moss, The Centre Cane Company; Kevin May, Forest Enterprise; David Muncey, Forbo; Paul Newman, Timber Research & Development Association; Claire Potter, Amorim Benelux BV; Maria Waters, Friends of the Earth; John Williams, Blackdown Horticultural Consultants; Verdant Works, Dundee.

Part 5: Plants that cure us
Colin Abbott, Felton Grimwade & Bickford Pty Ltd; Nicholas Archer, McFarlan Smith; Ben van Baarle, R-S Information Centre for Natural Rubber; Sandra Costigan, Medical Devices Agency; Deborah Dermont, Vermont Witch Hazel Company; Sheri Doran, Ethicon Endo-Surgery; Heather Gillanders, Colon Cancer Concern; GlaxoSmithKline; Eva Kakyomya, World Health Organization; Stefan Kraan, Irish Seaweed Centre; Dr Nick Lampert, Register of Chinese Herbal Medicines; Samuel K. Lee; Chris Leon, Chinese Medicinal Plants; Authentication Centre, RBG, Kew; Dr Gwil Lewis, Royal Botanic Gardens, Kew; Liz McConachie, Schering; Des McGrahan, Thomas Swan & Co. Ltd; Mike Murray, Association of the British Pharmaceutical Industry; Reckitt Benckiser; Fiona Robbins, Serono; Christine Ross, Smith & Nephew; Ernie Ruvalcaba, Lymphoma Research Foundation; Professor Monique Simmonds, Royal Botanic Gardens, Kew; Peter Soukas, Public Health Service, USA; Bronwen Tomas, Medicines Control Agency; Amie Tompkins, CMS One Stop Information Shop; Andreas von der Weth, Smith & Nephew.

Part 6: Plants that transport us
Carol Abel, British Marine Federation; Keith Addison, Journey to Forever; R. D. Allen, NASA (Langley Research Centre); Geoff Bishop, Shell Aviation Ltd; Steve Boyd, Uniqema; John Brant,

ConocoPhillips Lubricants; Geoff Burrows, Bentley Motors Ltd ; Albert Cho, Production Sheng Chan Pharmaceutical; Jenny Coleback, Wigglesworth & Co. Ltd; Professor R. M. Cripps, Royal National Lifeboat Institution; Christopher Dobbs, Mary Rose Trust; Tom Dougherty, Shipbuilders and Shiprepairers Association; Bill Farquhar, Macduff Shipyards Ltd; Fuchs Lubricants (UK); Sarah Goldstone, Robbins Timber; Peter Goodwin, Keeper and Curator, *HMS Victory*, HM Naval Base; Guy Gratton, British Microlight Aircraft Association; Janette Green, Bentley Motors Ltd; Professor Mike Guiry, University of Ireland; John Hamer, Minimax Club, UK; Dr Alan Hogenhauer, Loyola Marymount University, Los Angeles; Roger Kemp, Alstom Transport; Morris Minor Owners Club; Dock No, International Rubber Study Group; Andy Parr, Master Ropemakers Ltd; Professor Sir Ghillean Prance; Chris Reid, Baltek UK; Karen Russell, Horticulture Research Institute; Damien Sanders; Tracey Satchwell-Smith, Royal Caribbean International; Nigel Saw, British Marine Federation; Felix Schmid, University of Sheffield; Paul Schulha, Castrol; John Scofield, Marlow Ropes Ltd; Dr Dave Simpson, Royal Botanic Gardens, Kew; C. A. Stark, NASA (ASRC Aerospace); Sharon Styles, Alternative Crops Technology Interaction Newsork; Sonae Tafibra (UK) Ltd ; Swift Group Ltd; US Department of Energy (National Alternative Fuels Hotline); Roger Walker; Matthew Wharmby, London Bus Page; John Winfield, Castrol.

Part 7: Plants that entertain us
Tim Belcher, WHSmith; Charles Berolzheimer, Calcedar; Jerry Bix; Blue, MacMurchie Bagpipes Paul Boyle, PB Pro Shop; *British Journal of Photography*; Dr Charlie Clarke, Sappi Forests Research; Dunlop Slazenger International; Mark Fairbrass, The Beacon Press; Mike Gilbert, Natural History Museum; Sue Girard, The Paper Federation of Great Britain; Emily Green, Green Futures; Dr Mike Guiry, National University of Ireland; Barbara Hakim, Kolstein Music, Inc.; Steve and Susie Hammett, Recycled Paper Supplies; Walter van Hauwe, Stichting BlokFluit; Joel Hercek, Hercek Fine Billiard

Cues; Alun Hughes, Boosey & Hawkes Musical Instruments; Dr David Hughes, School of Oriental and African Studies, University of London; Rieka Hughes, Staedtler UK Ltd; Japanese Embassy ; Eric Kilby, The Paper Federation of Great Britain; Patrick Knill, Japan Society; Peter de Koningh; Dr G. Lugert, Faber-Castell Germany; George Manning, Power Bilt; Chris Mayo, Kodak; Sarah Miller, Winsor & Newton; Barbara Murray, Cumberland Pencil Company; Basha Nazir, PIRA International; Sara Oldfield ('Soundwood'), Fauna & Flora International; Chris Pearce, Fuji Photo Film UK Ltd; Daniella Peloquin, Taiga Rescue Network; Dr Terry Pennington, Royal Botanic Gardens, Kew; Professor Donald Pigott, Cambridge Botanic Garden; Troy Pluckett, Jr, Cayman Golf Company Inc.; John Purse, Prima Information; Fritz Reuter, Fritz Reuter & Sons, Inc; Matthew Rigby, T. S. Hattersley & Son Ltd; The Royal Photographic Society; Andrew Simmons, PPS Ltd; Daniel Smith; Justin Stead, University of British Columbia, Vancouver, BC; Andrew Stevens, Crown Green Bowler; Tim Synnott, Forest Stewardship Council, Oaxaca ; Hazel Tachtatzis, Calcedar; Eric Tay, Teh Cane Trading Company Pte Ltd; Clare Taylor; Marsha Taylor, Kestrel Reeds; Randy Teitloff, Ebonite; Gail Thacker ('Soundwood'), Fauna & Flora International; Christopher Torerro, Paperbase Abstracts; David Watson, International Hemp Association; Derek Weightman, Sappi Saiccor; George Wood, Wood Malletts Ltd, New Zealand; David Woodbridge, Institute of Wood Science; Beck Woodrow, Forest Stewardship Council (UK); Jeremy Williams, Boosey & Hawkes Publishing; Martin Wright, Green Futures; Roger Wright, Hawkins Wright.

Picture credits
All photographs by Edward Parker with the exception of the following, whom the publishers would like to thank: Bryan and Cherry Alexander for the pictures on pages 241 and 269. The Hutchinson Picture Library for the photograph on page 69. Thank you, too, to June and Mike Harrison for their help with the source material for the leaf illustrations.

304